D1154979

PROPERTY OF
Gilford Information Center
Chiron Diagnostics

Mixing in the Process Industries

Butterworths Series in Chemical Engineering: internationally acknowledged specialists provide authoritative treatment of topics of current significance in chemical engineering.

Series Editor

J W Mullin
Professor of Chemical Engineering, University College, London

Published titles:

Solid–Liquid Separation, Second edition
Liquids and Liquid Mixtures, Third edition
Fundamentals of Fluidized-Bed Chemical Processes
Diffusion in Liquids
Introduction to Electrode Materials

Related titles:

Heterogeneous Reactor Design
Phase Equilibria in Chemical Engineering

Mixing in the Process Industries

Editors: **N Harnby, M F Edwards, A W Nienow**

Butterworths

London   Boston   Singapore   Sydney   Toronto   Wellington

All rights reserved. No part of this publication may be reproduced or transmitted in any form or by any means, including photocopying and recording, without the written permission of the copyright holder, applications for which should be addressed to the Publishers. Such written permission must also be obtained before any part of this publication is stored in a retrieval system of any nature.

This book is sold subject to the Standard Conditions of Sale of Net Books and may not be re-sold in the UK below the net price given by the Publishers in their current price list.

First published 1985
Reprinted 1987, 1989

© Butterworth & Co (Publishers) Ltd, 1985

British Library Cataloguing in Publication Data

Mixing in the process industries.
  1. Chemical engineering   2. Mixing
  I. Harnby N.   II. Edwards, M.F.
  III. Nienow, Alvin W.
  660.2′84292   TP156.M5

  ISBN 0-408-11574-2

Library of Congress Cataloging in Publication Data

Main entry under title:

Mixing in the process industries.

  Includes bibliographies and index.
  1. Mixing.   I. Harnby, N.   II. Edwards, M. F.
III. Nienow, A. W.
TP156.M5M54   1985   660.2′84292   85-11034
ISBN 0-408-11574-2

Typeset by Mid-County Press, 2a Merivale Road, London SW15 2NW
Printed and bound by Hartnoll Ltd, Bodmin, Cornwall

Preface

'But you've no idea what a difference it makes, mixing it with other things—such as gunpowder and sealing wax' ... *Lewis Carroll, Through the Looking Glass.*

Whilst we don't actually mix gunpowder and sealing wax in this book, Lewis Carroll evidently had a presentiment of some of the unlikely mixing problems which would face process engineers today. Liquids, solids and gases have to be mixed in all combinations to satisfy a very variable process or product quality requirement. Long gone are the days when process mixing was left exclusively to the experienced worker who had a 'feel' for the product; and when academic interest was limited to the power consumption of paddle mixers as a fine example of the application of dimensional analysis.

There has been a post-experience course for the last ten years with the same title as that of the book. It has been organized at the University of Bradford in association with the Institution of Chemical Engineers. The course has been responsive to change in the process industries and it now reflects the balance of interest shown in the topic by industrialists over this period from both the UK and western Europe. The book has grown from the course and therefore is based on a well-tried and successful formula. That is a big advantage. Another advantage is that each chapter is by a specialist from either industry or academia. A slight disadvantage is that this inevitably leads to some variation in style between chapters. However, we feel that this is a small price to pay for in-depth expertise in each topic and the use of the Institution of Chemical Engineers Recommended Nomenclature for Mixing wherever possible has considerably increased the level of uniformity.

With so many authors involved, production problems were inevitable with some authors producing a chapter at the drop of a telephone and others requiring a long gestation period before a satisfying mix emerged. Our apologies go to the 'hares' for the long delay in publication, our thanks go to the tortoises for completing the course.

We think and hope that the resulting volume satisfies the demand by industry and academia, both undergraduate and postgraduate, for a text on mixing which gives a relatively comprehensive treatment of solid, liquid and multiphase mixing processes.

The editors

Contents

List of Contributors

Dr D. Boland, Imperial Chemical Industries P.L.C., Teesside

Professor J. R. Bourne, PhD, DSc, FIChemE, Swiss Federal Institute of Technology, Zurich, Switzerland

Dr R. Donaldson, Unilever Research Laboratory, Port Sunlight, Merseyside

Professor M. F. Edwards, BSc, PhD, FIChemE, CEng, Schools of Chemical Engineering, University of Bradford

Dr J. C. Godfrey, Postgraduate School of Chemical Engineering, University of Bradford

Dr N. Harnby, BSc(Chem Eng), ACGI, PhD, MIChemE, CEng, Schools of Chemical Engineering, University of Bradford

Dr R. King, BHRA Fluid Engineering, Cranfield, Bedford

Dr J. C. Middleton, BSc, PhD, MIChemE, CEng, New Science Group, Imperial Chemical Industries P.L.C., Runcorn

Professor A. W. Nienow, BSc(Eng), PhD, DSc, FIChemE, CEng, Department of Chemical Engineering, The University of Birmingham

Professor G. D. Parfitt, BSc, PhD, DSc, FRSC, formerly of the Department of Chemical Engineering, Carnegie-Mellon University, Pittsburgh, Pennsylvania, U.S.A.

B. K. Revill, Heavy Chemicals New Science Group, Imperial Chemical Industries P.L.C., Runcorn, Cheshire

Chapter 1

Introduction to mixing problems

A W Nienow

Department of Chemical Engineering, The University of Birmingham

M F Edwards and N Harnby

Schools of Chemical Engineering, University of Bradford

Mixing operations are encountered widely throughout productive industry in processes involving physical and chemical change. Although much of our knowledge on mixing has developed from the chemical industry, many other sectors carry out mixing operations on a large scale. Thus mixing is a central feature of many processes in the food, pharmaceutical, paper, plastics, and rubber industries. As a result, the financial investment in both the capital and running costs of mixing processes, when viewed on a national scale, is considerable.

It is therefore unfortunate that very few scientists and engineers receive a sufficiently thorough grounding in the fundamentals of mixing processes. Even degree-level courses in chemical engineering rarely leave the graduate in a position from which he can design with confidence equipment to satisfy industrial mixing duties[1]. What is more, he will even find difficulty in the selection of appropriate mixer types for a given application. To compound this problem, there are no widely accepted design codes associated with mixing, as in the case of shell and tube heat-exchangers for example.

A final problem is the performance analysis of existing installations. Generally there is very little instrumentation on mixing plant. Thus, although failure to mix adequately is obvious because of unsatisfactory product quality, there is often no means of detecting over-design. Thus, lack of knowledge of mixing processes can be hidden by overdesign and this may then go undetected if assessed by product quality. However, in terms of capital and operating costs, throughout one company's operations, the penalty for ignorance of mixing processes can be severe.

It is hoped that this book will help to provide a scientific underpinning of the mixing operations carried out in a wide range of industries. At present there are several problems which are still inadequately understood. However, the following chapters set out the present extent of our ability to design mixing systems.

1.1   Range of problems

In attempting to classify mixing operations it is helpful to consider the phases (solid, liquid or gas) involved in a particular process. Following this approach, it is possible to define operations which are common to several industries. For example, mixing duties from the food and pharmaceutical industries may be brought together in this way. Thus the general classification becomes independent of a particular product or a given industry.

1.1.1   Single-phase liquid mixing

In many operations it is necessary to mix together miscible liquids, e.g the blending of petroleum products. This is sometimes regarded as a simple mixing duty since it involves neither chemical reaction nor interphase mass transfer. It is necessary only to reduce the non-uniformities, i.e. variations in concentration to some acceptable level. However, such blending operations can be difficult to achieve when the liquids have widely different viscosities or densities. Also, problems can be encountered if one of the liquids to be mixed forms only a small volume fraction of the final mix. In Chapters 8 and 11 consideration is given to miscible liquid blending operations in mechanically agitated vessels, while blending using jet mixing devices is treated in Chapter 9. Some of the theory appropriate to the blending of high-viscosity materials is presented in Chapter 12.

When chemical reactions occur between miscible liquids it is necessary to bring together the reactants at the molecular level by mixing before the reaction can occur. The important interaction between mixing and reaction is developed in detail in Chapter 10.

1.1.2   Liquid–liquid (immiscible) mixing

When two immiscible liquids, often of low viscosity, are agitated a system is created having dispersed liquid droplets in a continuous liquid phase. Such a situation is often created in solvent extraction units where a high interfacial area between the two immiscible liquid phases is necessary to achieve interphase mass transfer. Thus agitation is used to create conditions favourable for mass transfer and if stirring is stopped the two phases will separate, leading to a greatly reduced interfacial area. In this instance the term 'mixing' is seen to include mass transfer considerations.

Another very common industrial mixing process involving immiscible liquids is emulsification. This is frequently encountered in the food and pharmaceutical industries when very small liquid droplets are created in a second liquid phase. In these cases the resulting mixture is often stable and will separate only after long periods of time. Furthermore, the stable emulsion will usually be relatively viscous and will often exhibit non-Newtonian rheological characteristics. The dynamics of emulsification is treated in Chapter 15.

1.1.3 Solid–liquid mixing

In operations such as crystallization or solid catalysed liquid reactions it is necessary to suspend solid particles in a relatively low viscosity liquid. This can be achieved in mechanically agitated vessels where the mixer is used to prevent sedimentation of the solids and to provide conditions suitable for good liquid–solid mass transfer and/or chemical reaction. If agitation is stopped the solids will settle out or float to the surface, depending upon the relative densities of the solid and liquid phases. The suspension of solids in mixing vessels and the design of mixing vessels for solid–liquid reactions are treated in Chapters 16 and 18 respectively.

At the opposite extreme it may be required to disperse very fine particles into a highly viscous liquid. For example, the incorporation of carbon black into rubber is such an operation. Here, as with emulsification in liquid–liquid mixing, the product is stable, highly viscous and may well exhibit complex rheology. Such processes often involve surface phenomena and physical contacting only, in contrast to the mass transfer and chemical reactions described in the previous paragraph. The dispersion of fine particles in liquids is considered in detail in Chapter 6.

1.1.4 Gas–liquid mixing

Several major industrial operations, e.g. oxidation, hydrogenation, and biological fermentations involve the contacting of gases and liquids. It is the objective of such processes to agitate the gas–liquid mixture thus generating a dispersion of gas bubbles in a continuous liquid phase. Mass transfer then takes place across the gas–liquid interface which is created. In some instances chemical reactions may also accompany the mass transfer in the liquid phase.

These gas dispersing duties are similar to the crystallization processes described in solid–liquid contacting and solvent extraction in liquid–liquid contacting. That is, the term 'mixing' covers a mass-transfer process. Furthermore, all the mixtures are unstable and separate if agitation is stopped. Gas–liquid contacting involving mass transfer and chemical reaction in mixing vessels is considered in Chapter 17.

In some instances gases and liquids are mixed to provide a stable batter or foam. This contacting is of a physical nature and the resulting product will often exhibit non-Newtonian flow characteristics.

1.1.5 Three-phase contacting

In some operations, e.g. hydrogenation, froth flotation, and evaporative crystallization, it is necessary to achieve contact between three phases. This is an area where little other than preliminary studies have been made[26] and more work is necessary to generate data which are of general use for design[27].

1.1.6   Solids mixing

In the previous categories of mixing processes the liquid phase is always present. However, many industrial operations involve solid mixing only in the absence of any liquid. Such processes are central to a very wide range of traditional as well as new process industries. Commonly the operation is batchwise but industries with a requirement to mix large throughputs of powders to give a constant composition product represent an expanding sector.

A feature which tends to be present only in solid mixing is segregation. This is the tendency of particles to separate out according to size and/or density. However, the 'unmixing' of powders due to segregation can be likened in some way to the tendency of solid–liquid, liquid–liquid and gas–liquid systems to separate once agitation is stopped. This analogy breaks down when it is remembered that segregation of solids can be caused by agitation.

Various aspects of powder mixing are treated in detail in Chapters 2–5.

1.1.7   Heat transfer

While many mixing devices are designed on the basis of blending, mass transfer, and chemical reaction, in some cases heat transfer is the controlling mechanism. Thus, for instance, the rotational speed of an impeller in a mixing vessel may be selected to satisfy a given thermal duty and the agitation conditions so generated may be more than sufficient for the mixing duty. No attempt is made to cover heat transfer in the text; however, the reader may find reviews of the literature to be of use[2,3].

1.1.8   Overmixing

There are a number of situations where excessive mixing in some form or other is not only wasteful of energy but also counter-productive. Thus, in the mixing of biological materials, excessive power inputs or impeller speeds may damage suspended micro-organisms especially if they have a morphology that is elongated with hyphae connected to them. Some polymer solutions whose desired rheological characteristics result from structured molecular forms may have them broken down and not reformed[29]. In crystallizers, high mixer speeds to enhance mass transfer are often counter-productive to the overall process because they cause a huge increase in secondary nucleation and therefore a smaller product. This point is covered in Chapter 18. In the mixing of solids, excessive mixing may mean mixing for too long in a batch mixer. Thus, while a good mix may be produced after a relatively short period of mixing, segregation may set in after this time and reduce mixture quality. It is important to appreciate therefore that more mixing, i.e. mixing more energetically or for more time, may do more harm than good.

1.2   Mixing mechanisms

As indicated in section 1.1, mixing is carried out usually to reduce inhomogeneities, especially in the mixing of particulate solids, or to enhance a rate
process, particularly in mechanically agitated vessels. An understanding of the
way that the solids and liquids move to achieve these requirements is an
essential prerequisite for a successful mixer design or selection. The different
forms of movement constitute the mechanisms of mixing.

1.2.1   Liquid mixing

In all liquid-mixing devices, it is necessary to have two elements. Firstly, there
must be overall bulk or convective flow so that no stagnant regions exist
within the device. Secondly, there must be an intensive or high-shear mixing
region which is capable of providing the reduction in inhomogeneities or rate
process enhancement required by the duty. Both these elements require
energy to sustain them. The proportion of energy going to each depends on the
particular application and, whatever its distribution, the mechanical energy is
finally dissipated as heat.

As in most fluid flow situations, the regimes may be characterized as laminar
or turbulent. In mixing vessels there is a substantial region between the two,
representing a transition from one to the other. It is convenient to describe
mixing mechanisms under either laminar or turbulent-flow conditions
because they are quite different in most respects.

1.2.1.1   Laminar mixing

Laminar flow is associated normally with high-viscosity liquids. At typical
rates of energy input, viscosities greater than about 10 Pas are required if the
flow is to be truly laminar. Fluids of such a high viscosity are often
rheologically complex, too. Under laminar flow conditions, inertial forces
quickly die out under the action of the high viscosity. Therefore, rotating
impellers must occupy a significant proportion of the vessel if adequate bulk
motion is to be achieved. Close to these rotating surfaces, large velocity
gradients exist. These are laminar regions of high shear rate which can cause
fluid elements to be deformed and stretched. As they do so, they continually
thin and elongate. Each time the elongated elements pass through such a high
shear rate region, they undergo a similar process but on a smaller scale. *Figure
1.1* shows these effects diagrammatically.

Associated with shear flows, in most practical situations, there are generally
extensional flows, i.e. flows where the velocity in a particular direction
increases in that direction. *Figure 1.2* shows a simple example of such a flow.
Again, the process leads to an increase in surface area and a reduction in
thickness of the element[4].

If there are solid agglomerates present within the high-viscosity fluid in
either of the above flow fields, they experience high shear stresses which can

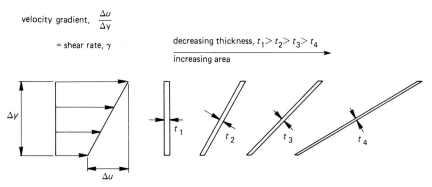

Figure 1.1   The thinning of fluid elements due to laminar shear flow

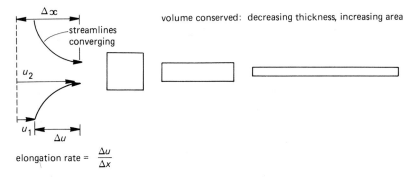

Figure 1.2   The thinning of fluid elements due to extensional flow

lead to agglomerate break-up and homogenization. Similarly, in the creation of emulsions, foams and batters, it is these stresses resulting from shear and elongation which bring about a reduction in droplet or bubble size.

Of course, molecular diffusion is occurring all the time. However, until the fluid elements are sufficiently small, their specific surface area is not great enough for the rate of diffusion to be significant. On the other hand, for ultimate homogenization of miscible fluids to occur, molecular diffusion is required. With highly viscous fluids, this is a slow process as molecular diffusivity is itself inherently slow in such cases.

In laminar flow, a similar mixing mechanism arises when laminar shear occurs in a concentric cylinder geometry; this is indicated diagrammatically in *Figure 1.3*. Thus each revolution leads to a further reduction in the thickness of the fluid element until once again molecular diffusion becomes significant.

Finally, there is mixing which arises from physically cutting the fluid into smaller elements and redistributing them; *Figure 1.4* indicates this process. Laminar mixing in in-line mixers, discussed in Chapter 13, is very largely a result of this mechanism. In addition, laminar shear and distributive mixing is dealt with in detail in Chapter 12.

In all the cases discussed above, the size or scale of the pure fluid elements

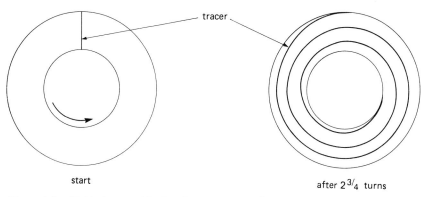

Figure 1.3   Fluid element thinning due to rotational shear

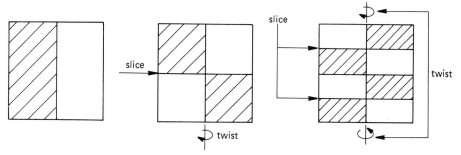

Figure 1.4   The concept of distributive mixing due to cutting and folding

decreases progressively as mixing progresses due to flow or re-distribution. At the same time, the difference in concentration between the different elements is reduced due to molecular diffusion, especially as the area available for diffusion is increased with the decreasing size of the element.

1.2.1.2   Turbulent mixing

For all practical purposes, the bulk fluid flow in mixing vessels containing rotating impellers is turbulent if the fluid viscosity is less than about 10 mPas. Here, the inertia imparted to the fluid by the rotating impeller is sufficient to cause it to circulate readily throughout the vessel and back to the impeller again. During the fluid's passage, turbulent eddy diffusion takes place, though it is at its maximum in the impeller region. Eddy diffusion leads to mixing which is much more rapid than the rate of mixing associated with the mechanisms of laminar flow. Once again, for homogenization to occur at the molecular scale, molecular diffusion must occur. However, in these low-viscosity fluids, molecular diffusion is much faster than in high-viscosity materials. So the overall mixing process right down to the molecular level in turbulent flow is much more rapid than it is in laminar flow.

The rate of mixing in turbulent flow is greatest close to the impeller. Here, there is a high shear rate due to the trailing vortices associated with disc turbine impellers[5] and in addition there are large Reynolds stresses[6] in the radial discharge stream. Furthermore, a high proportion of the energy introduced by the impeller is dissipated here[7]. Thus the rate of homogenization of miscible liquids is greatest in this region and gas and liquid–liquid dispersion occurs predominantly here.

The complexity of the turbulent flow in mechanically agitated vessels, particularly because of the three-dimensional flow field, makes an analysis extremely difficult. However, provided the Reynolds number of the main flow is high enough, Kolmogoroff's theory of local isotropic turbulence can be used to give some insight into its structure[8]. Turbulent motion can be considered as a superposition of a spectrum of velocity fluctuations and eddy sizes on an overall mean flow. The large primary eddies have large velocity fluctuations of low frequency and are of a size comparable with the physical dimension of the impeller, diameter D. They are anisotropic and contain the bulk of the kinetic energy. Interaction of the large eddies with slow-moving streams produces smaller eddies of high frequency which further disintegrate until finally they are dissipated into heat by viscous forces. There is a transfer of kinetic energy down the scale from larger eddies to smaller eddies, the directional elements of the main flow being progressively lost in the process.

Kolmogoroff argued that, for large Reynolds numbers, the smaller eddies are independent of the bulk motion and are isotropic. The properties of these eddies are firstly a function of the local energy dissipation rate/unit mass, ε_T. Below the eddy size, λ_k, at which viscous dissipation occurs, their properties also depend on viscosity. There is therefore an equilibrium established which contains a very wide range of eddy sizes, the universal equilibrium range. With the larger of these eddies ($D \gg \lambda_T \gg \lambda_k$), energy passes from bigger to smaller without dissipation. This is the inertial sub-range. The sizes below λ_k over which viscous dissipation occurs is the viscous sub-range. *Figure 1.5* shows this structure. An eddy Reynolds number, Re_k, is now defined for the Kolmogoroff length scale as

$$Re_k = \lambda_k V_k / \nu \qquad (1.1)$$

Conceptually the Reynolds number represents the balance of inertial to viscous forces; and within the spectrum of sizes the Kolmogoroff scale is the size where these two forces are in balance. Therefore, it is convenient to quantify eddy Reynolds numbers in general in relation to Re_k by the definition

$$Re_k = 1 \qquad (1.2)$$

From dimensional reasoning, the Kolmogoroff length scale is defined as

$$\lambda_k = (\nu^3 / \varepsilon_T)^{1/4} \qquad (1.3)$$

and the velocity scale v_k associated with it is

$$v_k = (\nu \varepsilon_T)^{1/4} \qquad (1.4)$$

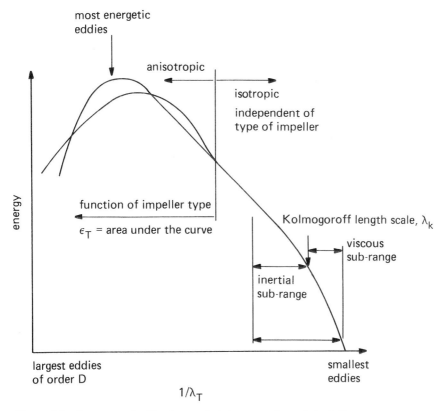

Figure 1.5   Spectrum of eddy sizes and their energy in turbulent flow

For an energy dissipation rate of 1 W/kg in water, equation (1.3) gives $\lambda_k \approx$ 30 μm.

By implication, therefore, processes which are particularly dependent on turbulent eddies and their associated forces are likely to be well correlated by energy dissipation rate. Bubble formation (Chapter 17) and micromixing phenomena (Chapter 10), for instance, fall into this category. However, processes which are dependent on the anisotropic main flows and for which the non-homogeneous nature of stirred tanks turbulence is significant, e.g. solid suspension and solid–liquid mass transfer (Chapters 16 and 18), are not well correlated that way.

Measurements have been made of turbulence structure by a number of workers, e.g. Molen *et al.*[9] using laser–Doppler methods and Rao and Brodkey[10] using hot-film anemometry. However, little progress has been made to date using this data for design purposes.

1.2.2   Solids mixing

The mixing of a particulate system differs from that of liquid systems in three important respects:

1. There is no particulate motion equivalent to the molecular diffusion of gases and liquids. The rate at which the randomization of the constituent particles occurs is entirely dependent on the flow characteristics or handling pattern externally imposed on the particles. There is no relative movement of the particles without an energy input to the mixture.

2. Although the molecules of a single-phase liquid system may differ, and may diffuse at different rates, they will ultimately achieve a random distribution within the confines of the system. Particulate and granular components do not usually have the constant properties of molecular species and can differ widely in physical characteristics. Thus a mixing motion which depends on identical particulate properties is unlikely to achieve its objective. More commonly such a 'mixer' would produce a grading or segregation of particles according to such characteristics as size, density, resilience, etc. Common household examples of the segregating effect of some 'mixing' actions are the grading of particles in a pepper pot or in a jar of coffee granules.

3. The ultimate element of the particulate mixture is several degrees of magnitude larger than the ultimate molecular element of the liquid mixture. In practical terms this means that samples withdrawn from a randomized particulate mixture will have a coarser texture or poorer mixture quality than would the equivalent samples taken from a gaseous or liquid mixture.

The industrial implications of these differences should be considered very carefully. Particles will change their relative positions only when subjected to movement. Once movement begins, the particles may randomize or segregate depending on both the type of movement imposed on the system and on the physical characteristics of the constituents. Thus, in marked contrast to the mixing of miscible liquids, the mixing of particles is often a readily reversible process and also a process in which the point of equilibrium can vary along a production line. A mixture of miscible liquids leaving a mixing unit will retain or even improve its 'mixedness' during the transport process to the next operation, while a well-mixed batch of particles can be separated almost completely at a subsequent process stage if incorrectly handled. If segregation or 'unmixing' is to be avoided the process designer has to be aware of the possibility of segregation occuring throughout the entire handling process and not only within the confines of a mixing unit.

The topics of powder handling and the mixing and segregation of powders are evidently very closely related. Powder handling is concerned with the delivery of powder at a defined rate and composition. If a process material has a strong tendency to segregate then the control of the delivery rate will be more difficult and the control of the composition impossible. A study of the mechanisms by which particulate mixing and segregation occur within a mixing unit will also indicate the best methods of handling such particles in the process as a whole.

A major influence on the mechanisms of mixing and segregation within a powder are the flow characteristics of that powder. A different mixing and handling approach is required for 'free-flowing' and for cohesive powders. If a free-flowing powder is poured from a vessel it will flow consistently, while the cohesive powder will flow intermittently, if at all. From a practical point of view a free-flowing powder will flow consistently but at the same time the particles have a high individual mobility and are presented with excellent opportunities for interparticulate segregation. A cohesive powder will present flow and storage problems but will not exhibit the gross segregation often encountered with free-flowing powders. As the size of the particle decreases the gravitational separation force between particles becomes less dominant and interparticulate bonding forces such as the van der Waals, electrostatic or moisture bonding forces can lead to the formation of agglomerates within the mixture. It is possible in some circumstances to take advantage of the preferential bonding of particles to produce a very-high-quality mixture.

Particle size is an important variable in determining the flow characteristics of a mixture and experience suggests that:

(a) materials with a size greater than 75 μm will segregate readily;
(b) the reduction of size below 75 μm will reduce segregation but it may still be detectable down to about 10 μm;
(c) below 10 μm no appreciable segregation will occur.

As a general rule, a process which suffers from serious flow problems will not suffer from segregation, and a process which has problems of segregation will not suffer from flow problems. In the transitional zone of powder flow it is, alas, possible to have both problems.

Mixture properties other than particle size can influence the flow characteristics of the mixture. The addition of quite small quantities of moisture can transform a strongly segregating mixture into a cohesive and non-segregating mixture. A mixture containing quite coarse particles can be surprisingly free of segregation if the major component has a rough or fibrous shape.

If one component in an otherwise coarse mixture is very fine, the fine particles can 'coat' the coarser particles. In this situation the very fine particles lose their freedom of movement and a high-quality, non-segregating mixture can result.

Perhaps the most obvious method of avoiding segregation is to ensure that the components of the mixture have the same particle size. Unfortunately, quite small differences in particle size can result in quite considerable segregation within a free-flowing mixture and outside the laboratory it is not usually possible to obtain an exact matching of particles.

In many industrial situations a free-flowing powder is a process requirement and in this situation segregation has to be minimized rather than eliminated. This can be done by the careful choice of the properties of the mixture and of the method of mixing.

1.2.2.1   Segregation

Most of the early work on solids mixing was carried out with two component systems whose particles were identical in all important properties, differing perhaps only in colour. Such a system lends itself to a relatively simple statistical analysis and if the mixing process goes on long enough random mixing will be achieved, and different types of mixer will vary only in the speed at which randomness is approached. The results of such idealized tests still affect attitudes to the selection of mixing equipment.

In industrial practice there is almost always some difference in the physical properties of the components to be mixed and this may lead to a tendency for the components to unmix or segregate. When this occurs the final state reached in the mixer is an equilibrium between mixing and segregation. For a given duty a mixer should be selected which gives the best mixing at equilibrium. It is essential to begin a study of solids mixing with a discussion of segregation. In fact, anyone concerned with the handling or processing of particulate solids should have an understanding of the causes of segregation. It has consequences in the collecting of samples from a powder mass for size or chemical analysis, in the feeding of materials to either tabletting machines, packaging machines or reactors and in many other powder-handling operations.

Segregation occurs within a mixer when differences in particulate properties cause a preferential movement of particles to certain regions of the mixer. Difference in most particulate properties can in certain circumstances cause a non-random movement of particles but those properties which chiefly cause segregation, in order of importance, are:

> difference in particle size
> difference in density
> difference in shape
> difference in resilience

Of these effects, size difference is by far the most important and almost always this is the most serious cause of segregation. The other effects are generally minor and unimportant.

If a stream containing particles of different size falls vertically the particles will tend to travel at different speeds and with different trajectories. In a continuous stream this will not cause segregation, as at any section of the stream there will be instantaneously representative particles of all the sizes contained in the stream flow. However, if the particles are projected in such a way that they have a horizontal component then appreciable segregation may occur as the horizontal component of velocity falls off very much more rapidly for a small particle than for a large particle. If a particle of diameter d_p and density ρ_s is projected horizontally with an initial velocity v_0, then the horizontal distance travelled, or stopping distance, is equal to $v_0 \rho_s d_p^2 / 18\mu$, where μ is the viscosity of the fluid.

When a mass of particles is disturbed in such a way that individual particles

have a freedom of movement, then a rearrangement in the packing characteristics of the particles can take place. The mobility of individual particles within
the mixture will be a function of their ability to fall into, or pass through, gaps
appearing amongst adjacent particles. Evidently mobility will be a function of
size, as it will be easier for a small particle to change position than for a large
particle. For this preferential movement to occur it is not necessary for the fine
particles to be so small that they can pass through the voids between the static
larger particles. A very small difference in size is enough for measurable
segregation to take place.

There are a number of industrial handling situations in which segregation
by percolation is promoted:

(a) *Segregation in pouring a heap.* When a mixture of particles is poured into
a heap the larger particles tend to run down to the edge of the heap. The scale
of segregation caused by such a pouring process is not as well known but is
certainly evident from the sectional view of a heap illustrated in *Figure 1.6*.
When material is poured on to the top of a heap a mobile layer of particles,
which is several particle diameters thick, is formed over the surface of the heap.
This mobile layer forms an effective screen through which all but the largest
particles are able to pass to the stationary region below. The ease of passage is

Figure 1.6   A mixture of multi-sized particles poured into a sectioned heap

Figure 1.7   Banded segregation resulting from the rotation of powder in a cylinder

a function of particle size and this leads to a very thorough size segregation within the heap.

(b) *Shear segregation.*   When a powder mass flows in such a way that a velocity gradient is set up, any one layer of particles will have a velocity relative to a neighbouring layer and, if the opportunity is presented, particles can drop or roll into a new particle layer. Smaller particles again have a greater mobility and segregation can result. An interesting analysis of the velocity gradients existing in a rotating horizontal cylinder was made by Donald and Roseman[11,12] and 'banded' segregation resulting from such rotation is illustrated in *Figure 1.7*.

(c) *Segregation due to vibration.*   The vibration of a powder mass creates voids into which fine particles can drop more readily than coarse particles. In addition to this percolation effect there is another mechanism of segregation which occurs when a mass of differently sized particles is vibrated. When one large particle is placed at the bottom of a bed of smaller particles and is vibrated it will rise to the surface. This can readily be demonstrated. The explanation is that a large particle causes consolidation among the particles immediately below it which prevents the large particle from sinking. At the same time any upward movement of the coarse particle enables fine particles to percolate beneath it and these in turn are consolidated as the coarse particle redescends. If the vibration is of a suitable type the large particles will rise to the surface.

1.2.2.2 Mixing

Mixing due to diffusion occurs when particles roll over a sloping surface of the powder. The random movement of particles on the free surface results in a redistribution of the particles. As was discussed in section 1.2.2.1, the movement of particles on a free surface can also give rise to segregation due to the percolation of fine particles if particles of different size are present in the mixture. Only when the particles within the mixture have identical properties can diffusional mixing be a truly randomizing process.

Mixing due to shear forces occurs when slip planes are established within the powder. Mixing takes place because of the interchange of particles between adjacent layers. As with diffusional mixing, this mechanism is not exclusive of segregation, as a size or density difference among the sheared particles can lead to a preferential particulate movement. Only for particles with identical properties will the mechanism be exclusively that of mixing.

Mixing due to convection involves the movement of comparatively large masses of particles from one location to another within the mixture. Particles do not have the same individual freedom of movement associated with shear and diffusional mixing and the resultant mixture is generally less prone to segregation.

1.2.2.3 The interaction of mixing and segregation

Mixing and segregating mechanisms cannot usually be separated within a powder-handling system. The final mixture quality will be determined by the relative importance of the two mechanisms but ideally segregation should be eliminated or at worst suppressed in favour of the mixing operation.

From the study of the mechanisms of segregation it is evident that segregation can only occur when particles possess an individual freedom of movement. Only when the particles retain independence can variations in individual particle characteristics influence particle movement and produce segregation. A 'free-flowing' mixture will generally permit individual particulate freedom while a 'cohesive' mixture generally has some inter-particulate bonding mechanism which permits particles to move only with an associated cluster of particles. There is, unfortunately, no hard boundary between a free-flowing and a cohesive powder, so there is no simple division into segregating and non-segregating mixtures.

1.3 Assessment of mixture quality

From the previous classification of mixing problems in section 1.1 it will be clear to the reader that the term 'mixing' covers much more than processes in which non-uniformities in composition are reduced. This latter type of operation, i.e. the blending of miscible liquids or powder mixing, are both duties for which a method of evaluation of mixture quality is necessary to assess the process result. Other 'mixing' processes, however, may include chemical reaction and/or mass transfer and the effectiveness of the operation is judged by factors such as reaction yield, extent of mass transfer, product rheology, etc.

In areas such as miscible liquid blending, the formation of emulsions, solid dispersions such as paints, and dry powder mixing, it is understandable, therefore, that several criteria have been developed to assess mixture quality[13,14,15]. It is unfortunate that of the many definitions presently available for mixture quality, in solid or liquid mixtures, none is universally applicable. In the case of powder mixing, further details may be found in Chapter 2, while for liquid mixing more information is presented in Chapters 8 and 10 for mechanically agitated vessels, Chapter 9 for jet mixers and Chapter 13 for in-line static mixers.

At this juncture it is worthwhile discussing two widely used measures of mixedness and to consider the problems in the experimental determination of these quantities. This can be done by considering the laminar shearing and distributive mixing processes illustrated in *Figures 1.1–1.4* where a coloured tracer element is seen to deform and/or change its configuration. However, although the tracer undergoes these geometrical changes which are interpreted as mixing, there is in the ideal situation represented in *Figures 1.1–1.4* no exchange between the tracer and bulk fluid and there are therefore no changes in concentration. Thus, ideally any point concentration will indicate pure tracer or pure bulk liquid. In such circumstances the progress of mixing can be assessed by the 'scale of segregation' which effectively measures the size of the unmixed regions. Thus as laminar shear or distributive mixing proceeds the scale of segregation is progressively reduced.

In any real liquid mixing process, the ideal behaviour illustrated in *Figures 1.1–1.4* will be accompanied by molecular diffusion and, as discussed earlier, this latter phenomenon will become more and more important as mixing approaches completion. This diffusional process causes changes in concentration to occur so that perfect segregation between the tracer and bulk fluid no longer exists. To allow for this diffusional effect an 'intensity of segregation' has been defined[25] which gives a measure of the concentration variations in the mixtures. Thus, as diffusional mixing proceeds the intensity of segregation is reduced.

In any attempt to measure these mixture parameters, i.e. scale and intensity of segregation, it is important to bear in mind that any probe to detect regions of varying composition in a mixture has a finite sampling volume, and the detector will be incapable of indicating any changes on a scale which is finer than this sample volume. Once the scale of segregation is reduced by the mixing process under consideration to a value which is of the same order as the sample dimension of the detector, it will not be possible to measure the scale of segregation. Any further improvements in mixture quality will not be detected accurately.

The above limitation of the detection volume is a major drawback in attempts to measure mixture quality, particularly towards the end of a mixing process when the scale of segregation can become very small. Thus, while the scale and intensity of segregation and the many other measures of mixture quality are simple to define and aid understanding of the mixing mechanisms, the experimental determination of these parameters is a major problem in mixing studies.

1.4   Rheology

Rheology is the study of material flow behaviour. The fluids referred to in the range of mixing problems discussed in section 1.1 cover an enormous variation of rheological properties. The possibilities are extended further when two- and three-phase mixtures, especially stable ones which are products of mixing operations such as emulsions, foams and dispersions are included. There are many books available which deal with this topic in some detail[22,23,24]. All that is presented here is sufficient to allow the reader to understand better those parts of later chapters which refer to rheologically complex fluids.

For fluids undergoing laminar shear as illustrated in *Figure 1.1*, the resistance to deformation depends on the dynamic viscosity. For a linear velocity gradient arising due to the movement of one parallel plate over another, the relationship between shear rate $\dot{\gamma}$ and shear stress τ, is given by

$$\tau = \mu \, du/dy = \mu \dot{\gamma} \tag{1.5}$$

where, for Newtonian fluids, the viscosity μ is constant, independent of shear rate $\dot{\gamma}$. However, for many of the fluids mentioned when discussing the range of mixing problems in section 1.1, the viscosity is a function of shear rate. This is generally the case when a fluid contains large complex molecules in solution, such as polymers or biological materials; or if stable two- or three-phase dispersions such as suspensions, emulsions and foams are being mixed.

To determine the relationship between shear stress and shear rate a range of viscometers is available. Two of them try to simulate the ideal flow represented by parallel plates; one has a cone and plate configuration and the other concentric cylinders. *Figure 1.8* illustrates these in diagrammatic form. Another simple configuration is the capillary viscometer. Full details and the advantages and disadvantages of the various types are given in the books referred to above.

The range of shear rates which can be covered is extremely large. This is vitally important because matching the shear rate in the mixer, of which there is a wide range as indicated in Chapter 7, is necessary if meaningful viscometric data are to be obtained. Viscometry is also sometimes used in quality control of products from mixing operations, especially of material such as emulsions, pigments, etc. In this case, the shear-rate range should match the use to which the product will be put or that it may have to experience during its shelf life.

Data from viscometry experiments are given normally in the form of a flow curve, a plot of τ against $\dot{\gamma}$. Apart from Newtonian fluids, three other major

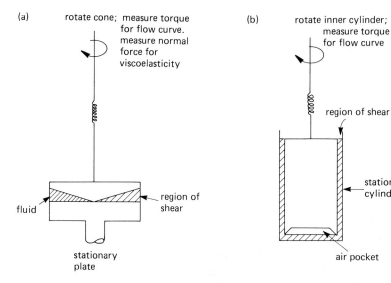

Figure 1.8   Two rotational rheometers

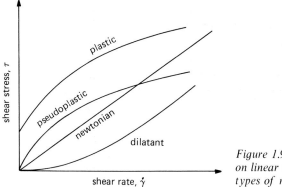

Figure 1.9   Examples of flow curves on linear co-ordinates of different types of rheological behaviour

categories can be defined, as indicated in *Figure 1.9*. These categories and their implications for mixing operations may be classified as follows:

(*a*) *Pseudoplastic or shear thinning.* This property is exhibited by an extremely wide ranges of material in the food, biological and polymer industries. An apparent viscosity μ_a is defined as

$$\mu_a = \tau/\dot{\gamma} \tag{1.6}$$

and, with this type of fluid, viscosity decreases with increasing shear rate. Thus, with high-speed rotating mixers, the viscosity tends to be low close to the impeller even though it is high elsewhere. In this case, good mixing may occur in the impeller region with near-stagnant regions in the remainder of the vessel.

(*b*) *Plastic.* Such materials are characterized by a yield stress. This is found particularly with stable dispersions where a minimum stress is required to break down the structure sufficiently before any movement will occur. In this case, the well-mixed region close to the impeller which has been called a cavern, may be accompanied by a totally stagnant fluid elsewhere. This extreme mixing problem is exhibited in *Figure 1.10*[16].

Figure 1.10   Flow visualization of a cavern, found with shear thinning plastic fluid

(*c*) *Dilatant or shear thickening.* This phenomenon is not very common but does occur with some concentrated suspensions, e.g. china clay. In this case, it is considered that the flow causes particle–particle alignment to develop and, from this, inter-particle bonds due to the presence of surface charges. Very little work has been done on the mixing of such fluids though the region close to the impeller must increase in apparent viscosity as compared with the remainder of the vessel.

Using the concept of apparent viscosity as indicated in Chapter 11, there is often no need to express flow curves (i.e. experimental data relating shear stress to shear rate) in algebraic form or to develop mathematical models. Nevertheless, it is often convenient to do so and the power law model is the most commonly used:

$$\tau = K\dot{\gamma}^{n}$$

(1.7)

where K is the consistency index and n the flow behaviour index. Almost all flow curves can be fitted quite well by such an equation over a limited range of shear rates but it is dangerous to extrapolate outside the range of measurements. If $n<1$, the fluid is pseudoplastic; if $n=1$, it is Newtonian and $K=\mu$; if $n>1$, the fluid is dilatant.

For plastic fluids which have a yield stress, one simple equation which is commonly used is the Herschel–Bulkely equation:

$$\tau = \tau_y + K_{HB}\dot{\gamma}_{HB}^n \tag{1.8}$$

where τ_y is the yield stress. Another is the Casson equation which has been found to be a good fit to the data from a wide range of fermentation broths[29]:

$$\tau^{0.5} = \tau_y^{0.5} + K_c\dot{\gamma}^{0.5} \tag{1.9}$$

It is important to realize that the yield stress, τ_y, is extremely difficult to determine unequivocally. The flow curve of a highly shear thinning fluid on linear co-ordinates very often indicates a finite yield stress. However, this may be due to the bunching of the data at low shear rates. Extending the low shear rate range graphically on log–log paper often suggests that the power law alone is sufficient. Suffice to say that, while yield stresses exist, they are irrevocably linked with the philosophical question: How low a controlled value of shear rate can be applied and τ still be measured? or, if a low value of shear stress is applied: How low a value of shear rate $\dot{\gamma}$ can be detected?

All the above categories of non-Newtonian behaviour are classified as time-independent, i.e. the shear stress is a unique function of the shear rate and does not depend on the time of shearing. Thus the material responds instantaneously to changes in shear rate. However, some pastes, foods, paints, etc., exhibit marked changes in rheology as the time of shearing increases. Such materials are time-dependent and two types of behaviour are possible:

1. Thixotropic liquids, e.g. paints, tomato ketchup, salad cream. Here the apparent viscosity reduces with time as the material is sheared at a constant shear rate (*Figure 1.11*). Usually after a large time of shearing, an equilibrium apparent viscosity is reached. If the shearing is now stopped the apparent viscosity gradually increases. Some implications of thixotropic behaviour for mixing have been discussed by Edwards *et al.*[17,18].

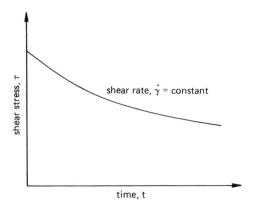

shear rate, $\dot{\gamma}$ = constant

shear stress, τ

time, t

Figure 1.11   The change of shear stress with time for a thixotropic fluid

2. Rheopectic liquids. These are rare. Here the apparent viscosity gradually builds up with time as the liquid is sheared.

In addition to the above, some materials behave in part as a viscous liquid and in part as an elastic solid, i.e. they can exhibit elastic recovery. These liquids are known as viscoelastic fluids. Although rather little has been done to establish the effect of viscoelasticity on mixing phenomena in a quantitative way, some effects are clear.

Firstly, when sheared in a cone and plate device, large normal forces as well as large shear forces may be developed in viscoelastic liquids (see *Figure 1.8(a)*). When such fluids are agitated by rotating stirrers, the fluid will climb up the shaft due to the Weissenberg effect[19]. This phenomenon can be so great that fluid may reach the gearbox or motor. In the laminar flow region, viscoelasticity enhances the power drawn by an agitator[20] and it can also lead to flow reversal, i.e. flow into, rather than away from, the turbine[16]. In dilute polymer solutions, on the other hand, the viscoelasticity associated with polymers may lead to drag reduction, i.e. the reduction in the power drawn by the rotating agitator[21].

When fluids are subjected to extensional flows of the type indicated in *Figure 1.2*, the resistance to deformation depends on their elongational viscosity. For Newtonian fluids, the elongational viscosity is three times the shear viscosity. This is known as Trouton's rule. However, for non-Newtonian fluids, especially highly viscoelastic ones, the ratio can be two or three orders of magnitude greater than this. Little has been published yet on the role of elongational viscosity in real mixers and its measurement is very difficult. There is need for considerably more work on this topic and the reader is referred to the book by Walters[22] and the review article by Cheng[4].

Notation

D	impeller diameter (m)
d_p	particle size (m)
K	consistency index (Pa s^n)
n	flow behaviour index, dimensionless
Re_k	Reynolds number of an eddy of size λ_k, dimensionless
u	velocity in laminar flow (m s^{-1})
v_0	initial velocity of a particle (m s^{-1})
v_k	velocity scale of size λ_k (m s^{-1})
x	distance parallel to the flow direction (m)
y	distance normal to the flow direction (m)
$\dot{\gamma}$	shear rate (s^{-1})
ε_T	energy dissipation rate (W kg^{-1})
λ_k	Kolmogoroff microscale of turbulence (m)
λ_T	scale of a turbulent eddy (m)
μ	dynamic viscosity, Pa s
μ_a	apparent viscosity, Pa s

v kinematic viscosity (m^2 s^{-1})

ρ_s density of a solid particle (kg m^{-3})

τ shear stress (Pa)

τ_y yield stress (Pa)

Subscripts

HB   Herschel–Bulkley

c   Casson

References

1   NIENOW, A. W. (1982) Inaugural Lecture, 'Stirred, Not Shaken; Mixing Studies', University of Birmingham, 20 pp.

2   EDWARDS, M. F. and WILKINSON, W. L. (1972) *Chem. Engr.*, no. 264, 310.

3   EDWARDS, M. F. and WILKINSON, W. L. (1972) *Chem. Engr.*, no. 265, 328.

4   CHENG, D. C-H. (1979) *Proc. 3rd Eur. Conf. Mixing*, BHRA, Cranfield, p. 73.

5   RIET, K. van't and SMITH, J. M. (1973) *Chem. Engng. Sci.*, **28**, 1031.

6   RUSHTON, J. H. and OLDSHUE, J. Y. (1953) *Chem. Engng. Prog.*, **49**, 161 and 267.

7   CUTTER, L. A. (1966) *A.I.Ch.E.J.*, **12**, 35.

8   LEVICH, V. (1962) *Physico-Chemical Hydrodynamics*, Prentice-Hall, New Jersey.

9   MOLEN, K. van der and MAANEN, H. R. E. van (1978) *Chem. Engng. Sci.*, **33**, 1161.

10   RAO, M. A. and BRODKEY, R. S. (1972) *Chem. Engng. Sci.*, **27**, 137.

11   DONALD, M. B. and ROSEMAN, B. (1962) *Brit. Chem. Engng.*, **7**, 749.

12   DONALD, M. B. and ROSEMAN, B. (1962) *Brit. Chem. Engng.*, **7**, 823.

13   SCHOFIELD, C. (1974) *Proc. 1st Eur. Conf. Mixing*, BHRA, Cranfield, p. C1–1.

14   STRIEFF, F. A. (1979) *Proc. 3rd Eur. Conf. Mixing*, BHRA, Cranfield, paper C2, 171.

15   UHL, V. W. and GRAY, J. B. (1966) *Mixing: Theory and Practice*, Academic Press, New York.

16   SOLOMON, J., ELSON, T. P., NIENOW, A. W. and PACE, G. W. (1981) *Chem. Eng. Comm.*, **11**, 143.

17   GODFREY, J. C., YUEN, T. H. and EDWARDS, M. F. (1974) *Proc. 1st Eur. Conf. Mixing*, BHRA, Cranfield, p. C3–33.

18   EDWARDS, M. F., GODFREY, J. C. and KASHANI, M. M. (1976) *J. Non-Newt. Fluid Mech.*, **1**, 309.

19   ELSON, T. P., SOLOMON, J., NIENOW, A. W. and PACE, G. W. (1982) *J. Non-Newt. Fluid Mech.*, **11**, 1.

20   NIENOW, A. W., WISDOM, D. J., SOLOMON, J., MACHON, V. and VLCEK, J. (1982) *Chem. Eng. Comm.*

21   RANADE, V. R. and ULBRECHT, J. J. (1978) *Proc. 2nd Eur. Conf. Mixing*, BHRA, Cranfield, p. F6–83.

22   WALTERS, K. (1975) *Rheometry*, Chapman & Hall, London.

23   WILKINSON, W. L. (1960) *Non-Newtonian Fluids, Fluid Mechanics, Mixing and Heat Transfer*, Pergamon Press, London.

24   SKELLAND, A. H. P. (1967) *Non-Newtonian Flow and Heat Transfer*, Wiley, New York.

25   DANCKWERTS, P. V. (1952) *App. Sci. Res.* (Hague), **A3**, 279.

26   CHAPMAN, C. M., NIENOW, A. W. and MIDDLETON, J. C. (1983) *Trans. I. Chem. E.*, **61**, 71, 82, 167, 182.

27   CHAPMAN, C. M. (1981) Ph.D. Thesis, University of London.

28   BAILEY, J. E. and OLLIS, D. F. (1977) *Biochemical Engineering Fundamentals*, McGraw-Hill, New York.

29   METZ, B., KOSSON, N. W. and SUIJDAM, J. C. (1979) *Adv. Biochem. Eng.*, **11**, 104.

Chapter 2

Characterization of powder mixtures

N Harnby

Schools of Chemical Engineering, University of Bradford

2.1 A qualitative approach

When is a mixture well mixed? A traditional, but insensitive, reply to this question is that a mixture is well mixed when it is good enough for its duty. A pigment is well mixed with a bulk product when the eye can no longer detect colour variations within the bulk material. This does not mean that the mixture could not be improved but only that it passes an arbitrary test of quality. The use of a magnifying glass to inspect the mixture could well mean that the previously satisfactory mixture becomes unsatisfactory.

It is a 'go/no go' test which gives no information on the relative quality of two 'satisfactory' mixtures. This is a grave disadvantage from the point of view of process control. Based on this test alone the process manager would have no forewarning if his process was slowly moving to an 'unsatisfactory' condition and a comparatively minor change in process conditions could produce an embarrassing amount of off-specification product.

The characterization can be improved by providing a better qualitative description of the state of the mixture. Danckwerts[1] defined several useful descriptive terms. The 'scale of segregation' of a mixture is a measure of the size of regions of segregation within the mixture. The smaller the scale of segregation the better the mixture.

Consider again the problem of dispersing the pigment in the bulk product. The scale of segregation in that case is an area on the inspected surface which does not have the mean composition of the bulk of the mixture. The divergence from the mean composition could vary from all of the particles in the area of segregation being of one type to a very small composition divergence. The 'intensity of segregation' is a measure of this divergence.

Alternatively, the intensity of segregation can be regarded as the amount of dilution that has occurred within the segregated areas. The lower the intensity of segregation the better the mixture. Both the scale and the intensity of segregation will affect the mixture quality. High-quality mixtures will have a small scale of segregation and a low intensity of segregation, whilst poor mixtures will have a large scale of segregation and a high intensity of segregation. Evidently, the function of a mixer is to reduce both the scale and intensity of segregation within the mixture.

When a specific mixture is considered, the assessment of the quality of that mixture depends on how closely it is examined. In the case of the pigment dispersion a microscopic examination of the mixture surface could well reveal a lack of colour homogeneity which was not visible to the naked eye. In the limit, if the mixture was examined on the scale of the individual particle then no mixing at all is possible and the mixture would always be unsatisfactory. There is evidently a critical scale of examination of a mixture at which it becomes unsatisfactory.

Danckwerts defined a 'scale of scrutiny' for a mixture as 'that maximum size of the regions of segregation in the mixture which would cause it to be regarded as imperfectly mixed'. The scale of scrutiny provides a vital link between product specification and the state of mixedness of the mixture. For a particular product it fixes the scale or sample size at which the mixture should be examined. If the eye is able to distinguish colour variations in areas greater than, say, 1 mm^2, then this becomes the scale of scrutiny for the pigment dispersion. (Great care has to be taken in fixing the scale of scrutiny for a product.) If nutrient is fed to animals in the form of cake mixture and the assumption is made that the daily nutrient intake should be nearly constant then the 'scale of scrutiny' of the nutrient mixture would be that of the animal's daily consumption.

In the case of the animal nutrient the scale of scrutiny would depend on the consumption of the animal being fed. A suitable feed for a sheep would require a smaller scale of scrutiny than that for a cow. If it was decided that a weekly, rather than daily, intake of nutrient was critical then the scale of scrutiny would increase. It could well be that in this case the scale of scrutiny would be determined not so much by the animals' requirements but by the marketing requirement of having a product of uniform appearance. An almost identical analysis of scale of scrutiny could be carried out on the mixture requirements for a pharmaceutical tablet.

An identification of the scale of scrutiny for a product fixes the size of the sample to be taken from the mixture and assessed for mixture quality. It determines how closely the mixture will be examined. The relationship between mixture quality and scale of scrutiny is difficult to predict and as a result a theoretical scaling up or down of mixture quality with a scale of scrutiny should be avoided. It is much safer to re-sample the mixture at the new scale of scrutiny.

2.2 A quantitative approach

Once a scale of scrutiny and sample size have been fixed a mixture can be sampled and the individual sample compositions compared. These values can be plotted on a histogram or distribution curve.

The mean composition value, $\bar{y}$, usually has an economic significance in that it is used to control the overall content of the key or expensive component.

The standard deviation of the distribution, s, is a measure of the quality of

the mixture. A high-quality mixture will show very little variation in composition between samples and will have a low standard deviation. If 'n' samples are withdrawn from the batch, then

$$\bar{y}=\sum_1^n y_i/n \quad \text{and} \quad S=\sqrt{\frac{\sum_1^n (y_i-\bar{y})^2}{n-1}}$$

The square of the standard deviation is the variance. This is widely used to typify particulate mixture quality because of its additive properties.

It is beyond the scope of this chapter to discuss in detail the problem of obtaining representative samples from a mixture though it must be emphasized that the validity of the conclusions on mixture quality is entirely dependent on the unbiased nature of the sampling process. Generally, the sampling problems increase as the quality of the mixture decreases. Biased sample selection is avoided if potential samples are selected by the use of random numbers. Biased sample retrieval from the mixture is generally reduced if the sample is taken from a moving stream of the mixture rather than a static mass of particles. On this basis it is usually better to sample the outflow from a powder mixer rather than the '*in situ*' mixture[2].

The experimental variance obtained from these samples is of little value unless it can be related to limiting variance values. If the assumptions are made that the mixture is a two-component system, and that the component particles differ only in colour, then these limiting values can be predicted with the help of the binomial theorem. The binomial theorem is applicable for the particular cases where the value of y_i is either 0 or 1. This condition is applicable in two extreme cases.

(i) When the components of the mixture are completely segregated it can be assumed that any sample withdrawn from the mixture will be entirely composed of a pure component. In such conditions Lacey[3] showed that

$$S_0^2=p\,\partial\,q \qquad\qquad (2.1)$$

where S_0^2 is the variance of the completely segregated mixture and p and q are the proportions of the two components estimated from the samples.

(ii) When any sample is withdrawn from a fully randomized mixture every *particle* within that sample can be considered as randomly representative of the mixture as a whole. If a sample contains A particles, then

$$S_R^2=\frac{pq}{A} \qquad\qquad (2.2)$$

where S_R^2 is the variance of a fully randomized mixture. This value of S_R^2 is normally the minimum variance which can be obtained within a mixture and represents the best attainable mixture. If a mixture is 'ordered' then the sample variance can be reduced to zero. Such a mixture would have a regular distribution of the components and every sample would contain exactly the same number of particles of each component. In some particulate systems fine particles can adhere preferentially to coarse particles and an approach to ordered mixing can be obtained.

Using the two datum values of variance of equations (2.1) and (2.2) and the experimental value of variance calculated from the samples a variety of 'Indices of Mixing' can be calculated.

The Lacey index[4] is defined as

$$M_1 = \frac{S_0^2 - S^2}{S_0^2 - S_R^2} = \frac{\text{How much mixing has occurred}}{\text{How much mixing could occur}} \qquad (2.3)$$

The index will have a zero value for a completely segregated mixture and increase to unity for a fully randomized mixture. A criticism of the Lacey index is that it is insensitive to mixture quality. Even a very bad mixture will have a variance value much closer to S_R^2 than to S_0^2, and as a result practical values of the Lacey index are restricted to the range of 0.75 to 1.0.

Poole, Taylor and Wall[5], omitted the value of S_0^2 from their index which is defined as

$$M_2 = \frac{S}{S_R} \qquad (2.4)$$

As the mixture approaches the random state the index will approach unity. Non-random mixtures will have a mixing index greater than unity.

As well as providing essential limiting variance values, equations (2.1) and (2.2) indicate how the mixture quality will be affected by the scale of scrutiny, or sample size. The variance, and hence the quality, of a completely segregated mixture is independent of the scale of scrutiny. If a fully randomized mixture is attainable then the quality of that mixture is inversely proportional to the number of particles in the sample. In this case a reduction in the particle size of the mixture will increase the number of particles in each fixed weight sample and produce a predictable improvement in the attainable mixture quality. Between these extremes of mixture types there is no infallible guide as to the relationship between scale of scrutiny and mixture quality.

The equations (2.1) and (2.2) defining the limiting variance values S_0^2 and S_R^2 were based on the assumption that a two-component, equi-sized particulate system was being analysed. Real systems are unlikely to be so obliging!

For the ideal system the measured proportions of the components will be constant no matter which particle characteristic is measured. The fraction of particles measured by weight will be exactly the same as that measured by volume, number of particles, surface area or any other characteristic of the mixture. For a real system the values of the limiting variance will be dependent on the characteristic measured. An additional problem for multi-sized particulate systems is that the assumption made in defining S_R^2 that each sample will contain the same number of particles is no longer valid and an estimate of the mean number of particles within the sample must be made.

Based on the work of Stange[6], Poole, Taylor and Wall[5], showed that for a two-component multi-sized particulate mixture

$$S_R^2 = p \cdot q \left/ \frac{W}{q(\sum f_a W_a)_p + p(\sum f_a W_a)_q} \right. \qquad (2.5)$$

where W is the total sample weight, p and q are the proportions by weight within the mixture and f_a is the size fraction of one component of average weight W_a in a particle size range.

The denominator in equation (2.5) is the estimate of the mean number of particles in a sample and is directly comparable with the denominator value A of equation (2.2). In order to estimate the limiting variance by equation (2.5) the size analysis of the components is required along with a knowledge of particle shape and specific gravity. The use of equation (2.5) is illustrated in section 2.5.

This limiting equation for random mixtures has been extended to cover multi-component mixtures by Stange[7] so that

$$S_R^2 = \frac{p^2}{W} \left\{ \left[\frac{1-p}{p} \right]^2 \cdot p(\sum faWa)_p + q(\sum faWa)_q + r(\sum faWa)_r + \text{etc.} \right\} \qquad (2.6)$$

In this expression one component is regarded as the 'key' component. If the variance of more than one component is regarded as critical in a process it could be necessary to monitor the state of mixedness of these components independently.

Equations (2.2), (2.5) and (2.6) are extremely important in that they provide a method of estimating the best theoretically attainable limit of mixture quality for any particulate mixture. The effect of altering the size distribution of the ingredients on the theoretically attainable mixture quality can be calculated readily. In practice this theoretically attainable mixture might prove to be unobtainable but at least the boundaries of possibility in a randomizing process are established.

2.3 Statistical inference

Samples are withdrawn from a mixture in order to estimate the quality of that mixture. This estimate is of little value if it is imprecise. The precision of a mixture quality estimate is usually expressed in terms of the confidence with which an estimate can be stated to lie within specified limits, and it can generally be increased by requiring less confidence in the estimate or by basing the estimate on more samples.

Without a knowledge of the precision of the estimate of mixture quality no meaningful comparison of mixtures can be made and no process quality control established.

A single set of sample compositions can be plotted as in *Figure 2.1* and an estimate of the mixture characteristics calculated from these values. If this process was repeated a large number of times then a range of values of mean composition and variance would be obtained. These values in turn could be

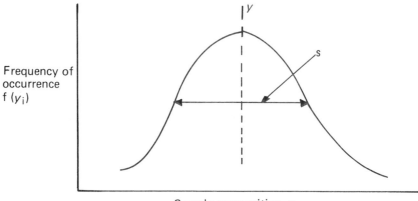

Figure 2.1   Distribution curve for sample compositions

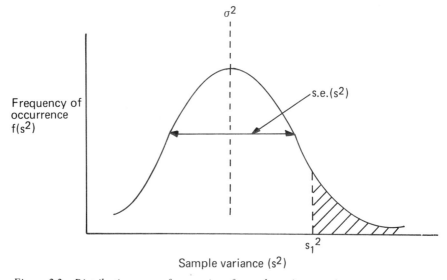

Figure 2.2   Distribution curve for a series of sample variance estimates

expressed in the form of a second distribution curve with a mean value and a variance as shown for the case of sample variance estimates in *Figure 2.2*.

If sample bias has been avoided the second distribution curve will peak at the true sample variance of the mixture, σ^2, and the standard deviation of the estimates of variance, or standard error of the variance, s.e. (s^2), can be calculated. The area under this second curve can be used to express the probability of certain variance values occurring when the mixture is sampled. Thus the probability that a variance value greater than s_1^2 will be obtained can be expressed as the ratio of the hatched area to the total area under the curve.

Such a procedure would require a very large analytical effort and the assumption is usually made that the distribution curve at this second stage can be described either as a normal curve or as a χ^2 curve. The statistical validity of this second stage estimate of precision is based on the form of the first-stage distribution curve, i.e. *Figure 2.1*. An estimate of the precision of mixture quality is only possible if the sample distribution curve described in *Figure 2.1* is that of a normal distribution. The test for normality can be carried out by an adaptation of the χ^2 test and is illustrated in section 2.4.

If the assumption is made that the compositions of samples withdrawn from the mixture are normally distributed then the precision of the mixture variance can be estimated by two methods:

(i) When more than about fifty samples are withdrawn from the mixture. In this case it can also be assumed that the second-stage distribution of estimates of variance of *Figure 2.2* will also be normal. This second normal curve can be 'standardized' in terms of a single variable 't' such that

$$t = \frac{s^2 - \sigma^2}{\text{s.e. } (s^2)}$$

or

$$\hat{\sigma}^2 = s^2 \pm t \cdot \text{s.e. } (s^2) \tag{2.7}$$

where $\hat{\sigma}^2$ is the best estimate of the true variance.

For high precision the values of t and of the standard error of variance should be minimized. The value of t will be obtained from Student's t distribution tables in this case and its value will largely be determined by the confidence level required of the estimate. If a 95% confidence level is required then the corresponding Student's t value for 60 samples is 2.00. The 95% confidence level infers that on average one out of twenty estimates of mixture variance fall outside the stated precision limits. If a higher confidence level was required then the value of Student's t would increase and the precision of the estimate would decrease.

The value of the standard error of the sample variance is estimated by

$$\text{s.e. } (s^2) = s^2 \sqrt{\frac{2}{n}} \tag{2.8}$$

where n is the number of samples withdrawn from the mixture. The increase in precision is proportional to the square root of the number of samples taken.

(ii) When fewer than about fifty samples are withdrawn from the mixture it is no longer safe to assume that the second-stage distribution of estimates of variance will be normal and the curve is better described as a χ^2 distribution. As with estimates of precision based on normality, the precision can be increased by requiring a lower confidence level in the precision estimate and by taking a larger number of samples. Unlike estimates based on normality, the limits of precision will not be symmetrical. The precision of the estimate will decrease as the number of samples decreases. For the χ^2 distribution

$$\hat{\sigma}^2_{(lower)} = s^2 \frac{(n-1)}{\chi^2_{(lower)}}$$

and

$$\hat{\sigma}^2_{(upper)} = s^2 \frac{(n-1)}{\chi^2_{(upper)}} \tag{2.9}$$

The values of $\chi^2_{(upper)}$ and $\chi^2_{(lower)}$ for a given confidence level can be obtained from χ^2 distribution tables. An example of the use of the χ^2 distribution to determine precision limits is given in section 2.4.

If it cannot be assumed that the samples withdrawn from the mixture are normally distributed then no estimate of precision can be made. Agricultural surveys reveal errors in the standard error of the variance of twice that calculated assuming a normal distribution, and the estimate cannot be improved by taking a large number of samples.

In practical terms a non-normal distribution is likely to result from a strongly segregated mixture and as the mixture improves so will the distribution curve approach normality. This means that reliable comparisons can be made between good mixtures but not between bad mixtures.

2.4   A typical mixture analysis

Some of the quantitative techniques outlined in sections 2.2 and 2.3 will now be applied to a typical mixture analysis problem. Let the problem be to assess the state of mixedness of coloured plastic chips with neutral plastic chips in a sheet of plastic.

The first step is to define the scale of scrutiny, and hence the sample size, required for the product. In the case of the plastic sheet this will be an area and, once it has been determined, the entire sheet of plastic can be subdivided into potential sample areas. Evidently the smaller the scale of scrutiny becomes the greater the number of potential sample sites. Assume that an analysis for every potential site has been carried out and the result expressed as a percentage of the coloured chips. The results are shown in *Figure 2.3*. In this case there are 900 potential sample locations each of which can be identified by two coordinates. A three-dimensional mixture analysis would require a third coordinate to identify potential samples.

In practice, only a fraction of these potential samples would be selected and analysed and the mixture quality estimated from these samples. To avoid sample selection bias, and to enable precision estimates to be carried out, a simple random selection of sample sites should be made. A simple random selection is one in which every *unselected* site has an equal chance of selection. A site can only be chosen once. *Table 2.1* provides a series of random numbers of range 1–30 from which random coordinates can be chosen for the mixture of *Figure 2.3*. The selection process is evident.

	1	2	3	4	5	6	7	8	9	10	11	12	13	14	15	16	17	18	19	20	21	22	23	24	25	26	27	28	29	30	
1	62	61	60	59	60	58	58	58	59	54	53	52	54	47	51	59	59	59	62	54	51	52	55	55	60	60	60	54	58	59	1
2	60	64	61	63	59	57	56	56	55	55	54	54	51	52	50	57	56	59	46	58	58	57	59	58	59	59	56	60	57	50	2
3	59	59	59	60	56	61	61	61	57	57	52	53	53	50	52	50	56	59	55	58	56	55	55	58	58	51	57	60	60	57	3
4	63	62	58	62	58	55	57	58	57	53	55	44	38	49	54	48	49	49	48	47	51	54	46	46	53	51	56	56	54	54	4
5	57	59	61	57	55	59	58	55	42	48	41	39	52	51	50	52	52	48	49	43	44	44	43	49	49	50	61	53	53	54	5
6	58	58	56	55	59	59	58	55	51	45	42	50	52	53	45	46	50	44	46	44	41	43	45	44	46	50	48	53	61	55	6
7	61	61	59	61	57	42	37	40	39	42	41	51	51	53	52	51	50	47	47	45	48	41	56	41	44	49	50	53	55	56	7
8	62	58	58	58	44	54	40	41	48	49	43	43	50	48	52	48	48	49	44	38	41	42	42	41	40	42	50	58	53	55	8
9	61	60	69	60	51	42	40	40	45	38	46	49	47	48	49	38	51	49	44	49	38	44	45	45	41	40	42	56	55	53	9
10	57	60	62	56	46	56	43	45	51	43	39	43	44	53	48	50	39	50	42	45	45	46	44	42	40	38	43	59	52	57	10
11	58	63	57	51	45	45	49	42	52	39	37	52	43	49	53	54	47	43	48	49	43	46	45	46	44	43	39	54	55	60	11
12	64	62	61	58	59	45	43	40	50	46	44	57	52	51	49	47	49	47	46	46	42	42	42	43	46	54	54	54	56	52	12
13	58	60	56	46	40	47	40	49	43	46	40	51	50	50	51	49	51	49	44	40	49	41	41	41	42	46	39	52	54	59	13
14	59	63	58	54	41	41	41	43	44	43	40	53	52	51	55	49	50	50	44	46	46	44	41	43	43	41	43	52	58	51	14
15	61	61	60	41	41	43	40	55	36	40	43	45	51	55	50	51	55	56	45	44	43	43	47	39	39	43	53	57	54	54	15
16	62	60	62	42	39	41	48	41	39	37	50	42	53	55	48	48	46	48	45	48	40	40	43	47	41	42	43	53	62	53	16
17	60	58	57	39	41	45	43	43	39	42	58	53	52	50	53	50	49	48	55	38	45	46	46	40	41	42	40	41	40	51	17
18	56	57	55	38	40	42	40	45	43	45	56	51	42	51	46	48	45	45	43	50	42	42	42	43	40	43	37	40	57	51	18
19	61	62	49	43	44	42	47	45	39	38	53	56	55	49	38	52	49	43	47	48	46	47	47	42	45	42	42	39	52	53	19
20	60	59	63	58	42	41	40	43	44	44	47	53	52	42	47	47	39	47	46	46	43	44	44	44	43	41	43	50	56	52	20
21	59	56	61	53	43	39	40	42	39	40	40	50	50	51	48	51	47	49	45	46	46	43	41	39	41	44	40	52	56	51	21
22	58	61	58	50	43	38	44	59	37	36	41	42	53	42	42	49	50	45	41	48	44	45	45	42	42	41	41	54	58	53	22
23	59	60	59	52	41	40	41	38	42	40	49	53	48	41	49	53	46	48	42	43	44	39	39	44	45	42	48	48	48	53	23
24	62	57	57	60	44	40	49	56	42	43	42	41	41	39	52	50	49	51	50	47	42	42	42	42	38	43	52	54	52	54	24
25	59	61	59	48	55	43	43	40	42	43	41	40	40	40	40	40	47	47	45	44	44	44	43	43	43	41	53	56	57	55	25
26	60	59	60	56	46	42	41	57	37	55	51	41	40	39	50	48	50	50	44	45	44	56	56	45	44	56	51	57	60	56	26
27	61	60	57	47	58	56	57	55	55	52	53	47	47	57	49	48	50	60	44	47	45	54	56	59	54	56	57	61	54	54	27
28	62	59	58	58	58	57	58	55	55	50	39	48	47	57	49	52	48	47	49	53	54	58	55	45	59	55	60	49	60	57	28
29	56	58	59	57	57	57	56	56	54	55	55	54	57	58	56	57	53	53	57	58	50	50	56	43	59	57	59	58	58	54	29
30	60	62	60	57	59	59	58	58	58	58	57	53	54	61	60	60	58	55	62	55	55	54	57	53	55	59	53	57	53	60	30

Figure 2.3   Two-dimensional array of potential sample sites and percentage compositions

Table 2.1   Table of random numbers (1–30)

20	17	12	28	23	17	29	6	8	1	2	10	6	7	14	13	15	5	4	19	15	25	18	27
26	10	21	25	10	22	14	25	14	19	4	19	19	17	25	18	21	6	12	23	30	17	30	30
3	4	10	3	23	10	11	24	18	3	19	30	30	19	2	27	6	12	28	8	10	3	3	28
19	27	8	1	8	7	23	12	29	5	10	28	12	18	2	7	19	10	21	29	3	15	30	12
18	11	7	18	18	4	6	27	22	15	18	15	28	24	25	13	7	22	19	10	16	12	10	13
9	24	2	22	2	24	15	12	24	2	1	7	5	20	30	2	4	24	5	30	27	25	20	28
8	7	12	29	12	18	27	30	30	25	27	20	14	18	30	19	17	15	28	10	7	19	16	24
27	28	21	19	6	12	23	10	15	4	17	6	11	14	8	20	26	24	30	1	28	9	27	24
14	15	22	0	25	3	23	21	27	25	26	22	15	9	28	23	21	5	9	14	5	2	19	16
7	14	19	29	9	30	18	19	6	17	4	12	1	29	3	19	24	28	4	19	5	29	13	9
29	26	8	21	25	27	16	23	2	19	17	29	29	26	4	7	19	5	13	14	18	9	13	28
29	5	27	24	30	3	7	26	24	23	6	4	2	3	6	25	26	16	7	30	21	8	4	17
20	5	4	5	5	22	12	10	21	25	4	1	17	7	27	29	21	10	12	1	14	28	28	25
17	28	23	4	22	7	2	10	17	12	3	1	26	23	8	16	15	18	13	7	8	26	26	27
6	26	24	11	22	17	28	3	17	18	8	13	1	28	11	29	18	7	12	30	11	5	28	30
19	3	4	29	7	12	21	9	6	11	20	21	27	21	9	1	15	23	6	21	6	28	14	15
4	21	30	9	8	21	12	2	19	30	7	26	7	1	28	28	19	30	1	14	25	5	1	15
18	1	23	19	7	7	8	19	3	27	17	22	13	15	15	4	18	4	23	28	2	7	23	15
3	1	26	6	14	3	18	28	21	7	3	10	29	5	16	9	12	20	10	3	16	5	7	4
6	26	3	16	8	27	29	3	30	19	28	14	11	10	30	8	14	24	1	21	16	30	2	3
21	4	9	30	15	22	3	22	20	1	19	3	19	14	19	28	10	21	9	23	22	2	2	21
21	1	12	3	25	22	10	30	7	6	11	29	22	4	13	11	3	12	25	5	30	26	30	15
27	14	29	25	24	25	19	1	27	26	19	19	18	17	26	15	4	20	16	25	20	3	24	24
4	4	2	22	30	23	17	3	26	14	10	14	9	1	23	3	29	4	27	8	12	14	22	10
10	19	5	14	18	10	25	5	25	24	28	20	30	2	25	20	4	24	5	17	29	25	29	14
22	5	6	6	4	26	5	23	19	14	7	25	7	21	16	17	24	27	7	9	25	24	7	3
26	3	12	23	1	28	29	28	9	3	6	13	11	11	14	30	20	29	2	16	16	27	4	21
22	20	26	22	13	28	8	19	25	8	12	17	11	17	16	23	29	13	16	12	13	17	3	2
5	5	3	26	7	25	26	19	18	27	17	13	28	30	29	3	29	22	17	23	1	8	3	30
9	16	8	19	22	18	24	25	15	26	30	17	20	8	12	11	25	10	4	21	5	20	1	3
8	12	27	16	15	13	11	27	7	5	20	30	9	5	2	24	21	15	28	29	24	23	25	19
25	22	21	4	7	19	13	2	17	21	1	12	29	29	29	7	16	4	19	22	22	29	4	17
4	11	16	22	2	15	19	2	20	25	7	27	11	6	24	8	25	6	23	29	12	1	20	24
1	10	10	6	25	16	12	24	2	6	14	9	10	24	4	11	8	23	5	27	23	10	7	24
20	12	10	5	19	28	7	25	3	15	18	28	28	18	23	11	11	11	22	19	7	15	24	22
11	23	14	24	28	18	23	6	15	25	25	3	29	26	4	10	13	30	8	8	26	1	5	6
12	19	22	29	1	11	9	21	10	6	5	26	18	16	22	11	18	1	30	30	3	18	27	3
11	18	2	7	28	25	8	6	24	12	3	25	6	9	15	21	29	12	4	19	23	3	16	26
19	24	5	9	4	26	8	19	21	25	26	18	27	28	12	18	21	5	18	27	10	24	30	24
2	29	21	22	26	13	13	24	15	30	29	27	22	16	26	24	11	1	28	21	7	20	12	21
6	27	8	15	10	10	30	9	5	28	18	14	14	28	3	13	1	2	28	9	19	24	26	27
3	13	28	25	7	11	18	17	25	2	22	21	23	28	24	27	17	6	20	5	11	17	29	27
25	29	18	30	20	17	20	26	29	27	14	23	4	28	2	13	15	29	3	8	9	24	24	14
14	6	20	20	19	26	22	11	3	17	17	23	12	23	7	29	6	27	13	17	4	15	19	5
20	33	2	2	19	29	3	15	26	7	14	9	1	11	4	1	15	19	8	30	27	29	15	28
27	16	4	26	4	24	25	8	27	2	8	10	19	8	17	25	9	11	30	26	19	7	10	23
20	4	24	25	3	1	20	20	13	23	18	11	29	30	15	22	27	5	26	21	16	29	7	30
30	28	10	30	28	16	30	21	15	10	6	29	9	5	5	24	19	29	7	12	12	12	21	24

| | Random number for | | Analysis of selected sample |
	abscissa	*ordinate*	%
	20	17	38
	12	28	39
	23	17	46
	29	6	53
	etc.		etc.

Twenty samples were selected by this means and a mean composition, $\bar{y}$, and a sample variance, s^2, calculated in *Table 2.2*.

Table 2.2 Table for estimating the sample mean composition and variance based on a simple random selection

y_i	$\delta y_i = (y_i - 50)$	$\delta y_2 = [y_i - \bar{y}]$	$(\delta y_2)^2$
38	-12	7.65	58.5225
39	-11	6.65	44.2225
46	-4	0.35	0.1225
53	$+3$	7.35	54.0225
58	$+8$	12.35	152.5225
60	$+10$	14.35	205.9225
41	-9	4.65	21.6225
50	0	4.35	18.9225
50	0	4.35	18.9225
43	-7	2.65	7.0225
40	-10	5.65	31.9225
47	-3	1.35	1.8225
38	-12	7.65	58.5225
44	-6	1.65	2.7225
40	-10	5.65	31.9225
40	-10	5.65	31.9225
53	$+3$	7.35	54.0225
43	-7	2.65	7.0225
50	0	4.35	18.9225
40	-10	5.65	31.9225

$$\bar{y} = 50 + \frac{\sum \delta y_1}{20} = 50 - \frac{87}{20} = 45.65\%$$

$$s^2 = \frac{\sum (\delta y_2)^2}{n-1} = \frac{852.55}{19} = 44.87$$

If based on a fractional rather than a percentage composition, $s^2 = 0.004487$

If it can be assumed that the mixture under consideration is ideal, in that the component particles differ only in colour, then the limiting variance values for the mixture can be estimated using the values of $\bar{y}$ and s^2. From equation (2.1),

$$s_0^2 = p \cdot q = 0.4565 \times 0.5435 = 0.2481$$

This is the worst possible mixture variance for these components. From equation (2.2), assuming that each sample contains 100 particles, i.e. $A = 100$,

$$S_R^2 = \frac{p \cdot q}{A} = \frac{0.4565 \times 0.5435}{100} = 0.002481$$

In most mixing systems this is the best possible mixture variance for these components. The experimentally determined value will lie between these extremes and can be used to calculate a mixing index for that particular mixture. Thus from equation (2.3)

$$M_1 = \frac{S_0^2 - S^2}{S_0^2 - S_R^2} = \frac{0.2481 - 0.00449}{0.2481 - 0.00248} = 0.992$$

Figure 2.4 is a typical plot of Mixing Index, M_1, against time for a segregating mixture. Even though the Mixing Index after 30 minutes was approximately 0.85 this mixture visibly exhibited quite strong segregation. The appearance of an optimum mixture after a very short mixing time is a common occurrence when handling a segregating mixture.

The precision of the estimate of the experimental variance is expressed in terms of confidence in which an estimate lies within specified limits. Let us assume that in this case we require 95% confidence that our estimate will lie between specified limits, i.e. on average one in twenty experimental values will lie outside these limits. Then, on the assumption that we are sampling from a normal population and by using the χ^2 test, equation (2.9) gives

$$\hat{\sigma}^2_{(upper)} = s^2 \frac{(n-1)}{\chi^2_{(upper)}}$$

$$= 0.00449 \frac{(19)}{8.91} = 0.00957$$

$$\hat{\sigma}^2_{(lower)} = 0.00449 \frac{(19)}{32.85} = 0.00260$$

where 8.91 and 32.85 are the upper and lower values respectively of χ^2 at the 95% probability level.

On average we would obtain an experimental variance value outside the range 0.00260 to 0.00957 only one time in twenty. These limiting variance values could be used to compare the mixture quality within a mixer at different mixing times or to compare the performance of two competing mixers. An *F*-test would give a more refined statistical comparison of two mixture qualities as described by variance values.

A basic assumption in the application of the χ^2 curve to predict the precision limits for the variance is that the population of potential samples within the mixture is normally distributed. Such a test requires a larger number of samples to be withdrawn from the mixture than would normally be the case for assessing mixture quality and for this reason the test is applied occasionally or when the process undergoes a significant change. It should not, however, be

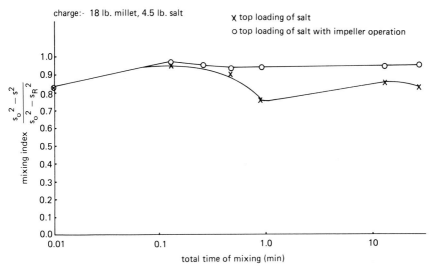

charge:- 18 lb. millet, 4.5 lb. salt

X top loading of salt

o top loading of salt with impeller operation

Figure 2.4   The variation of Lacey's 'mixing index' with time for the Rotocube

omitted if meaningful precision limits are to be estimated for a mixture. Such a test is carried out in *Table 2.3* based on a simple random selection of 100 samples from the two-dimensional array of *Figure 2.3*.

The basis of the test is that a comparison should be made between the distribution of the samples withdrawn from the mixture and the distribution of a normal curve having the same mean value and standard deviation as the sample distribution. The steps in the test are:

(i) Sample the mixture.
(ii) Calculate a mean value and standard deviation based on this sample.
(iii) Postulate class boundaries for the distribution ensuring that there are more than six classes and that each class contains more than five samples.
(iv) Convert the class boundaries to '*t*' values.
(v) Calculate the area of the normal curve corresponding to each class.
(vi) Compare the observed frequency and expected frequency by a χ^2 test and deduce the probability of the sampled population being normal.

The χ^2 value corresponding to the differences between the expected and the observed sample frequencies is given by:

$$\chi^2 = (8.8-10)^2/8.8 + (12.6-20)^2/12.6 + (19.8-13)^2/19.8$$
$$+ (22.0-14)^2/22 + (18.6-19)^2/18.6 + (11.1-19)^2/11.1$$
$$+ (7.1-5)^2/7.1$$
$$\chi^2 = 16.0063$$

In this case χ^2 has $7-3=4$ degrees of freedom since the total frequency, the mean and the standard deviation of the fitted distribution are made equal to

Table 2.3   Test for normal distribution of samples

From *Figure 2.3* the range of values is 36 to 69%

Select class boundaries as 4% apart and take 100 simple random selections

Class boundaries, x_i	Frequency of occurrence
$0 \rightarrow 40.5$	10
$40.5 \rightarrow 44.5$	20
$44.5 \rightarrow 48.5$	13
$48.5 \rightarrow 52.5$	14
$52.5 \rightarrow 56.5$	19
$56.5 \rightarrow 60.5$	19
$60.5 \rightarrow 100$	5

By computation the mean composition of this sample is 50.06 with a standard deviation of the samples of 7.08. These values can be taken as the true mean and standard deviation of a normal distribution. The corresponding t values for the class boundaries x_i are given by:

$$t = (x_i - \bar{x})/s$$

or

$$t = (x_i - 50.06)/7.08$$

Class boundaries, x_i	Value of t	Area under normal curve	Area for each class	Expected frequency	Observed frequency
0	$-\infty$	0.5000			
40.5	-1.35	0.4115	0.0885	8.8	10
44.5	-0.79	0.2852	0.1263	12.6	20
48.5	-0.22	0.0871	0.1981	19.8	13
52.5	0.34	0.1331	0.2202	22.0	14
56.6	0.91	0.3186	0.1855	18.6	19
60.5	1.47	0.4292	0.1106	11.1	19
100.0	$+\infty$	0.5000	0.0708	7.1	5

those of the observed distribution. From χ^2 tables:

$$\chi^2_{0.975}(4) = 0.484$$

$$\chi^2_{0.025}(4) = 11.143$$

at the 5% confidence level. As the experimental χ^2 level lies outside these probability limits it seems unlikely that the distribution of samples is normal and there is a corresponding doubt as to the estimates of precision of variance.

Table 2.4   The calculation of the random variance of a non-ideal mixture

$$S_R^2 = p \cdot q \left/ \frac{W}{q(\sum f_a W_a)_p + p(\sum f_a W_a)_q} \right.$$

Particle		Average weight in size range, W_a	Fraction undersize, f	Range fraction, f_a	Mean particle weight of component
Size (μ)	Weight (μg)				
210	13.044		1.000		
		10.629		0.095	1.010
180	8.214		0.905		
		6.484		0.535	3.469 Spherical
150	4.754		0.370		component A
		3.752		0.250	0.938 (SG. 2.69)
125	2.751		0.120		
		2.190		0.040	0.088
105	1.630		0.080		
		1.328		0.040	0.053
90	1.027		0.040		
		0.810		0.040	0.032
75	0.594		0.0		
					$5.590 = (\sum f_a W_a)_p$
210	12.850		1.000		
		10.471		0.084	0.880
180	8.092		0.916		
		6.387		0.697	4.452 Spherical
150	4.683		0.219		component B
		3.696		0.182	0.673 (SG. 2.65)
125	2.710		0.037		
		2.158		0.013	0.028
105	1.606		0.024		
		1.309		0.020	0.026
90	1.012		0.004		
		0.798		0.004	0.003
75	0.585		0.0		
					$6.062 = (\sum f_a W_a)_q$

2.5   Non-ideal mixtures

Most industrial mixtures fail to conform to the statistically ideal pattern of equi-sized particles distinguishable only by colour. It is for this reason that equations (2.5) and (2.6) are so important in establishing the best attainable limits of mixture quality. The statistically precise work of Stange[6] was applied to real powder mixtures by Poole, Taylor and Wall[5]. This application involved assumptions and estimations which in view of the central value of the equation are worthy of investigation.

Equation (2.5) is applied to a typical two-component mixture of ingredients A and B. The components are in equal proportions by weight and the weight of sample withdrawn is 2 g. The average particulate weight of each component can be *estimated* from a knowledge of a sieve size determination and an assumption as to the shape of the particles. In this case the particles were assumed to be spherical. The calculation is set out in *Table 2.4*.

The mean particle weight of each of the components is calculated from a sieve analysis in *Table 2.4*. Substituting the summation values in equation (2.5) gives:

$$S_R^2 = 0.5 \times 0.5 \bigg/ \left\{ \frac{2 \times 10^6}{0.5 \times 5.590 + 0.5 \times 6.062} \right\}$$

$$= 7.2 \times 10^{-7}$$

This mixture quality might be improved if some ordering of fine particles occurs amongst the coarse particles, but commonly it represents the limiting mixture quality of the system.

It is generally desirable to keep this value as small as possible. This can be done by keeping the denominator of equation (2.5) as large as possible by choosing a large scale of scrutiny or by keeping the mean particle weight of the components small. *Table 2.4* indicates that even small proportions of relatively coarse particles have a disproportionate effect on the mean particle weight and hence the attainable mixture quality.

References

1   DANCKWERTS, P. V. (1953) *Research* (London), **6**, 355.
2   HARNBY, N. (1975) *Proceedings of the Third International Powder Technology and Bulk Solids Conference*, Powder Technology Publication Series No. 6.
3   LACEY, P. M. C. (1943) *Trans. Instn. Chem. Eng.*, **21**, 53.
4   LACEY, P. M. C. (1954) *J. Appl. Chem.*, **4**, 257.
5   POOLE, K. R., TAYLOR, R. F. and WALL, G. P. (1964) *Trans. Instn. Chem. Eng.*, **42**, T166.
6   STANGE, K. (1954) *Chem., Ing. Tech.*, **26**, 331.
7   STANGE, K. (1963) *Chem., Ing. Tech.*, **35**, 580.

Chapter 3
The selection of powder mixers

N Harnby

Schools of Chemical Engineering, University of Bradford

3.1 The range of mixers available

Powder mixers are relatively simple machines of low capital cost. For this reason they are manufactured in an apparently bewildering variety by a large number of equipment manufacturers. Despite the many patented designs on the market, most mixers can be categorized relatively simply according to the mixing mechanism used.

3.1.1 Tumbler mixers

'Tumbler' mixers are the simplest and the most common type. A totally-enclosed vessel is rotated about an axis causing the particles within the mixer to 'tumble' over each other on the mixture surface. In the case of the horizontal cylinder, rotation can be effected by placing the cylinder on driving rollers. In most other cases the vessel is attached to a drive shaft and supported on one or two bearings. Common vessel shapes include the cube, double-cone, V and Y mixers.

A V-mixer is illustrated in *Figure 3.1*. The capacity of the vessel ranges up to 50 m³ and the mixer will normally operate at half the total capacity. Rotational speeds depend on capacity but, as a rough guide, the recommended speed will normally be about half the critical speed of the mixer. Critical speed is here defined as that speed at which the centrifugal force acting on a particle just balances the force due to gravity.

[handwritten margin note: dependent on particle size]

The addition of an internal impeller operating on the same axis of rotation as the shell and in a counterdirection extends the application of the tumbler mixer to lightly agglomerating mixtures. The impacting action of the impeller breaks down the agglomerates for subsequent redistribution.

The provision of internal baffles is often claimed to improve the rate and/or equilibrium quality of mixing in the tumbler mixer. An equilibrium mixture quality is usually reached in 10 to 15 minutes. *[handwritten: but is equilibrium good enough]*

3.1.2 Convective mixers

The choice of mixers within the convective group is much wider. In most convective mixer designs an impeller operates within a static shell and groups

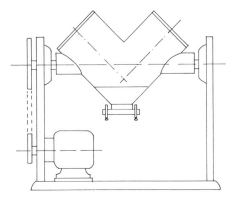

Figure 3.1   The V-mixer

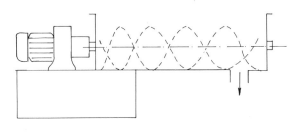

Figure 3.2   The ribbon blender

of particles are moved from one location to another within the bulk of the mixture. Generally the tumbling mixing mechanism is not important in such mixers.

The ribbon blender is perhaps the most widely used convective mixer (*Figure 3.2*). A ribbon rotates within a static trough or open cylinder and the particles are relocated by the moving ribbon. Rotational speeds are normally within the range of 15 to 60 rev/min. A large variety of ribbons and blades are available for different duties; smoothly contoured and highly polished cast blades are frequently used when cleanliness is an important process requirement. Capacities of these mixers range up to 20 m³, or even larger in special cases.

However, several mixers operate at much higher impeller speeds. The containing vessel in this case is commonly a vertical cylinder (*Figure 3.3*), and the impeller a relatively simple rotating blade. With rotational speeds of the order of 1000 rev/min the energy input to the mixture is large and commonly results in a 'fluidized' particulate system in which the temperature can rise appreciably.

In the ribbon blender the entire mixture volume is actively swept by the passage of the impeller and the power requirements are directly related to the

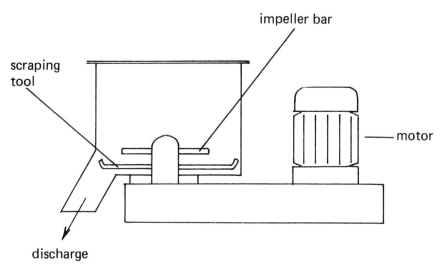

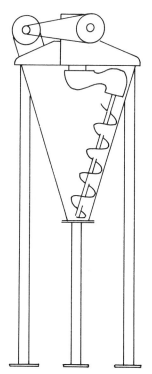

Figure 3.3   A high-speed impeller mixer

working capacity of the mixer. In several mixers the impeller actively agitates only a small proportion of the mixture volume at a given time and either the impeller progressively sweeps the entire mixture volume over a period of time or natural circulation is used to ensure that all the mixture passes through the impeller zone. In the case of the 'Nautamix' (*Figure 3.4*) an Archimedian screw

Figure 3.4   The Nautamix

lifts powder from the base of a conical hopper to the powder surface whilst at the same time progressing around the hopper wall. With this design a central conical volume of powder relies on natural circulation to sweep powder into the mixing zone. A modified driving action permits the screw to describe an epicyclic path moving towards the central axis of the hopper and provides a more complete sweep of the mixture volume. In the vertical screw mixer (*Figure 3.5*) the natural circulation zone is larger and the risk of static zones existing is increased, particularly if the powder is of a cohesive nature. The central fixed screw provides a more robust engineering design.

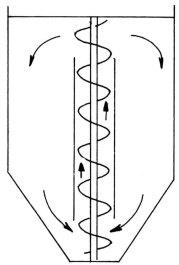

Figure 3.5   The vertical screw mixer

3.1.3   Fluidized mixers

Fluidization is caused by the passage of a gas through a bed of particles. In such a system the bulk density of the powder is reduced and individual particle mobility is potentially increased. If the gas flow rate is sufficiently large there will be considerable turbulence within the bed and the combination of turbulence and particle mobility can produce excellent mixing. A constant danger in the fluidized mixer is that if turbulence is not complete then the constituent particles can readily segregate due to variable settling or projection rates.

Very few commercially available fluidized mixing units exist. A typical example is the 'Airmix', see *Figure 3.6*. The system necessarily has more units than the tumbler and convective mixers. A blower or compressor is required to generate the gas flow and, downstream of the mixing vessel, some device is required to separate entrained particles from the gas stream. The Airmix operates on a closed circuit so that any gas can be used as the fluidizing medium. The gas is introduced into the mixing chamber through a conical

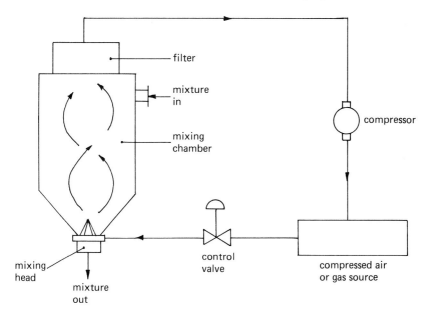

Figure 3.6   The 'Airmix'

mixing head which produces a swirling, turbulent, action within the mixing chamber.

Aeration of the mixture takes place over a period of 15 to 30 seconds and a similar period is allowed for the particles to re-settle. The length and frequency of the cycles can be varied by the control valve setting. The high capital cost of the unit compared with the equivalent capacity tumbler or convective mixer is compensated by the very rapid mixing time. Typically, the 15 minutes required to reach equilibrium mixture quality in the tumbler and convective mixers is reduced to 1 to 2 minutes in the Airmix, thus enabling a smaller capacity unit to be used.

Perhaps the greatest attraction of the fluidized bed is not so much its ability to mix powders but its general versatility as a piece of process equipment. Within a single vessel it is possible to react, dry, coat, agglomerate or mix a powder system, thus avoiding transportation and contamination problems. Because of the diversity of its application the bed is usually designed for a specific process and is not available as a manufacturer's standard line product. Several fluidized beds are purpose-built for the pharmaceutical industry—the Manesty mixer being a good example.

3.1.4 Hopper mixers

When a hopper discharges, there is a pronounced velocity gradient across any horizontal section. If the outflow from the hopper is recycled then there will be considerable axial mixing. If a suitable distributor can provide radial mixing at some point in the cycle then, in principle, mixing can be carried out within a storage area. The recycle can be effected by either pneumatic or mechanical means.

If effective, this technique would provide a very neat solution to the problem of mixing powders for many processes and would be particularly attractive if very large volumes of material were to be mixed. It presupposes that all the material is in motion and that there is no hold-up of material within the cycle. Segregation is likely to be a problem both on the free surface of the hopper and within the bulk of the material due, in both cases, to percolating segregation.

In some designs, material is withdrawn simultaneously from several locations within the hopper and these flows recombined at a common outlet for either recycle or delivery. Such mixers are likely to be applicable only to very-free-flowing powders.

3.1.5   Size-reduction equipment

It is frequently desirable from both a practical and statistical point of view to comminute the constituent particles of a mixture. In carrying out this comminution appreciable mixing can occur. The 'muller' mills are particularly useful in that the intensive shearing action which causes particle fracture also produces intensive mixing.

For particles with better flow characteristics the 'pin' mills can be used for both mixing and comminution. Such mixers are particularly useful when one or more components of the mixture have strongly adhesive characteristics.

3.2   Selection based on process requirements

Only seldom can a mixer selection be made solely on the basis of the mixture quality achieved in a series of isolated tests. Generally, a compromise has to be reached between mixture quality and the compatibility of the different mixer types with a process. An initial selection eliminates those mixers which are incompatible.

In many process industries product purity has an overriding importance. If contamination between batches is to be avoided the mixer must be thoroughly cleaned after each usage. To meet this condition, smoothly contoured, highly polished mixer surfaces are required with easy access for cleaning purposes. Bearings and glands should not be in contact with the mixture, or be positioned above the mixture, if potential lubricant contamination is to be avoided. The tumbler mixers meet all these requirements, having a simple shape and no contact with rotating parts within the mixing vessel.

Low-speed convective mixers with complex impeller arrangements present cleaning problems and the whole class of convective mixers potentially have lubricated surfaces in contact with the mixture. Closed-circuit hopper mixers and fluidized mixers are not generally suitable if frequent changes of product specification require a complete flushing of the circuit, and gas-circulated systems often have the problem of lubricant entrainment in the gas stream.

As well as providing a high-purity product it is often essential that the escape of powder to the surroundings be minimized for health, safety and economic reasons. The total enclosure of the powder in a closed mixing vessel is evidently an advantage in these circumstances though the enclosure has to be broken for loading and unloading operations. Open mullers and convective mixers can effectively disperse fine powder to the surrounding atmosphere.

Mixers can also be selected on the basis of the avoidance or encouragement of particle size degradation. The 'gentle' operation of the tumbler mixers and the low-speed convective mixers does not generally cause appreciable comminution. Comminution is encouraged in those mixers which have high-speed impellers impacting the particles or in which the mixture is subjected to an intensive shearing action. Thus the high-speed impeller mixers, the mullers and any convective mixers in which there is a small clearance between impeller and mixer wall should be avoided if comminution is to be minimized.

Comminuting mixers with a high energy input can produce considerable temperature increases within the mixture. This can be undesirable if the product is heat degradable though, on occasions, as with the high-speed impeller mixers, this temperature rise is used to approach the melting point of a mixture constituent and ensure a physical bonding of mixture components. For very abrasive particles the problem might well be the degradation of the process equipment rather than that of the particles. Bearings can be stripped even by contact with airborne dust from such a mixture. In such circumstances the tumbler mixers are very attractive as the mixture is contained away from lubricated surfaces and the simple walls of the vessel can be clad with wear-resisting liners.

The enclosed nature of the tumbler mixers becomes a disadvantage if controlled additions of ingredients are to be made over an extended period. With the rotating shell, the mixer has either to be stopped in order to make additions, or a mechanically difficult axial feed provided to the rotating vessel. Similar problems are apparent if it is required to heat or cool the mixture during the mixing process.

The tumbler mixers are usually simple to construct as a monoshell vessel but difficult to jacket. After construction there remains the mechanical problem of providing a constant flow of heat transfer fluid to the rotating vessel. The continuous visual or mechanical monitoring of the mixture quality presents the same problems of access. The static shell, low-speed, convective mixers, on the other hand, give every opportunity for interference in the progress of the mixing process.

Additions can readily be made and monitored whilst mixing is in progress and the vessel shells are generally of a shape that can easily be jacketed. The convective mixers which do not rely on natural circulation have an additional advantage in that they can mix moist powders. Moisture addition to a tumble mixer tends to create large agglomerates unless it is fitted with an internal impeller.

If a mixing process is to be continuous, then frequent product specification changes are unlikely and the ability to clean the equipment becomes less important. The essential process feature of a continuous mixer is that it should be readily capable of continuously charging and discharging the process mixture.

Enclosed mixing systems such as the tumbler mixers cannot be readily adapted to continuous operation, for this reason, and convective mixers such as the ribbon blender or fluidized mixers are generally favoured. The design criteria for a continuous mixer are quite different to those for a batch mixer. Depending on the control system used for the feed streams to the mixer it could be desirable that a substantial amount of backmixing occurs within the mixer. A fluidized bed could backmix most effectively whilst the performance of a ribbon blender would be a function of the ribbon geometry.

3.3   Selection based on mixture quality

The effectiveness of different mixers can be assessed by comparing their performance using a standard mixture. The only reasonable basis of comparing mixer efficiencies is to consider the quality of the equilibrium mixture. The times taken to reach this equilibrium condition will vary from mixer to mixer.

Before considering some of the experimental work on mixer efficiency, it is possible to make predictions on mixer performance based on the discussions of the interaction of mixing and segregation of Chapter 1. Mixers which have a predominant mixing mechanism of diffusion or shear mixing can be classified as segregative whilst those which rely on convective mixing are likely to be less segregative. Any mixer which relies on tumbling or stirring, such as the rotating shells or vertical stirred mixers, can produce considerable segregation whilst scooping impeller mixers, such as the ribbon blenders and the 'Nautamix', are likely to minimize segregation.

The importance of the segregating mechanism in a given mixer will largely be a function of the properties of the components. If the materials are not free flowing because they are fine or moist, then any type of mixer will produce a mixture which avoids gross quality variations and selection will probably be based on cost and process consideration. If, however, the materials are free flowing, and have a size differential, then it is desirable for the best mixing results to choose a mixer from the group of non-segregating mixers.

For cohesive powders the rate at which mixing can be achieved is often more important than ensuring that a very-high-quality equilibrium mixture is attained. In a tumbler mixer radial mixing is relatively fast whilst axial mixing is a slow and rate-controlling process. In such a case the equilibrium state mixture would probably be the same irrespective of the component loading pattern adopted but the time to reach this equilibrium value could be significantly different. For the free-flowing mixture individual particle mobility ensures that the original loading pattern is less important.

When assessing the performance of a mixer the fact that the mixer has to be emptied is often overlooked. The emptying operation often involves particles diffusing down an inclined surface and always involves a shearing action within the bulk of the mixture. As these are mechanisms which promote segregation, it is possible that a mixture which was of high quality within the mixer can be destroyed by the discharging process. Evidently quite different conclusions could be reached on relative mixer performance if the discharge rather than *in situ* qualities of mixture were compared. A qualitative assessment of some of the major characteristics of a range of mixers is given in *Table 3.1*. It should be emphasized that these comments should be used as a guide and are not a substitute for practical quality tests carried out using a specific process material.

The work of Adams and Baker[1], is very useful in that it represents one of the few cases in which the performance of a range of mixtures is compared using

Table 3.1   Summary of mixer characteristics (Williams)

Type of mixer	Batch or continuous	Main mixing mechanism	Segregation (suitability for ingredients of different properties)	Axial mixing	Ease of emptying	Tendency to segregate on emptying	Ease of cleaning
Horizontal drum	B	Diffusive	Bad	Bad	Bad	Bad	Good
Lödige mixer	B	Convective	Good	Good	Good	Good	Bad
Slightly inclined drum	C	Diffusive	Fair	Bad	Good	Good	Good
Steeply inclined drum	B	Diffusive	Bad	Good	Bad	Bad	Good
Stirred vertical cylinder	B	Shear	Bad	Good	Good	Bad	Good
V mixer	B	Diffusive	Bad	Bad	Good	Bad	Good
Y mixer	B	Diffusive	Bad	Bad	Good	Bad	Good
Double cone	B	Diffusive	Bad	Bad	Good	Bad	Good
Cube	B	Diffusive	Poor	Good	Good	Bad	Good
Ribbon blender	B	Convective	Good	Slow	Good	Fair	Fair
Ribbon blender	C	Convective	Good	Fair	Good	Good	Fair
Air jet mixer	B	Convective	Fair	Good	Good	Good	Fair
Nauta mixer	B	Convective	Good	Good	Good	Good	Bad

standard segregating components. Their work was concerned with the blending of small quantities of 3 mm cubes of black master batch polythene with 4 mm cubes of natural polythene. The specific gravity of the two polythene types were 1.2 and 0.92 respectively. The two properties of size and density are reinforcing in this case and should emphasize segregating tendencies. Three tumbler mixers and one convective mixer were tested. A major criticism of their work is that the size, speed and degree of fill of the mixers tested is not indicated and there is no indication as to whether these conditions have been optimized. As the proportion of the black master chips in the mixture was small it was possible to assume a Poisson distribution of black particles, and on this basis they were able to place 10% probability limits on the number of particles expected to be found in a sample withdrawn from a random mixture. Results obtained for a Rotacube, V-Mixer and Ribbon blender are given in *Figure 3.7*.

Rather unusually Adams and Baker analysed the discharge from the mixer rather than the more conventional *in situ* analysis. Results are thus given in sequential form and despite the scatter of the individual samples it is possible to identify trends of segregation. Sample size is not quoted, but as only twenty-five or so black particles were counted in each sample it is likely that the sample is 'coarse grained' and will reflect quite wide variations due to local 'patchiness' of the black polythene component.

'End to end' segregation was observed for the ribbon blender, and under the test conditions the results obtained from this convective mixer were inferior to those from the Rotocube and V-mixer. Adams and Baker attributed this result to a lack of efficient axial distribution within the ribbon mixer tested, and indicated that the initial distribution of the minor component had an important bearing on the quality of the mixture. Testing a 'double cone' blender they found extreme segregation of the smaller and denser particles to the centre of the rotating particulate mass and a very poor quality of discharged mixture resulted. Surprisingly, the quality of mixture discharged from the other two tumbling mixers tested was entirely satisfactory from the point of view of the process. In view of this apparent contradiction in performance of the tumbler mixers it is particularly unfortunate that the authors do not give details of mixer operating conditions. Such details could well provide the basis for an explanation of this apparent anomaly.

The results of Adams and Baker are to some extent contrary to the general conclusions summarized in *Table 3.1* and emphasize the need to carry out practical tests on any proposed mixing equipment using the process materials.

During the discussion following the presentation of Adams and Baker's paper, Sherwin briefly mentioned tests carried out on a ribbon blender, a rotating drum and a concrete mixer in which sand ($\frac{1}{2}$ mm to 1 mm) and gravel ($1\frac{1}{2}$ mm to 3 mm) were mixed. In this case, the quality of mixture obtained from the ribbon blender was much higher than that from the tumbler mixers. Although not specific about operating conditions, this reference does mention that the quality of mixture obtained is dependent on the degree of fill of the mixer and the proportions of the components.

(i)   *Ribbon Blender*

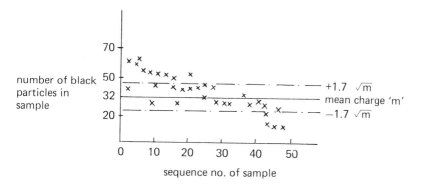

number of black particles in sample

sequence no. of sample

(ii)   *Rotacube* (320 turns)

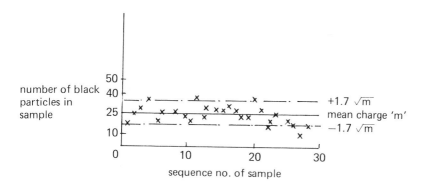

number of black particles in sample

sequence no. of sample

(iii)   *V-Mixer* (100 turns)

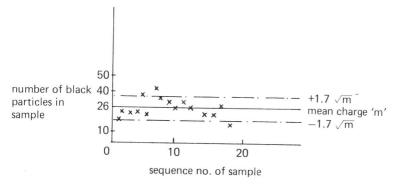

number of black particles in sample

sequence no. of sample

Figure 3.7   Results of Adams and Baker

Greathead and Simmonds[2] confirmed that the point of addition of a minor component can influence the equilibrium quality of a mixture in a ribbon blender. Most of the work they report should be treated with caution, as the results are based on only six samples taken from the surface of the mixture.

One of the few significant works on a high-speed mixer with a predominantly convective action is due to Poole, Taylor and Wall[3]. They tested a Morton 'Lodige' type of mixer, with a nominal working capacity of 0.125 ft[3], and used mixtures of copper/nickel spheroidal powders as well as ceramic oxide powders of differing shape and agglomerative characteristics. Working with non-cohesive systems and with particle-size ratios of the order of 2:1 they found that the mixture sampled from the vertical drum approached the theoretical random mixture after less than one minute of operation. They also found that the mixing time required to obtain a near random mixture increased markedly both with the presence of agglomerates and with the increase in the proportion of one of the minor constituents. Unfortunately there was no comparative work done using the same mixture in other mixers.

Ashton and Valentin[4] compared the performance of an Airmix, Nautamix, Oblicone and Z-blade mixer using equi-sized particles of sand and calcite. The mixture quality achieved approached that of the random mixture quality in all four mixers but this condition was achieved most rapidly in the Airmix followed by the Oblicone, Nauta and Z-blade mixers. When testing the effect of mixer variables on the quality of mixing achieved it was found that the tumbler mixer permance was very dependent on rotational speed.

Tests were also carried out on the Oblicone and the Airmix using sand/calcite mixtures with a size differential. There was little to choose between the qualities of mixture produced by these two mixers except that the Airmix produced a very-high-quality mixture when operating with a fine-sand/coarse-calcite charge. Test work carried out on the Airmix showed not only a rapid rate of mixing but also a longer-term de-mixing tendency which, if allowed to proceed, resulted in a low-quality mixture. Working with the Nautamix, Ashton and Valentin came to the conclusion that the ultimate quality of the mixture was independent of the speed of rotation of the impeller.

Using a highly segregating mixture and analysing mixture quality at the point of discharge, Harnby[5] tested Nautamix, ribbon blender, Rotacube and twin-shell mixers using a standard mixture. The Nautamix and ribbon blender both produced appreciably better equilibrium mixtures than either the Rotacube or the twin-shell mixers. On an arbitrarily defined basis of efficiency the two convective mixers had a 90% efficiency whilst the two tumbler mixers had an equilibrium efficiency of only 20%. After a very short mixing time both tumbler mixers passed through an optimum mixture quality which deteriorated quite rapidly up to the equilibrium quality.

A similar general conclusion was reached by Williams and Khan[6] mixing fertilizer granules of varying size ratio in a V-mixer, a ribbon blender and a Nautamix. Their results are plotted in *Figure 3.8* and indicate the variation of the equilibrium quality of the mixture, for the size ratio of the granules. For all

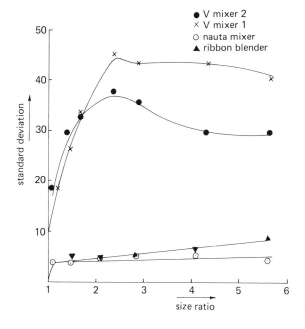

Figure 3.8 *The effects of varying particle size ratio on mixer performance*

size ratios the performances of the ribbon blender and the Nauta mixer are significantly better than those of the V-mixer.

To summarize, the quantitative assessments that have been carried out on the mixing performance of a variety of mixers generally support the qualitative assessment of *Table 3.1*. It would, however, be unwise to use such an assessment without the support of some practical test work. In the case of the tumbler mixers the mixing performance is certainly a function of the operating conditions and an optimization programme could be necessary in order to make a true mixer assessment.

3.4   Selection based on mixing costs

If more than one mixer can be found that satisfied both the process and the mixture quality requirements then it is likely that a final mixer selection will be based on the unit costs of mixing.

Costs should be kept in perspective. Generally, powder-mixing costs represent only a very small percentage of the total product manufacturing costs. Mixing becomes expensive only when production time is lost due to a failure to meet product specification and it is therefore more important that the mixer meets mixture quality specifications and integrates fully into the process than a small economy be made on running costs.

The costs of mixing can be split into three components:

(i) Depreciation on capital cost;
(ii) Power requirements;
(iii) Labour costs.

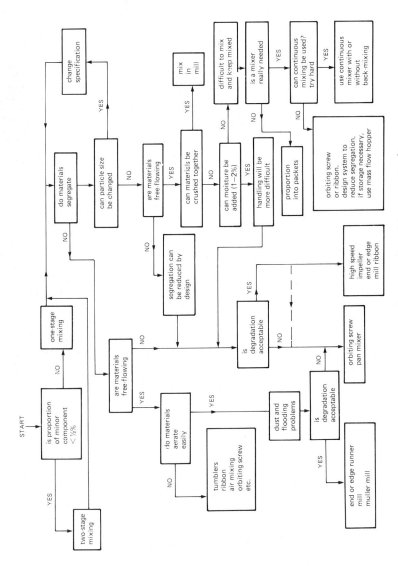

Figure 3.9   Mixer selection decision chart (based on Schofield and Miles[8])

An analysis of powder batch mixing[7] suggests that there is little difference in the unit running costs of conventional convective and tumbler mixers. Labour costs were the dominant cost component in determining the overall costs of mixing in the majority of the cases examined and any reduction in labour utilization produces a marked improvement in unit costs.

One method of reducing the labour content is to reduce the frequency of operating the mixing cycle by increasing the installed capacity of the mixer beyond the minimum flowsheet capacity. This has the effect of offsetting savings in labour with an increase in depreciation costs and an optimization of the cycle time can be carried out. In the cases considered overcapacity was justifiable for the lower mixing rates. The practical advantage of decreasing the cycle frequency is that labour is released for longer periods and can be used for other operations.

A second method of minimizing the labour content is to operate a continuous mixing system. In this case labour utilization is very low and the function of the mixer is chiefly to act as a buffer vessel to eliminate surges in feed rate of the components. A continuous system does not, therefore, have the disadvantage of increasing depreciation costs and labour costs are reduced. Whilst such a system minimizes the mixing unit costs, it should be emphasized that the costs of ancillary equipment will increase as the emphasis is placed on the controlled flow of solids from a storage system.

3.5   Selection decision chart

As has been seen in sections 3.2 and 3.3, the choice of a mixer is governed by the process materials, process requirements and the efficiency of the mixer. In *Figure 3.9* an attempt is made to order the various decisions to be made into logical sequence[8]. Again it must be emphasized that the chart is a guide only and is not a substitute for carrying out tests on a selected mixer.

References

1   ADAMS, J. F. E. and BAKER, A. G. (1956) *Trans. Inst. Chem. Eng.*, **34**, 91.
2   GREATHEAD, J. A. A. and SIMMONDS, W. H. C. (1957) *Chem. Eng. Prog.*, **53**, 194.
3   POOLE, K. R., TAYLOR, R. F. and WALL, G. P. (1964) *Trans. Instn. Chem. Eng.*, **42**, T166.
4   ASHTON, M. D. and VALENTIN, F. H. H. (1966) *Trans. Instn. Chem. Eng.*, **44**, T166.
5   HARNBY, N. (1967) *Powder Technology*, **1**, 94.
6   WILLIAMS, J. C. and KHAN, M. L. (1973) *The Chem. Eng.*, **269**, 19.
7   HARNBY, N. (1963) *Chem. and Proc. Eng.*, **49**, 53.
8   SCHOFIELD, C. and MILES, J. E. T. (1968) *Proc. Eng.*, Sept., 3.

Chapter 4

Powder mixing in gas fluidized beds

D Boland

Imperial Chemical Industries, p.l.c., Teesside

Over the last 25 years the industrial utilization of gas fluidized beds has increased markedly, and their application now extends into many diverse fields. Gas fluidized beds have proved of particular use in the cracking and reforming of hydrocarbons, in synthesis reactions used for producing organic products and in physical processing, such as drying and freezing. The fluidized state also provides a convenient means of transporting and mixing particles.

This chapter is concerned with the phenomenon of particle mixing and the associated effect of particle segregation. A comprehensive but brief record of the mechanics and capabilities of fluid bed mixers and segregators is provided and the techniques which are available for assessing the effectiveness of such units are also discussed.

4.1 Fundamentals of fluidization

When gas is injected into the base of a bed of particles, the magnitude of the drag force exerted on the particles is generally about 70–80 times that experienced by an isolated particle subjected to the same gas velocity. If the velocity of the gas is raised to a sufficiently high level, then the drag force balances the weight of the particles and the condition known as fluidization occurs. Incipient fluidization occurs at a gas velocity considerably below the terminal value.

4.1.1 Characteristics of fluidized beds

  (i) The bed behaves like a liquid and it is possible to propagate wave motion. The normal pressure–depth relationship of liquids exists and bubbles of gas can be seen to rise in a similar manner to bubbles in a liquid.
 (ii) The surface area afforded by fluidized particles is large.
(iii) Particles can be made to move about the bed in a rapid manner.

A secondary characteristic of bubbling fluidized beds is that no temperature gradient exists within the mass of the fluidized particles. This isothermal property results from the intense particle activity which occurs in such systems; the continual movement of particles from one bed zone to another prohibits the establishment of localized temperature gradients. Heat transfer between the bed and immersed surfaces is also enhanced by the rapid solids motion; the continual flow of particles around the heating surfaces ensures that a high temperature gradient is maintained.

The large surface area available has made the fluidized bed very popular for catalytic reactions, particularly those involving the critical control of temperature or large heat transfers.

4.1.2 Minimum fluidization velocity

Incipient fluidization occurs when the pressure drop across the bed is equal to the weight per unit area of the particles in the bed; i.e.

$$\Delta p = \frac{W}{A}$$

The gas velocity prevailing at this condition is known as the minimum fluidization velocity, (U_{mf}). It is usual to quote this quantity as a superficial linear velocity, i.e. the gas velocity achieved in the empty bed shell. If the pressure drop is plotted against gas velocity for a bed of closely graded particles, a characteristic curve (*Figure 4.1*) is obtained. The initial straight-line gradient represents the velocity range over which a fixed bed exists while the crest of the 'kink' at the end of the fixed bed section represents the point of minimum fluidization (U_{mf}). At velocities in excess of U_{mf}, the bed pressure drop remains substantially constant.

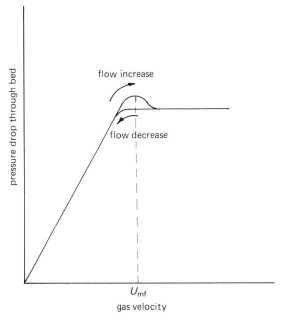

Figure 4.1 The variation of pressure drop with superficial gas velocity through a fluidized bed for a narrow particle size distribution

It should be noted that the kink is only seen on the flow-increase cycle. A slightly higher pressure drop is established during the transition from fixed to fluidized state as a result of the expenditure of energy in overcoming interparticle bridging forces which do not exist during defluidization.

It can be shown[1] that the minimum fluidization velocity is a function of the square of the particle diameter and the difference between particle density and gas density. It follows that the quantity of gas required for incipient fluidization rises markedly as particle size and/or density increases.

Various correlations are available for predicting U_{mf} (refs. 1 and 2), but as their accuracy is often only about $\pm 50\%$ a practical determination is recommended whenever possible.

4.1.3   Gas bubbles

When the fluidizing velocity is greater than U_{mf}, bubbles of gas rise through the bed. Generally speaking, the volume of gas passing through the system in the form of bubbles is approximately equal to the excess volumetric flow-rate; i.e.

Volume of bubbles $= A(U_t - U_{mf})$

where U_t is the total superficial gas velocity.

This generalization does not hold for fine, low-density systems (say, $<100\ \mu m$ mean size and <1.5 gm/cc density) which tend to accept considerably greater amounts of gas into the emulsion phase than that corresponding to U_{mf} without forming bubbles[3].

Gas bubbles are spherical capped and possess an indentation at their rear, known as the 'cusp' (*Figure 4.2*). The cusp is associated with the bubble wake which is discussed in section 4.4.1.

For gas velocities above U_{mf}, small bubbles are formed just above the gas distribution plate. These small bubbles tend to join together or coalesce as they rise through the bed with the result that the bubbles present in the upper

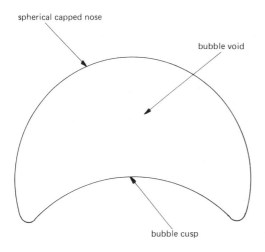

spherical capped nose

bubble void

bubble cusp

Figure 4.2   Gas bubble shape

regions of the bed are considerably larger, but less numerous than those near the base.

4.2   Application of fluidized bed to mixing

Particle mixing is usually easy to achieve in a fluidized bed. The only likely exceptions are when the particles are so fine that they are difficult to fluidize ($< 50 \mu m$) or when strong interparticle forces (usually of an electrostatic nature) are set up during fluidization. However, not all systems involving differences in particle size and/or density can be treated successfully by fluidization and proposals to mix systems of these types should be supported by preliminary practical tests at an early stage in the project.

Although it is difficult to achieve good fluidization and mixing of fines, the fluidization of large particles, up to about 1 cm or more in diameter, is quite possible. However, a practical limitation may be imposed by the large quantity of fluidizing gas required.

4.3   Mechanisms of particle mixing

The fundamental mechanisms of particle motion are now well known, largely as a result of research carried out at Harwell by Rowe and his co-workers. They were able to show that the causes of particle mixing are:

(i) Particle transport in the wakes of gas bubbles;
(ii) Bulk displacement of particles by rising bubble voids.

4.3.1   Bubble wakes

In order to demonstrate the effect of the bubble wake, Rowe *et al.*[4] used X-ray photographs of bubbles rising through a bed composed of a lower layer of lead glass beads and an upper layer of soda glass beads.

It was pointed out that as a bubble rose through the upper layer it took with it a wake composed of material from the lower layer. Further tests involved examining, by dissection, sections of a bed initially formed from layers of different colours of glass beads. It was found that most of the initial wake material gathered at the base of the bed was not carried all the way up to the top of the bed but was deposited in discrete amounts at intermediate points up the bed. The experiments further suggested that wake deposition was balanced by a continuous acceptance of material from the emulsion phase.

Rowe[5] suggested that the three-dimensional bubble wake is composed of a rotating torus into which particles are continuously accepted from the emulsion phase. The continuous growth of the torus provokes periodical instabilities, and these result in the deposition of discrete quantities of wake material to the emulsion phase.

Subsequent work[6] carried out in a two-dimensional bed (a bed in which two faces are separated by only a small gap, say, 1 cm) has indicated that particle motion around a rising bubble is basically of the form suggested by Rowe. A tracer technique was used to highlight particle motion around the essentially two-dimensional bubbles formed between the two closely spaced surfaces. Analyses of ciné sequences revealed that the streamlines of particle motion in the vicinity of a rising bubble are as shown in *Figure 4.3*.

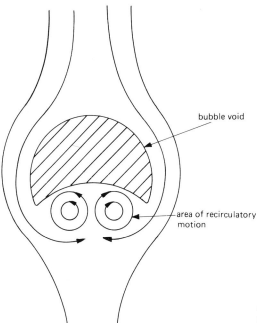

bubble void

area of recirculatory motion

Figure 4.3   Streamlines of particle motion in the vicinity of a rising bubble

Flow around the nose is as expected from potential flow theory while, at the rear of the bubble, there exist areas of recirculatory flow resembling a vortex pair. Material which has flowed around the sides of the bubble void also flows around the perimeter of the wake zone and then converges towards the vertical axis of the bubble in such a way that particles actually enter the wake by moving centrally upwards. As the wake accepts an increasing amount of material and grows progressively larger, the downflow of the emulsion phase, relative to the bubble, causes the vortices to become elongated. Eventually the vortices become so large and distorted that they are shed.

Thus, the information revealed by the ciné sequences appears to be in agreement with Rowe's earlier work and the overall conclusion to be drawn is that bubble wakes provide a means of elevating discrete masses of particles from low to high bed regions.

4.3.2   Drift profile effect

Darwin[7] has shown that a drift profile is produced when a solid body moves through an incompressible fluid; elements which initially form a straight line at right-angles to the line of motion of the body are drawn out in a spout behind the body due to its passage across the line of the elements. The two-dimensional effect is shown in *Figure 4.4* for the case of a cylindrical object.

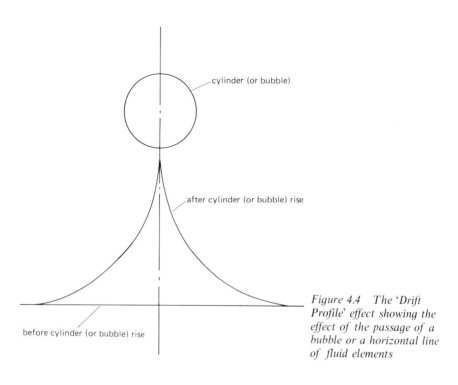

Figure 4.4   The 'Drift Profile' effect showing the effect of the passage of a bubble or a horizontal line of fluid elements

The reason for the formation of the spout-like drift profile can be understood by reference to *Figure 4.5*. As the cylinder moves through the fluid it tends to push elements of the fluid in front of it. Eventually these same elements 'slide' round to the back of the cylinder and are displaced further from their starting point as they fill the space behind the moving cylinder. The nearer the element is to the centre line of the path of motion of the cylinder, the further it is displaced.

A phenomenon of this type was observed by Rowe in the course of his X-ray experiments on beds of soda and lead glass beads. (See also section 4.4.1.) As a bubble crossed the lead/soda interface it was seen to draw up a spout of material from the lower layer behind it. Rowe pointed out that the passage of a bubble through a fluid bed must therefore give rise to a bulk displacement of particles up the bed in the region immediately surrounding the path taken by the bubble.

NUMBERS INDICATE TIME SEQUENCE

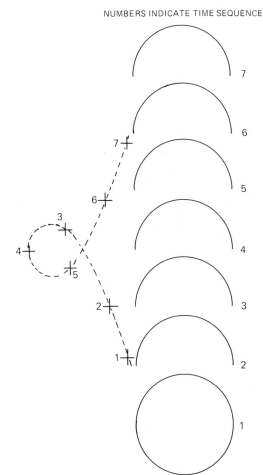

Figure 4.5   Incremental movement of a discrete particle during the passage of a bubble

4.3.3   Overall significance of bubbles

The exact contribution to particle mixing caused by an individual bubble is difficult to assess. Rowe[5] has estimated that bubbles displace two-thirds of their volume (the sphere volume) upwards through one to two bubble diameters; 75% of the displacement is caused by the drift profile, 25% by the wake. Woollard and Potter[8] have carried out a quantitative investigation into this matter and have reported that only 30–40% of the void (not sphere) volume is displaced by the drift and wake. A complication is that the number of bubbles required to stir up a bed and effect a good mix is greatest for those mixtures which are most difficult to attain.

As bubbles rise in a fairly random manner, the possibility of attaining random mixing of particles is good.

4.3.4   Mixing in the absence of bubbles

Mixing does not generally occur when the bed is fluidized at very low or zero excess gas velocities because insufficient bubbles are formed to cause bulk displacement of the particles. The only exception to this statement is for the case of fine mixtures with a mean particle size of less than about 100 μm.

Rowe *et al.*[5] have pointed out that fluidized fines are separated from their neighbours by greater distances than are their coarser counterparts and that this is reflected in the characteristically large bed expansion exhibited by fine systems during the transition from the fixed to fluidized state. It was suggested that a consequence of the high voidage was that interparticle penetration takes place giving rise to a diffusive type of particle mixing. Evidence in favour of this theory was provided by X-ray photographs of bubbles in fine mixtures; these showed that, although the wake and drift profile mechanisms occurred, the overall effect was modified from that for coarser mixtures due to a swirling eddy diffusion type of motion.

It should be stressed that there is little direct evidence to suggest that diffusive mixing, on the scale of individual particles, takes place for mixtures with a mean particle size of much above 200 μm.

4.4   Mechanisms of particle segregation

Before proceeding with a detailed discussion of the mechanisms of segregation it is worthwhile to give a summary of the conditions which can exist in an unbaffled bed containing a wide size distribution. (See also ref. 6.)

If the flow rate is very high then complete mixing is approached. However, if the flow is reduced somewhat, but is still relatively high in comparison with U_{mf} for the mean particle size, then the bed will consist of a strongly bubbling region, in which complete mixing of fine particles occurs, and a lower incipiently fluidized region of segregated coarse particles. Further flow reduction can lead to the formation of a fixed bed of coarse particles near the distributor with the incipient segregated region occupying the middle section of the column and the bubbling mixed section the top portion.

4.4.1   Drag forces

Sutherland and Wong[12] have demonstrated that segregation is dependent on differences in drag between the segregating particles as follows.

For a stable fluidized particle, drag and gravitational forces can be imagined to balance and the following expression can be written.

$$F_D = F_G$$

$$K \cdot f(\varepsilon)d^n U^m = \pi/b \cdot g \cdot (\rho_s - \rho_f)d^3$$

$$K \cdot f(\varepsilon)U^m = \pi/b \cdot g \cdot (\rho_s - \rho_f)d^{3-n}$$

where K is a constant; $f(\varepsilon)$ is some function of voidage, U is Re superficial velocity and d is the particle diameter.

If the mean size of the particles is d_{mean} then the drag on finer constituents of the distribution (size d_f) is

$$F_D = Kf(\varepsilon)d_f^m \cdot U^m$$
$$= \pi/b \cdot g \cdot (\rho_s - \rho_f) \cdot d_{mean}^{3-n} \cdot d_f^n$$

This force will not be balanced by gravity and the net upward force will be

$$F_{NET} = \pi/b \cdot g \cdot (\rho_s - \rho_f) \cdot d_{mean}^{3-n} \cdot d_f^n - \pi/b \cdot g \cdot (\rho_s - \rho_f)d_f^3$$
$$F_{NET} = \pi/b \cdot g \cdot (\rho_s - \rho_f) \cdot d_f^n(d_{mean}^{3-n} - d_f^{3-n})$$

For the laminar regime $n = m = 1$.

If the coarser particles of the distribution are considered, a net downward force is indicated. Also, examination of systems possessing particles of different density would demand that, for equi-sized particles, less dense material rises, whilst more dense material sinks towards the distributor. An expression of the form

$$F_{NET} = \pi/b \cdot g \cdot d^3(\rho_{sin} - \rho_s')$$

is obtained for this case.

For mixtures of varying size and density, those particles possessing the highest drag per unit mass will rise; this could be the large or small particles, depending on the relative densities.

For segregation by density, the net drag varies with the density difference. It seems likely that systems of variable size (net drag varies with diameter cubed) might be the more susceptible to segregation effects.

4.4.2 Conditions within the bed

Most researchers (e.g. refs. 2, 8 and 13) have considered segregation to occur as a result of the fluidizing gas transporting fine particles to the upper regions via channels formed within the bed. Thomas *et al.*[13] thought that transport of fines (by the fluidizing gas) through the interstitial spaces between particles might also be possible and Desphande *et al.*[10] stated that if the gas velocity was only sufficient to fluidize the component of a two-component mixture possessing the lower critical velocity, then the particles of this constituent would diffuse through channels in the fixed bed composed mainly of the heavier component.

The possible importance of bubbles was not considered by any of the authors mentioned above, although in each case segregation was reported for fluidization conditions under which the occurrence of some bubbles was very likely. Boland[6] looked for a possible inter-relationship between bubbles and segregation. In one experiment a mixture of equal volumes of 300 μm blue and 100 μm white glass beads was fluidized in a two-dimensional bed; disengagement between the blue and white beads could be followed visually and it was found that the rate of stratification appeared to be optimized by allowing the system to bubble gently.

It may well be that the fluidization condition required for segregation to occur at any section of a three-dimensional bed is also one of gentle bubbling. Such a state might induce particles to disengage as a result of disturbances caused by their flow around bubbles. If the bubbling were too vigorous it could be expected that the disengagement effect would be imperceptible in comparison to the bulk mixing effects caused by the bubbles. On the other hand, if the bed were bubble-free and the particles were just at U_{mf}, then they might be locked together in such a way that no segregation could occur.

4.4.3 Step and jump segregation

Experiments[6] indicate that the degree and nature of segregation differs according to whether the flow rate to the bed is reduced from the value for reasonably complete mixing to the selected segregation velocity in one sharp jump (jump segregation) or in a series of incremental steps (step segregation). Tests were carried out in a 15 cm diameter column using 15 cm depths of various segregation prone distributions of sand particles. Samples withdrawn from the bed were analysed by sieving.

4.4.3.1 Step segregation

For the step-segregation tests the procedure was to reduce the velocity from that at which the bed was completely mixed to the desired final velocity in small increments. Four minutes was allowed to elapse between each incremental reduction in order that particle disengagement might occur and that the system might regain equilibrium. The form of the segregation curves can be described and explained by reference to *Figure 4.6*, where the trends are exaggerated for the purpose of clarity. The vertical sections of the curves represent the bed region in which complete mixing occurs whilst the sloping sections indicate regions of segregation. Although the segregation regions are represented here by straight lines, there is no reason why in practice these regions might not be described by a curve.

If the bed is subjected to the minimum velocity at which complete mixing occurs, then its composition will be uniform at all heights. This situation is represented by curve AB. If the velocity is reduced to some slightly lower value, then the system will no longer be capable of maintaining the full distribution in the well-mixed condition and a small accumulation of coarse material will appear near the distributor. For each stage in the flow reduction, more coarse material will leave the well-mixed sector of the bed and some of this will become defluidized; the condition shown by curve CDPQ will be achieved at some intermediate flow rate. The section of the curve between D and Q represents the region of segregation.

As it is unreasonable to expect that interconnection of the vertical and sloping sections (C–D and D–Q respectively) represents a sudden change from

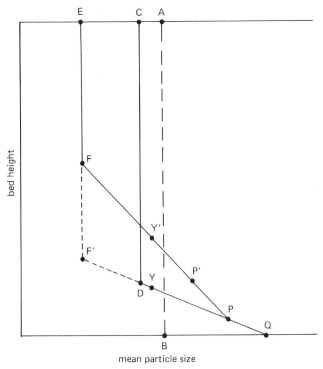

Figure 4.6   The form of step segregation curves

the fixed to the well-mixed fluidized state, it can be deduced that a transition region must exist. It is suggested that this takes the form of an incipiently fluidized region containing only a few small bubbles (insufficient to cause mixing) and in which segregation is evident. This is referred to hereafter as the 'incipiently segregated zone'. The section D–P represents this zone for the conditions of curve CDPQ, whilst P–Q represents the fixed segregated zone.

It might be expected that incremental reduction in flow from that giving the conditions shown by CDPQ would result in curve EF′DPQ. This prediction would be based on the idea that, as soon as the gas flow was reduced, part of the incipiently segregated zone would 'freeze' into a fixed-bed region. However, practical data suggest that in fact flow reduction causes a slight vertical displacement of the segregation curve such that point Y on curve CDPQ corresponds to point Y′ on EFP′PQ.

In effect, this means that flow reduction causes material in the incipiently segregated zone to re-locate itself so that the height at which a given mean size occurs is increased. No doubt, this is attributable to the downward migration of coarser fractions, which leads to the addition of further material to the fixed-bed region and to a modification of the particle distributions present in the incipiently segregated zone.

Figure 4.7 represents the form which is taken by a family of curves representing different end-points. The line passing through P_0, P_1, P_2, P_3

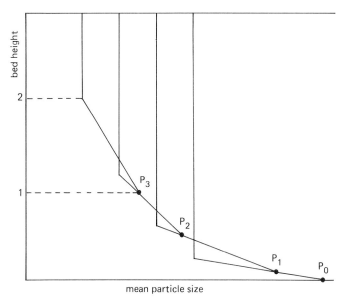

Figure 4.7 A family of step segregation curves

represents the progression of the fixed-bed segregation region. For the final condition shown (lowest flow rate) fixed-bed segregation occurs between the distributor and height 1, while incipient segregation exists in the region from height 1 to height 2.

4.4.3.2 Jump segregation

Jump-segregation tests were performed by reducing the flow from a velocity which gave reasonably good mixing to the end-point velocity in one sharp jump. The results of the practical tests revealed that, for any given distribution, a critical end-point velocity exists. The form taken by the segregation curve is markedly different for end-point velocities above and below the critical velocity. Typical jump segregation curves are presented in *Figure 4.8*; curve (a) typifies the high-velocity end-point and curve (b) that for the low-velocity case.

The form of curve (a) is identical to that expected for the step segregation system. Although the amount of coarse material disengaging per unit time must be much greater than that which does so in an incremental reduction, the similarity of the form of the curves indicates that the bed conditions corresponding to curve (a) are basically the same as those for step segregation, i.e. a bubbling well-mixed zone and incipient and fixed segregated zones. Curve (b) in *Figure 4.8* can be seen to exhibit an inversion point. Presumably this inversion results from large quantities of comparatively coarse material suddenly being rendered defluidized.

When the flow is cut back, the coarser cuts of particles fall out from the fluidized state almost instantaneously and as they descend they tend to entrain finer material. Because the entrained fines disengage to some extent during the

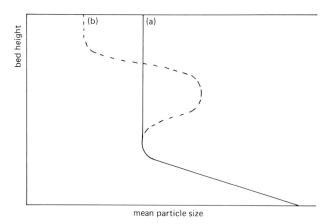

Figure 4.8 '*Jump*' *segregation curves for high and low end velocities.* (a)*Curve for high end-point velocity;* (b) *curve for low end-point velocity*

segregation/entrainment process, a size gradient is observed in the bottom section of the bed. The situation at this stage is stylized in *Figure 4.9*.

The upper zone contains well-mixed fine- and medium-sized fluidized particles, whilst the lower zone contains defluidized coarse with entrained medium and fine particles mainly situated in the upper regions of the zone. The conditions shown in *Figure 4.9* are also represented by curve (a) of *Figure 4.8*.

After the high-speed defluidization effect has occurred, a secondary segregation process takes place in the upper fluidized regions. This second slower phase results in the formation of a stable well-mixed bubbling region near the surface and a segregation region between this surface band and the original lower fixed bed region. This secondary segregation region is probably one of mainly incipient segregation, but, in order that an inversion zone might exist between the primary and secondary segregation zones, it is necessary that at least the lower portion of the secondary region is fixed. If it were not, then the uppermost portion of the primary zone, which is of lower mean size, would be expected to be fluidized and the two zones would merge.

The situation at this point is represented by *Figure 4.10*, which shows no change in the conditions of the lowest bed zone from *Figure 4.9*. However, a second segregation region composed of mainly medium-sized particles is shown to be formed, whilst the particles remaining in the upper well-mixed zone are mainly of a comparatively fine size. Curve (b) of *Figure 4.8* indicates the significance of this second phase of the segregation in terms of a bed height plotted against mean particle size.

Summarizing, the inversion effect of the jump system can be considered to result from the sudden flow reduction causing a high-speed drop-out of coarse particles followed by a slow segregation process occurring in the fluidized bed which remains after the primary fixed bed has been formed.

It should be noted that the effect only takes place if the jump in gas velocity is great enough to cause a sufficiently great mass of material to be deposited

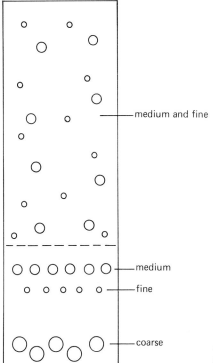

medium and fine

medium

fine

coarse

Figure 4.9   Initial stage in jump segregation

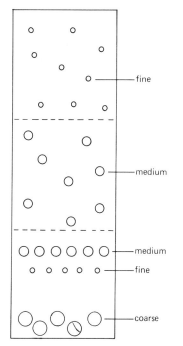

fine

medium

medium

fine

coarse

Figure 4.10   Second stage in jump segregation

instantaneously. It is thought that some degree of instantaneous coarse deposition occurs even when no inversion is perceptible. Due to entrainment of fines, the pureness of the segregated layer is not comparable with that for the step system.

4.5   Pressure drop/velocity curves for segregating systems

It has been shown[6] that it is possible to ascertain whether a system is segregation prone by examining the pressure drop/velocity curve.

The form taken by a segregation-prone mixture is indicated in *Figure 4.11*. During the flow increase cycle the whole of the bed will be defluidized until the velocity is increased to U_p. At this point, it is thought that most of the bed will

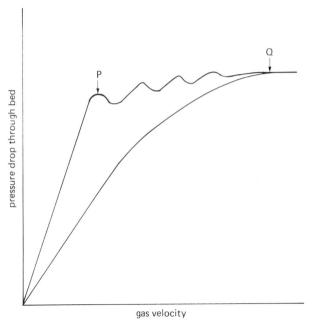

Figure 4.11   Pressure drop/velocity curve for segregating particulate system

be close to being fluidized but only the upper bed region will have actually reached the critical condition. When the velocity is increased by a further small amount, the surface regions will become sufficiently well fluidized for segregation to occur and coarse material from this locality will leave the fluidized state by moving downwards to the surface of the, as yet, fixed bed. As the flow is increased in successive increments, the basic process will be repeated as progressively lower layers of the bed are subjected to fluidization and segregation.

This repetitive process of fluidization and segregation of individual sections results in the pressure-drop curve for the flow-increase cycle exhibiting a characteristic undulating form. The form of the curve can be accounted for in the following manner.

As the flow is progressively increased, the pressure drop fluctuates according to whether the governing mechanism is fluidization or segregation. Every time segregation occurs, the depth of the fixed bed will be increased and consequently the overall bed pressure drop will be decreased. However, the next increase in the flow rate will probably institute incipient conditions in some lower bed region so that the pressure drop will rise again. This rise and fall will be cyclic.

The initial pressure drop decrease of any section will also be partly attributable to the reduction in fluidization energy which occurs after interparticle contacts have been broken. However, subsequent undulations at higher flows cannot be associated with this phenomenon, and must be attributed solely to segregation effects.

During the flow decrease cycle, particle disengagement is more complete. In this instance, stratification does not occur in the layer-by-layer fashion described above; rather, for a given flow reduction, coarse particles are able to drop simultaneously out of the fluidized state from a greater range of bed heights than for the previous case. The result is that for any flow rate, in the range P to Q, the fixed and incipient segregation regions are composed of purer coarse material than for the flow increase path. As a consequence the voidage of these regions is higher for the flow decrease path and the pressure drop is therefore lower.

The reason that, during the flow increase cycle, surface regions fluidize first is probably related to packing effects in the fixed bed. Particles near the surface may fluidize at a slightly lower velocity than those near the distributor where they must be more tightly packed due to the effects of the weight of the particles in the upper bed regions. Gas expansion is unlikely to be of importance in such shallow beds, but the variation in U_{mf} values is likely to be accentuated by the deposition of coarse material from fluidized upper layers to fixed lower layers. The presence of this coarse material may well increase the U_{mf} value of the fixed-bed layers.

For a given mean size of particles, it appears that the interval P–Q increases as the standard deviation (and therefore the segregation tendency also) of the mixture is increased.

4.6   Effects of critical parameters on mixing

4.6.1   Excess gas velocity

The most important single factor in a fluidized system is, without doubt, the gas velocity. More particularly, it is the gas which passes through the system in the form of bubbles which determines the degree of mixing or segregation. This quantity can be expressed in terms of an excess gas velocity as $(U_t - U_{mf})$. Virtually all investigators have mentioned the importance of the gas velocity.

It has already been mentioned that segregation is optimized by the presence of small bubbles in the bed by the use of a small excess gas velocity and the transition between the conditions for segregation and good mixing is therefore not clear cut. It cannot be assumed that because a system appears to be bubbling reasonably strongly complete mixing is taking place and more positive information should be obtained from experimentation.

4.6.2   Mean particle size

The ease of mixing appears to vary with particle size. The general trend seems to be that systems of higher effective bed viscosity require more energy in the form of bubbles to effect a good mix. According to viscosity data (e.g. ref. 26), this generally means that the larger the particle size, the greater must be the excess gas velocity for mixing. The situation is made more complex by the effect of particle shape. Microspheroidal particles tend to exhibit lower viscosities than do more angularly shaped particles.

Kunii and Levenspiel[9] have shown by back calculation from a solids circulation equation that the wake fraction decreases with increase in particle size. This alone means that larger particles require more bubbles for mixing.

Extensive tests on the mixing and segregation of distributions of sand particles[6] has shown that, generally speaking, the excess gas velocity required to effect a complete mix increases as both the standard deviation and the mean size of the particles increases. However, the experimental data indicates that for mixtures with a mean size of less than about 150 μm, the degree of mixing effected by a given excess gas velocity decreases as the mean size tends to zero, i.e. the standard deviation which can be supported in the well-mixed state at a given excess gas velocity becomes progressively narrower as the mean size is reduced from 150 to zero μm. This effect could well be related to the tendency of fine mixtures to accept considerably more gas into the emulsion phase than that corresponding to U_{mf} (ref. 3); thus it may be that for any given excess gas velocity the number of bubbles produced tends to decrease directly with mean particle size.

4.6.3   Particle density

Only limited data (e.g. refs. 10 and 11) are available on systems of mixed density and this is an area where practical information is of great importance. However, the basic behaviour of fluidized particles can be deduced from a consideration of the drag forces exerted on the particles by the gas stream (see section 4.5.1). Thus, the results of Nicholson and Smith's experiments[11] on mixtures of different densities but similar sizes of particles are entirely expected.

These authors showed that for a bed initially composed of an upper layer of heavy particles and a lower layer of light particles fluidization occurs in such a way that the mixing rate constant first increases as the heavier particles fall through the coarser ones. However, this initial phase is followed by a decrease in the rate constant as the heavy particles continue their downward migration and, in effect, segregate out. For the reverse starting conditions of fines at the top, the rate constant is found to be very low except for the relatively high flow rates at which the lower layer is sufficiently well fluidized for its particles to be mobile.

4.6.4   Aspect ratio

The conditions which favour segregation are likely to be the opposite of those favouring mixing. In particular, high aspect ratios lead to segregation, low ones, to perfect mixing.

Thomas *et al.*[13] investigated segregation effects in a 10 cm I.D. bed. They found that the concentration of fines in the upper regions of the bed increased as the aspect ratio was increased from 3 to 9. Sutherland, working in conjunction with first Wong[12] and then Capes[17], found that the same effect prevails in beds (both continuous and batch), containing cylindrical mesh baffles, and with aspect ratios of up to 22.

Nicholson and Smith[11] found that their mixing rate constant varied inversely with the cube of the aspect ratio. They considered that this effect was caused by bubbles only transporting their wakes over short distances before depositing them. Thus, for high aspect ratios, bubbles did not provide a direct path for particle transport from the bottom to the top of the bed.

It is possible that the decrease in efficiency of mixing with increase in aspect ratio may be a reflection of the tendency of high aspect ratio systems to 'slug'. The mixing properties of slugs are not likely to be comparable with those of bubbles, because drift profile and wake effects are of little importance.

4.6.5   Baffles

From the literature it appears that the presence of baffles in a fluidized bed increases the possibility of segregation. Hall and Crumley[16] observed no marked segregation during fluidization of Fischer–Tropsch iron catalyst in a 120 cm-deep 2.54 cm-diameter bed. However, when horizontal baffles were introduced into the system, a size segregation effect was noted.

Sutherland and Wong[12] and Botterill and Kunte[14] used baffled beds in an attempt to improve segregation. The latter workers carried out a factorial analysis of the results of their experiments. This revealed that out of the parameters tested (particle distribution, baffle hole sizes, gas velocity, number of baffles), the degree of segregation was most significantly affected by the number of perforated plate baffles present in the system. No significant trend was detected for variations of baffle spacing within a range of 8.5 to 3.5 cm. (The column was 3.8 cm square.)

This contrasts with some of Hall and Crumley's tests on batch systems[16] which showed that segregation became stronger as the horizontal baffles were located closer together. Capes and Sutherland[17] compared the segregation tendencies of cylindrically baffled and unbaffled beds of 7:1 aspect ratio. They found that the former gave appreciably better segregation.

Presumably baffles enhance segregation because their ability to break up bubbles[9] results in a low level of particle mixing even at quite large excess gas velocities. Loosely packed cylindrical mesh baffles are likely to be particularly effective for breaking up bubbles.

4.6.6   Vibration

It is generally accepted (e.g. refs. 18 and 19) that vibration leads to a reduction in the gas velocity required for incipient fluidization and also to an improvement in the quality of fluidization. Recent tests[6] have shown that vibration can also be an effective method of increasing particle segregation. It was found that for a fluidized bed vibrating at approximately 50 cps and operating at a gas velocity above the reduced U_{mf} value, the following mechanisms affect particle movement:

(i) Bubbles promote particle disengagement or mixing according to the number and size present. Disengagement only predominates for gently bubbling systems.

(ii) Fines which are exposed to a greater drag force than necessary for fluidization rise up the bed whilst coarser cuts which are not supported by their U_{mf} velocity accumulate near the base.

(iii) Vibration causes disengagement of fines and coarse so that, unless the drag force afforded by the gas is sufficiently high, fines fall under the effect of gravity through the voids between the coarse particles. This phenomenon is known to occur in non-fluidized particulate systems.

The degree of particle segregation achieved in a vibratory system depends initially on the gas velocity. If this quantity is large, then mechanism (i) is found to dominate and good particle mixing occurs. However, if the velocity is reduced somewhat then the vibration induces good particle disengagement and fines are transported to the upper regions and coarse cuts to the base in accordance with mechanism (ii). Finally, at very low excess gas velocities, the drag on even fine particles is insufficient for transport to the surface of the bed and mechanism (iii) predominates.

It can be seen, therefore, that for vibratory systems optimum segregation occurs at an intermediate excess gas velocity. Practical tests also revealed that when the flow was reduced to this optimum velocity in a stepwise manner, then the degree of segregation which occurred was significantly greater than the maximum step segregation attainable under normal conditions. However, the degree of segregation recorded for lower end-point velocities with mechanisms (iii) dominating was less than that achieved in a normal system.

4.6.7   Pulsation

If the air supply to the bed is pulsed on and off then the bed tends to rise and fall in a piston-like manner. It appears that this reciprocating motion enhances the disengagement between different types or sizes of particles[6].

Although only a limited number of tests have been carried out, the indications are that segregation is greater at a pulse frequency of 3 cps than at higher frequencies up to 25 cps. Also, the degree of segregation attainable by stepwise reduction of the mean velocity (from that required for complete mixing to conditions close to U_{mf}) appears to be greater than the maximum segregation which can be attained by step flow reduction under either normal or vibratory fluidization conditions.

4.7   Practical arrangements

4.7.1   General considerations

The design of the bed will depend on whether the unit is to be used for mixing or segregation. For large-scale units it is advisable to restrict the aspect ratio to a maximum of 1.5:1 if good mixing is required. High aspect ratios of 10:1 or even more are used in fluid-bed segregators.

In order to minimize the carry-over of fines from the surface of the bed, it is necessary to incorporate a disengagement zone above the surface of the fluidized particles. The diameter of this section is sized so that the superficial velocity of the gas passing through it is sufficiently low to preclude extensive entrainment of fines. The zone should be deep enough to eliminate the carry-over of any particles which might be shot high above the surface by the localized expulsion of gas from a bursting bubble or possibly even a channel. The depth required will depend on the fluidization conditions and on the particles themselves.

It is common practice to distribute the gas to the bed using either a perforated plate or a bubble cap plate. Perforated plates suffer from the drawback that particles are able to fall through the perforations when the bed is defluidized. Bubble cap types are limited in their effectiveness on account of their tendency to become choked by the fluidized particles. In order to ensure even distribution of gas between the holes of the distributor it is necessary for the plate to offer an appreciable resistance to the flow of the fluidizing gas. Plates are usually designed to operate at a specific fraction of the pressure drop across the fluidized bed. A safe fraction to design for is 0.5.

It has already been mentioned that most baffling arrangements tend to break up bubbles and enhance segregation. Whilst randomly packed cylindrical mesh baffles are particularly effective and horizontally orientated surfaces are also good, vertical rods are not recommended in the present application as they tend to act as 'chimneys' up which bubbles are able to rise.

The apparent increase in segregation brought about by the use of pulsation and vibration of the fluidized bed is discussed in detail in sections 4.6.6 and 4.6.7, but the combined effectiveness of these measures has not been investigated. It seems likely that segregation could be maximized by the simultaneous use of both systems. However, it may be impractical to employ vibration on some large-scale units.

It is possible to successfully operate both mixers and segregators on a continuous basis. The optimum feed location for a segregator appears to be in the top half of the column, say, two-thirds of the way up the column[17]. The feed and take-off point for a mixer can be conveniently located at the surface of the bed as can the top take-off point for a segregator. Exit streams can be easily controlled by weir arrangements. The bottom take-off for a segregator will be located just above the distributor and in this case the flow must be controlled possibly by use of either a rotary or iris (e.g. Mucon) valve.

One major problem with fluid-bed mixers is that the exit stream flowing over the weir may tend to segregate during its removal from the bed and careful design of the take-off arrangement is therefore called for. A potential problem area in the design of a segregator is the selection of ancillary equipment capable of providing a controlled continuous flow from the bottom take-off.

It is important to note that, where appropriate, the potential of the fluidized bed as a mixer or segregator may be combined with its advantages of good heat transfer, easy temperature control, isothermal conditions and large particle surface area. Further information on general aspects of fluidization engineering is available in ref. 9.

4.7.2   Data on industrial processes

4.7.2.1   Mixing

Literature describing industrial fluidized mixing is fairly limited. D'Arcy-Smith[20] discussed aspects of such processes with particular emphasis on applications in the cement industry. He considered that mixing units could be made more efficient by introducing external lifter tubes to elevate the powder from several levels simultaneously. Also mixing could be improved by supplying different air velocities to different parts of the distributor. This technique results in the bed developing a spiral mixing action. He warned that not all powder combinations can be successfully mixed by fluidization. Kunii and Levenspiel[9] have also given details of mixers in which lifter tubes are used, but in their illustrations the tubes are located internally.

Leva[21] investigated the industrial potential of fluidized mixing using two-component systems of glass beads. He attempted to improve the mixing properties of stratification-prone combinations by introducing a stirrer into the bed. However, problems arose due to the occurrence of 'dead areas' of little or no particle activity. To overcome this, Leva adapted a conical-shaped fluidization chamber fitted with a stirrer. This was more satisfactory. However, a conical unit can be expected to stifle bubble nucleation and growth at elevated bed levels and as a consequence natural particle motion will be limited. It seems that the adapted design rejects one of the important advantages of fluidization, viz., the nucleation and growth of bubbles at elevated bed heights, and replaces it with an energy consuming stirrer.

Fomichev and Gvoziev[22] have described tests on a small-scale continuous mixer designed to investigate the industrial potential of such units. They found that they could successfully mix salt and tridiphenylamine mixtures in a continuous manner.

Levey *et al.*[23], on the other hand, described tests on tapered beds which showed that continuous plug flow of particles can be approached in such units. In their case, this was desirable because the reaction which they were studying (conversion of UO_3 to UF_4) demanded long residence times with no short circuiting.

4.7.2.2 Segregation

Capes and Sutherland[17] carried out a thorough investigation into the use of gas fluidization as a means of separation of ores. The main study was concerned with narrow cuts ($-20+45$ mesh) of germanium ore, consisting of coarse sandstone, dispersed with fine 'eyelets' of carbonaceous material. The germanium was contained in the carbonaceous material and the density of particles varied between 1.2 and 2.7 depending on the amount of the latter present. Experiments were performed in 3 and 5 cm-diameter columns using aspect ratios of up to 22. Batchwise tests revealed the same segregation tendency as reported in a previous Sutherland paper[12], although, in this present case, segregation was on a density basis rather than the size basis used previously. Further experiments, concerned with continuous systems, employed an overhead top product take-off, bottoms take-off near the distributor and the feed located at some intermediate point. Tests showed that, as for the batch system, optimum segregation occurred at just above U_{mf}, with a minimum at some slightly higher velocity. Data were presented showing that there is an optimum combination of top product/feed ratio, at which recovery of germanium is a maximum. The effect of feed rate was also studied using a 108 cm-deep, 5 cm-diameter bed with the feed supplied two-thirds of the way up. For selected test ratios of top product/feed rate it was demonstrated that the quality of the top product increased as the average particle residence time increased until an optimum value was reached, above which no further improvement in segregation could be attained. This optimum residence time was found to be approximately 7 minutes, which corresponds to a solids throughout of 300 g/min. Batchwise tests on a similar bed suggested that it reached the equilibrium degree of segregation in about the same time.

Tests on various feed locations indicated that the best position was two-thirds of the way up the bed. At this position, entering material came into contact with a region of low-density material in which particle motion was probably freer than lower down the column.

Further tests showed that the technique could also be applied to the upgrading of iron ores.

Other workers have investigated systems which are of some relevance.

Toynbee and London[24] described a fluidized separation process based on the air slide conveyor. They used a continuous system in which fine coals could be dedusted by the upward air flow. The technique relied more on elutriation than segregation. Kaye and Jackson[25] described a new procedure for separating powders of different sizes. The system was designed to improve on the normal stratification which the authors observed to occur in gas fluidization. By introducing a sieve mesh into the fluidization chamber, the starting mixture could be split into two cuts. The separation procedure involved extending the gas velocity beyond that normally associated with stratification, thus it would appear that this process also depends on elutriation. Fitch[26] has described the fluidized bed classification of ores. The ore is fed to the top of a column in which an upward flow of wash fluid is arranged. The fines are removed with the wash fluid, whilst the course is taken off from the base of the unit. Again, an elutriation effect is involved.

4.7.3   Commercial equipment

Up to the present time, fluidized beds appear to have found more outlets as mixers than as segregators. In many cases the normal fluidized-bed arrangement has been supplemented by the addition of devices to supposedly assist the natural mechanisms of mixing. One such addition is the use of lifter tubes (see also section 4.7.2). While it is quite true that these tubes do elevate powder to high bed levels and thereby aid mixing, there seems to be no good reason why the tubes should be more effective than bubbles. Bubbles, moreover, have the advantage that their rise path is quite random. Possibly lifter tubes may be of real value in systems which do not fluidize very well and tend to channel, but, under normal conditions, their cost may outweigh their advantages.

Also mentioned in the last subsection is the technique of fluidizing different bed sections at different velocities in order to promote a spiral mixing action. Again, there is no guarantee that this orderly spiral action promotes a better mix than achieved when the same total volumetric flow is introduced in the form of randomly located bubbles. In the Airmix mixer[27], a spiral fluidized motion is achieved by arranging angled nozzles around the periphery of the base of the bed. Intermittent pulses of gas cause the bed to rise, spiral and fall back repeatedly. In view of the segregatory effect produced by pulsations (see section 4.6.7) it seems unlikely that this particular arrangement can be very efficient.

In the Nautamix system[28], the normal arrangement for a fluidized bed is supplemented by a set of nozzles which are superimposed on the distributor plate so that localized high-velocity jets of gas can be injected into the bed. It seems that these jets induce some degree of channeling and this in turn causes orderly patterns of solids circulation to be set up. Again it must be stated that there is no reason why the orderly circulation patterns should promote more efficient mixing than would occur if the nozzle gas were to merely pass through the system as bubbles introduced via the normal distributor plate.

In the Henschel mixer[29], a fluidized condition is produced by a high-speed agitator which causes the particles to be forced against the wall of the mixer and to become aerated. Mixing is effected in a kind of forced fluidized vortex.

References

1 DAVIDSON, J. F. and HARRISON, D. (1963) *Fluidised Particles*, Cambridge University Press, Chapter 1.
2 LEVA, M. (1959) in *Fluidisation*, McGraw-Hill, New York.
3 DAVIES, L. and RICHARDSON, J. F. (1966) *Trans. Inst. Chem. Engrs.*, **44**, T293.
4 ROWE, P. N., PARTRIDGE, B. A., CHENEY, A. C., HENWOOD, A. G. and LYALL, E. (1965) *Trans. Instn. Chem. Engrs.*, **43**, T271.
5 ROWE, P. N. and PARTRIDGE, B. A. (1962) *Proceedings of the Symposium on Interaction between Fluids and Particles*, Inst. Chem. Engnrs, London.
6 BOLAND, D. (1971) Ph.D. Thesis, Bradford University, UK.
7 DARWIN, C. (1953) *Proc. Cambridge Phil. Soc.*, **49**, 342.
8 WOOLLARD, I. N. M. and POTTER, O. E. (1968) *A.I.Ch.E. Journal*, **14** (3), 388.
9 KUNII, D. and LEVENSPIEL, O. (1969) in *Fluidisation Engineering*, Wiley.
10 DESPHANDE, A. D., THARGAPPAN NADAR, R. and PIA, M. W. (1965) *Indian J. Tech.*, **3**, 111.
11 NICHOLSON, W. J. and SMITH, J. C. (1966) *Chem. Eng. Prog.* (Symposium Series), **62** (62), 83.
12 SUTHERLAND, J. P. and WONG, K. Y. (1964) *Can. J. Chem. Engng.*, **42**, 163.
13 THOMAS, W. J., GREY, P. J. and WATKINS, S. B. (1961), *Brit. Chem. Engng.*, **6** (3), 176.
14 BOTTERILL, J. S. M. and KUNTE, M. V. (1962) *Birmingham University Chemical Engineers*, **13**, 77.
15 SCHUGERL, K., METZ, M. and FETTING, F. (1961) *Chem. Eng. Sci.*, **15**, 1.
16 HALL, C. C. and CRUMLEY, P. (1952) *Journal App. Chem. (London)*, Supp. Issue 1, **2**, 547.
17 CAPES, C. E. and SUTHERLAND, J. P. (1966) *Ind. Eng. Chem.* (*Process Design and Development*), **5** (3), 330.
18 JUDD, M. R. (1965) Ph.D. Thesis, Cape Town University, South Africa.
19 YOSHIDA, T., KOUSAKA, Y. and YUTANI, A. (1966) *Kagaku Kogaku* (Ab. Ed.), **4** (1), 158.
20 D'ARCY-SMITH, F. R. (1965) *Brit. Chem. Engng.*, **10**, 652.
21 LEVA, M. (1962) *Proceedings of the Symposium on Interaction between Fluids and Particles*, Instn. Chem. Engnrs., London.
22 FOMICHEV, A. G. and GVOZDEV, V. D. (1964) *Inst. Chem. Engng.*, **4** (4), 809.
23 LEVEY, R. P., DE LA GARNA, A., JACOBS, S. C., HEIDT, J. M. and TRENT, P. E. (1960) *Chem. Eng. Progr.*, **56** (3), 43.
24 TOYNBEE, P. A. and LONDON, F. W. (1962) *Brit. Chem. Engng.*, **7** (6), 425.
25 KAYE, B. H. and JACKSON, M. R. (1967) *Powder Technology*, **1**, 43.
26 FITCH, B. (1962) *Ind. Eng. Chem.*, **54** (10), 44.
27 'Airmix', Newton Machine & Tool Co. Inc., 2401 Atlantic Avenue, Brooklyn, NY 11233, USA.
28 'Nautamix', P.O. Box 773, Haarlem, Holland, Telex 41167.
29 'Henschel', 400 Dusseldorf, Stein Strasse 23, Germany, Telex 8587375.

Chapter 5

The mixing of cohesive powders

N Harnby

Schools of Chemical Engineering, University of Bradford

5.1   Introduction

It was seen in an earlier chapter that the role of a mixer was to progressively reduce both the scale and intensity of segregation of the mixture ingredients. With free-flowing mixtures the scale of segregation within the mixture could be very large due to the preferential and segregating movement of individual particles.

In the case of cohesive powder mixtures, individual particles do not have the same freedom to relocate and most of the mechanisms of segregation discussed earlier are not relevant. Working with fine powder systems thus minimizes the chance of gross segregation occurring. At the same time the large number of particles within the system improves the statistically attainable mixture quality for a given 'scale of scrutiny' of the mixture.

The combination of these advantages has generally resulted in a mixture quality which satisfies most process industries and, if powder mixing was the only process consideration, the best advice would be to work wherever possible with cohesive systems. Because of the lack of mobility of the particles the mixing process is usually slower for a given mixer but the final equilibrium mixture is of a much higher quality.

The problems associated with the mixing of cohesive powders have been highlighted by those industries which demand a very high degree of uniformity within the bulk mixture. The pharmaceutical and pigment dispersion industries are good examples of industries with this requirement.

Commonly a minor component is to be dispersed within the bulk of the material and then mixture quality analysed at a very small scale of scrutiny. The cohesive mixture is usually of good quality when analysed at a large scale of scrutiny but at a small scale of scrutiny a high, and unacceptable, intensity of segregation can occur. This results from the agglomeration of constituent particles into small but stable groupings of like constituents. Both the form and the strength of these groupings are of fundamental importance in understanding the mixing process as the role of the mixer must be to repeatedly break them down to the scale of the constituent particles.

It has been seen in an earlier chapter that for a free-flowing mixture the art of the process engineer is to restrict the movement of individual particles. For the cohesive system the problem is reversed as the cohesive system has a natural structure which has to be repeatedly broken down in order to give individual particles within that structure an opportunity of relocating themselves.

The nature and the strength of the interparticulate forces acting within the cohesive powder system will determine the ease, or difficulty, likely to be experienced in re-locating individual particles within a mixture and will also determine the final scale and intensity of segregation.

5.2 Interparticulate bonding

If solid 'bridge formation' between particles is excluded the most important forces causing the structuring of a mixture are:

(a) due to moisture;
(b) due to electrostatic charging;
(c) due to van der Waals' forces.

5.2.1 Bonding due to moisture

At equilibrium conditions the form of the moisture retention in a powder is a function of the nature of the solids, the ambient humidity, the temperature and the pressure[1].

Moisture will be present as adsorbed vapour if the humidity is below a critical value and as liquid bridges for higher humidities.

The binding forces between two particles in these two situations is different and is a function of the actual humidity in both cases.

5.2.1.1 Adsorbed layer bonding

The bonding is caused by the overlapping of the adsorbed layers of neighbouring particles and its strength will be proportional to the tensile strength of the adsorbed film and the area of contact. In the simple case of two dissimilar spheres (*Figure 5.1*), the area can be calculated and a bond strength calculated. Thus the contact area S is given by[2]

$$S = 4\pi(\delta - y/2)R_2\left(\frac{m}{m+1}\right) \tag{5.1}$$

where δ is the thickness of the adsorbed layer, y is the distance between the surfaces of the spheres, and m is the ratio R_1/R_2 of the sphere diameters. Then, using as a unit force the weight of the smaller particle and assuming water as the adsorbate

$$f = \frac{3 \times 10^5(\delta - y/2)}{\gamma R_2^2}\left(\frac{m}{m+1}\right) \tag{5.2}$$

where δ and y are in Å, R_2 is in μm; the specific gravity of the liquid absorbate γ is in g/cm³, and m and f are dimensionless.

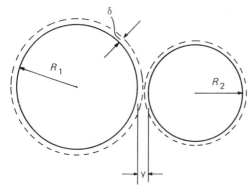

Figure 5.1   Adsorbed layer bonding between two dissimilar spheres

If one of the spheres is very large then the bonding force increases and from inspection of equations (5.1) and (5.2) the bonding force between a flat surface and a sphere would be twice as large as that between two equal-sized spheres. Evidently, there can be preferential bonding or segregation with such a system and particle shape will be an important variable.

Surface roughness is also an important variable. In most cases the surfaces of the contacting particles will not be perfectly smooth and surface roughness will reduce the area of contact and hence the bond strength. For rough particles the surface irregularities can certainly be comparable in size with the thickness of the adsorbed layer and can considerably reduce the effective contact area between particles. Surface debris on particles or the presence of fine 'coating' particles can have the same effect and keep particles apart[3].

The atmospheric relative humidity will influence the thickness of the adsorbed layer and hence the bonding area, but, more importantly, it will be a major variable in determining the transition point between adsorbed layer bonding and liquid bridge bonding. At this point there will be an incremental change in bond strength and potentially in the flow characteristics of the system. Turner[20] quotes from several authors values of this critical humidity ranging from 65% to 80%. Coelho and Harnby developed a theoretical approach for estimating the critical humidity, H_c, based on surface humidity of the particles.

Thus, for adsorbed surface moisture (*Figure 5.2*), the portion S_i covered by i layers of adsorbed moisture on the area S of solid leads to[2]

$$S_i = \beta \cdot S_0 \cdot (p/p_0)^i \tag{5.3}$$

where S_0 is the uncovered portion of S and $\beta = \exp\left[(Q - \lambda)/RT\right]$ is the B.E.T. constant which is characteristic of the solid-vapour system involved. p is the partial pressure of the adsorbate in the bulk of the gas and p_0 the vapour pressure of the adsorbate. Also

$$S = S_0 + \sum_{i=1}^{\infty} S_i$$

This is equivalent to dividing the space over the solid into elements one layer

Figure 5.2   *Adsorption region with surface films of variable thickness*

thick and in which the concentration of adsorbate decreases from the solid surface until it reaches the concentration of the bulk in the gas phase. This is likely to occur after a few decades of adsorbed layers. If it is assumed that at the *i*th layer the amount of adsorbate present is the sum of the liquid layer and gaseous adsorbate at the concentration of the bulk then at the first layer it has been shown[1] that the surface humidity

$$S.H. = H \left\{ \frac{1 - H + \beta}{1 - H + \beta H} \right\} \tag{5.4}$$

At the transition point to liquid bridge bonding the surface humidity calculated in this way should be the same as the surface humidity for the liquid bridge state. This relationship will be investigated further in the section on liquid bridging (see section 5.2.1.2).

The analysis of the contact areas for two spheres used in the derivation of equations (5.1) and (5.2) can be extended to allow for surface roughness[4]. The effect of surface roughness is to decrease the effective adsorbed layer thickness of the surface by an amount δ_0 equal to half the average peak-to-trough height. This implies not only a decrease in the binding force but also a minimum adsorbed layer thickness below which interactions are limited to the peaks of roughness and are therefore negligible. This minimum adsorbed layer thickness will occur at a minimum humidity value, H_{min}, below which there will be no significant interaction. For the dissimilar sphere system it may be shown[2] that:

$$H_{min} = 1 - \frac{\beta(\delta_0 + \varepsilon_0)/2\delta_0 - \sqrt{[1 - \beta(\delta_0 - \varepsilon_0)/2\delta_0]^2 + \beta - 1}}{\beta - 1} \tag{5.5}$$

where ε_0 is the thickness of an adsorbed layer (for water vapour, $\varepsilon_0 \simeq 3.5$ Å $= 3.5 \times 10^{-10}$ m).

If $\beta = 1$, equation (5.5) simplifies to

$$H_{min} = \frac{\delta_0}{\delta_0 + \varepsilon_0}$$

In the limiting case of perfectly smooth surfaces $\delta_0 = 0$ and $H_{min} = 0$.

In all actual cases $\delta_0 > 0$ and the value of H_{min} will depend on both the B.E.T. constant and on the surface roughness as illustrated in *Figure 5.3*.

When the surface roughness is small the value of H_{min} is strongly dependent

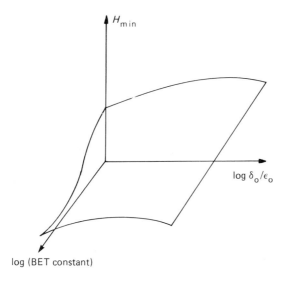

Figure 5.3   Values of minimum humidity plotted against the BET constant and the surface roughness of the particle

on the B.E.T. constant, β, and will decrease as β increases. As the surface roughness increases so does H_{min}, and the effect of the nature of the solid vapour interaction becomes less and less marked until it is negligible.

For sufficiently rough particles the value of H_{min} may be equal to or greater than the critical humidity H_c and thus no adsorbed layer interaction will occur for such powders.

5.2.1.2   Liquid bridge bonding

When the relative humidity reaches a critical point at which liquid bridges form between particles then there is an incremental increase in the bond strength between particles and an increased likelihood of the powder having a cohesive nature.

A liquid bridge will only be stable if its tendency to evaporate equals the tendency of water vapour to condense on it. This equilibrium can be expressed in terms of the Kelvin equation[5].

$$p = p_0 \exp\left[-\frac{M \cdot \Delta P}{\gamma R T}\right] \tag{5.6}$$

where p is the partial pressure of vapour over the bridge, p_0 the vapour pressure at the same temperature, M the molecular weight of the vapour, γ the density of the liquid, R the gas constant, T the absolute temperature, and ΔP the pressure difference across the liquid surface.

With a total atmospheric pressure of P the pressure inside the liquid bridge will be $P - \Delta P$. But the pressure inside the liquid bridge must be greater than or equal to its vapour pressure as the liquid would otherwise boil, so that

$$P - \Delta P \geqslant p_0$$

and

$$\Delta P \leqslant P - p_0$$

With the relative humidity defined as p/p_0 and denoting by *S.H.* the value of the humidity at the liquid surface then

$$S.H. \geqslant \exp\left[-\frac{M(P-p_0)}{\gamma R.T.} \right] \tag{5.7}$$

Liquid bridges will coexist with adsorbed films and thus equilibrium may only be established if at the edge of the bridge the partial pressure p given by the Kelvin equation is equal to the resultant partial pressure p_1 at the first adsorbed layer on the solid surface. To obtain equilibrium, mass transport of vapour between the bridge surface and the monolayer would occur and the bridge would either grow or evaporate until equilibrium was reached.

The critical humidity, H_c, below which it is no longer possible to have stable liquid bridges, can be obtained by elimination between equations 4 and 7 and gives:

$$H_c = 1 - \frac{1 - \exp\left[-M(P-p_0)/\gamma RT \right]}{2}$$

$$\times \left\{ 1 - \beta + \sqrt{1 + \beta^2 + 2\beta \cotan h \left[\frac{M}{2\gamma RT}(P-p_0) \right]} \right\} \tag{5.8}$$

The critical humidity is a function of the temperature, pressure and the nature of the solid and liquid. As would be expected, H_c decreases as P increases, as increasing the pressure favours condensation.

The vapour pressure is a little influenced by the overall pressure P but it changes much more markedly with temperature, especially near the boiling point. For water at ambient conditions the change in vapour pressure is small compared with P.

The value of H_c is sensitive to changes in the B.E.T. constant, β, which commonly has values ranging from 1 to 200.

A liquid bridge has a curved surface and therefore the pressure on both sides of the surface is not the same with the pressure difference ΔP being given by the Laplace–Young equation[5]:

$$\Delta P = \sigma\left[\frac{1}{R'} + \frac{1}{R''} \right] \tag{5.9}$$

where σ is the surface tension and R', R'' are the radii of curvature of the surface measured in two planes at right angles to each other. Thus the value of ΔP becomes larger and the possibility of a bridge existing becomes smaller as R' and R'' are reduced until a critical value of ΔP exists and the bridge evaporates. From equations (5.4) and (5.6) we may write that at equilibrium

$$\Delta P = \frac{\gamma R \cdot T}{M} \ln\left[\frac{1-H+H\beta}{H(1-H+\beta)} \right] \tag{5.10}$$

It follows from this relationship that for the same solid/liquid pair, at the same temperature, the size of a liquid bridge is a function of humidity alone. The geometry of the bridge in turn affects the magnitude of the binding force between the two surfaces.

The overall binding force has components due to the pressure deficiency and to the surface tension so that the total capillary force, F, is given by

$$F = F_p + F_t$$
$$= \pi r^2 \, \Delta P + 2\pi r\sigma \cos \alpha$$

where r is a circular approximation to the radius of the cross-sectional area of the bridge[6], σ is the surface tension, and α the angle at which the surface tension acts in relation to the axis of symmetry of the system.

For spheres, the component F_p decreases with increasing bridge size as the value of ΔP decreases faster than the cross-sectional area increases and at saturation it will have zero value. The component F_t will steadily increase with increasing bridge size. The total force varies much less markedly than each of the two components and the net effect is that for spheres a slow but steady decrease in force with increasing humidity until a sudden drop occurs at a value very near saturation. Recent work by Coughlin, Elbirli and Veraga-Edwards[22], suggests that the shape of the contacting surfaces is important and that in some cases the total capillary force, F, will increase with increasing relative humidity.

This treatment of capillary bonds assumes a two-particle system with no bridge interference. As humidity increases the area of the particle surface covered by the bridge increases and the independence of neighbouring liquid bridges becomes a function of porosity and of absolute particle size.

5.2.2   Electrostatic bonding

Two solids in rubbing contact will charge each other electrostatically. Electrons are transferred until at equilibrium a contact potential difference is established. Depending on the sign of the charge, particles will tend to bond to the mixer wall or to other dissimilar particles and produce a preferential movement or segregation.

Mixers generally provide excellent frictional contact between component particles and with the wall of the mixer and most workers in the powder-mixing field have had experience of electrostatic charging affecting their results. Generally such effects are unwelcome as they not only are unpredictable in magnitude but they also give the potential for a dangerous electrostatic discharge.

Several workers have attempted to harness the effect in a helpful way by giving opposite charges to two-component systems[7,8], and thus bonding dissimilar particles together preferentially. Such a system has not, as yet, an industrial application.

The estimation of the surface charge density of particles has been the object of a great deal of work, as it will largely determine the force of attraction between particles[9,10]. For conducting bodies the charges can migrate on the particle surface and produce a locally high surface-charge density whilst for non-conducting bodies the charge density will be relatively uniform over the complete surface. Evidently a complete range of semi-conductor solids is possible and the effect of surface impurities or irregularities of shape or charge density is not known.

If the surface charge exceeds the normal breakdown strength of the surrounding gas then discharge will occur. It would seem that a given particle system has a maximum stable charge and Lapple[11] and Harper[23] suggest that this will vary with particle diameter, with the smaller particles having higher values of maximum stable charge. As humidity increases the breakdown strength of air will be reduced and the maximum stable charge decreased.

Generally, estimates of bond strength quote a range of possible values ranging from zero to a value corresponding to the maximum surface charge density. Conventional capacitor equations suggest the adhesive force, F_{el}, is given by

$$F_{el} = \frac{\sigma^2}{2\varepsilon_0\varepsilon} \tag{5.11}$$

where σ is the surface-charge density on each of the two particles and ε_0, ε, are the static dielectric constants of the two materials.

Whilst equation (5.11) has to be qualified in many ways most workers are agreed on the proportionality between the adhesive force and the square of the surface charge density. The charge density is not easily determined. Using glass ballotini, Turner and Balasubramanian[20] found charge densities ranging from 0.1 nC/m² to 2 μC/m². Using Rumpf's equation for the mean bonding force[12], and the maximum charge density value of 2 μC/m², the same workers calculated interparticulate forces ranging from 0.24 to 0.71 nN, depending on particle size. Translated into a force per unit area of powder or an adhesive pressure, several authors quote pressures in the range 10^4 to 10^7 N/m². The relative magnitude of these forces of attraction with other bonding forces will be considered in a later section.

Unlike moisture bonding, contact is not essential to produce a bonding effect. The electrostatic forces remain constant over macroscopic distances and the effective area over which a force is applicable is large compared with the adhesional forces requiring physical contact.

5.2.3 Van der Waals' force bonding

Attractive forces exist between neutral atoms or molecules which are separated by a distance which is large compared with their own dimensions. These are known as van der Waals' forces and they decrease with the distance apart according to a power law.

They arise because of a transient polarization of the atom or molecule which will act on the surroundings to produce spontaneous fluctuations elsewhere.

The classes of such interaction have been extensively reviewed by Kauzmann[13] and Jehle[14]. The electromagnetic properties of van der Waals' forces were first shown by London in 1930 and are frequently referred to as the London–van der Waals' forces[15]. The extension of the theory of van der Waals' attractive forces from the atomic or microscopic scale to bulk powders on the macroscopic scale was first carried out by Lifshitz in 1955 (ref. 16).

Lifshitz suggested that the van der Waals' force of attraction between a sphere and a flat surface, F_{vdW}, could be calculated from the expression

$$F_{vdW} = \frac{\bar{hw}}{8\pi Z_0^2} \cdot R \tag{5.12}$$

where $\bar{hw}$ is the Lifshitz–van der Waals' constant for the system and is a function of the optical density of the system, R is the radius of the sphere, and Z_0 is the distance between the two solids. Equation (5.12) can be modified[7] to give the force between the two spheres so that

$$F_{vdW} = \frac{\bar{hw}}{8\pi Z_0^2} \left(\frac{R_1 R_2}{R_1 + R_2} \right) \tag{5.13}$$

and the force of the attraction is only half that between sphere and flat surface when the spheres are of equal size. As with adsorbed moisture bonding, there is preferential bonding to flat surfaces and the potential for non-random bonding exists. Particles do not have to touch in order that attraction occurs but it can be seen from equations (5.12) and (5.13) that the magnitude of the force decreases with the square of the distance between the surfaces. The range of attraction is significantly smaller than that for the electrostatic bonding and for a similar packing density the effective area of contact will also be smaller.

Using fused quartz Derjaquin, Abrikosova and Lifshitz compared experimental and theoretical results[18]. With gap widths varying from 10^{-5} to 10^{-3} cm they found measurable forces of from $1–2 \times 10^{-9}$ to 2×10^{-4} N, and they confirmed the power relationship between attractive forces and the distance apart of the surfaces of equations (5.12) and (5.13). Estimation of the Lifshitz–van der Waals' force for quartz enabled a theoretical estimation of the attractive force to be made and this was in good agreement with the experimental values. When these forces are compared with the electrostatic forces of section 5.2.2 it can be seen that the greatest surface charge densities produce attractive forces which are comparable with the least of the van der Waals' forces and that generally the van der Waals' forces are several degrees of magnitude larger.

The work of Rumpf[19] on bulk powders confirms the relative importance of the van der Waals' and electrostatic forces with theoretical strengths for the van der Waals' force in the range 2×10^7 to 3×10^8 N/m². These adhesive pressures are relatively constant with change in material whilst electrostatic pressures are highly sensitive to the material, its surface characteristics and impurities. The van der Waals' forces are, however, sensitive to the degree of compaction of the bulk powder, as this will affect the average distance apart of the particles.

A critical distance between particles will occur when the attractive force will just support the weight of one of the particles. This distance is the range of attraction, Z_R. At this critical point the weight of the sphere can be equated to the attractive force expressed by equation (5.12) so that

$$\frac{\bar{h}\bar{w}}{8\pi Z_0^2} R = \tfrac{4}{3}\pi R^3 \rho g$$

Then substituting typical particle values the range of attraction in Angstroms is,

$R\ (\mu)$	$\rho\ (1\ \mathrm{g/cm^3})$	$\rho\ (20\ \mathrm{g/cm^3})$
1	5000 Å	1000 Å
10	500 Å	100 Å
100	50 Å	10 Å

For the larger particles the range of attraction is certainly comparable with the scale of surface roughness and would prevent effective bonding.

5.2.4 Interaction of the bonding forces

Whilst various bonding forces have been identified, and to some extent quantified, real systems cannot be so conveniently compartmentalized and usually some combination of forces will apply.

In powder mixing the humidity of the atmosphere plays a crucial role in determining the relative importance of the forces. Adsorbed and liquid bridge bonding forces can dominate other possible bonding forces and also influence their magnitude. On a theoretical basis Rumpf plotted the theoretical tensile strength of agglomerates as a function of particle size[19] (see *Figure 5.4*).

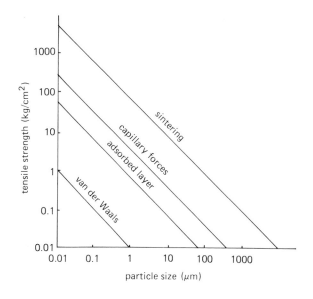

Figure 5.4 Theoretical tensile strength of agglomerates (Rumpf[12])

Whilst Rumpf interpreted this work for briquetting and tabletting processes, it also gives a good indication of the relative magnitude of the forces controlling the flow characteristics of a mixture. If the tensile strength at which a cohesive powder becomes free flowing is arbitrarily chosen as 0.01 kg/cm^2 then critical particle sizes can be identified for each of the bonding mechanisms. On this basis van der Waals' forces become important in the submicron range only, adsorbed moisture forces below about 80 μ, capillary forces below about 500 μ and sintered forces for substantial particles. These approximate values of critical particle size confirm the general relationship between particle size and the nature of powder flow suggested earlier in section 1.2.2.

Adsorbed moisture increases the van der Waals' forces as the adsorbed layers may be considered as part of the particle and a decrease in the interparticulate distance will occur. As was seen in section 5.2.3 for smaller particles, the range of attraction is of the same order of size as the thickness of the adsorbed layer. Rumpf suggests that the effect of such adsorbed layers can increase the contribution of van der Waals' forces to the cohesive strength of bulk powders by a factor of 100.

Electrostatic forces will decay rapidly as humidity causes air to be more conductive and therefore favour discharge from the particles. Adsorbed moisture also reduces the magnitude of mechanical forces such as friction and interlocking, as the adsorbed film acts as a lubricant, decreasing friction and reducing the chances of stable interlocking.

As the value of H_{min} is reached, the adsorbed film will be thick enough to cover most peaks of roughness and the opportunities of attraction between adsorbed layers should greatly increase. The value of H_{min} should not be considered as a sharp boundary, but rather as a turning point at which the contribution of the adsorbed layer attraction should be substantially increased. Rumpf[19] suggested that, whereas for particle sizes below 1 μm, van der Waals' forces should take over from adsorbed layer forces; for coarser particles van der Waals' interaction should be negligible and adsorbed layer bonding would be the dominant force. As the critical humidity H_c is reached another incremental change in bond strength occurs.

Some of the results of the varying bond strengths on powder mixtures can be seen in *Figure 5.5* (ref. 2). The two curves (a) and (b) illustrate the effect of moisture on the equilibrium mixture quality obtained after prolonged mixing in a V-mixer. In both cases a 50:50 mixture was mixed to equilibrium using a range of moisture contents. The ballotini/ballotini and sand/sand components were sieved to give a size ratio of 2:1 with the fine component having a sieve size of 70 μm. A high standard deviation indicates a poor mixture, free flowing characteristics and poor particle cohesion, whilst a low standard deviation indicates a good mixture with good particle cohesion.

Ballotini is an example of a relatively smooth, regular and spherical mixture. For very dry conditions the mixture quality is good and then deteriorates very rapidly within a very short range of moisture content and eventually improves again at high moisture content (*Figure 5.5(a)*). The good mixture quality in dry

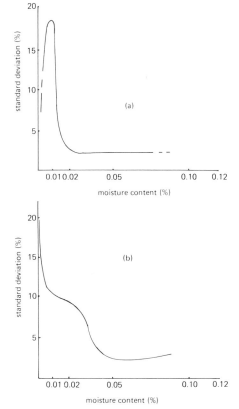

Figure 5.5   The effect of moisture on mixture quality for (a) ballotini mixture; (b) sand mixture

conditions is believed to be due to the interparticulate forces caused by electrostatic charging. The effect could be clearly seen. At higher moisture contents the charge is dissipated and the thickness of the adsorbed layer is not great enough to provide an effective contact area. This results in free flow conditions and poor mixture quality.

As moisture content continues to rise the critical humidity is approached, increasing numbers of strong liquid bridges are formed, and the mixture becomes cohesive and well mixed. The dominant forces in this case are thought to be due to electrostatic charging and to liquid bridge formation. The fine particles are thought to be too large to be effectively bonded by adsorbed moisture and by van der Waals' forces.

When using a sand mixture no statistical improvement in mixture quality was achieved in very dry conditions, though some electrostatic charging could be seen within the mixture. Surface charge is probably dissipated within the bulk of the powder due to the angular nature of the particles. The two-stage improvement in mixture quality is thought to be due to the build up of adsorbed layers followed by the formation of liquid bridges. Potential contact areas between particles are larger than for ballotini and the adsorbed moisture bonding regime is more significant.

Evidently, as the particle size of the fine component decreases to the submicron range the relative importance of the bonding forces will change again.

5.3   Selection of mixer

Within the mixing process the formation of a structural powder and perhaps agglomerates is not a bad feature as it will minimize large-scale segregation. The mixer used for such mixtures must be capable of repeatedly breaking down such structures and agglomerates to the scale of individual particles. If the mixer is not capable of doing this then a small scale of segregation with a high intensity of segregation is likely to result.

The most difficult powder mixtures to repeatedly restructure will have some submicron constituents, will have interlocking or plate-like shapes, will have some moisture present, and will have been subjected to compaction at some stage. To mix such a system the 'mixer' is likely to need high shearing or impaction characteristics and could well be a particle comminuter rather than a conventional 'mixer'.

Mixers using the impaction mechanism generally rely on a bulk circulation of powder to bring the mixture within the rotational volume of a high-speed impeller which breaks down particulate bonds. Bulk circulation can be effected in fluidized beds, tumbler mixers or convective mixers and is of greatest value when the mixture is not too cohesive. Impaction mixing is especially useful when a small proportion of a fine cohesive powder is to be added to a free-flowing bulk material. This is the so-called coating process.

If carefully chosen, the fine particles will preferentially bond to the coarser or 'parent' particles and can produce a mixture of very high quality. In this process the balance between bond strength and the impaction energy input is most important. Finer coating particles require a greater energy input to dislodge and a critical coating particle size, and bond strength will exist which the mixer will be incapable of breaking. If, on the other hand, an 'overkill' of impaction energy is provided, then there is the possibility that the parent particles are comminuted and that the very desirable free-flowing characteristics of the bulk powder will be lost.

With such mixers it is desirable that the impeller speed is variable so that the mixer can be tuned to the process requirements. It is also important that the size of the coating particle be controlled within as small a size spectrum as possible.

The impaction mixer is probably the best mixer if the mixture contains a very wide particle-size range. Such mixtures can exhibit the worst features of both the free-flowing and the cohesive powder systems. The coarser particles and some of the fine-particle agglomerates can be free flowing whilst the bulk of the material remains cohesive. The mixture does not have the smooth-flowing advantages of the coarser powder or the mixing advantages of the finer particles. Such 'fence-sitting' systems should be avoided if possible.

For very cohesive mixtures, bulk circulation to an impaction device becomes difficult, and there is the strong possibility that mixture will adhere to the mixer walls or form a dead spot within the circulating mixture. A better solution is then to look for shear mixing. A variety of runner mills are good examples of such mixers. Care should be taken not to overfill and to ensure that there is an efficient scraping system which returns sheared material into the path of the mill. Comminution is likely to occur as mixing proceeds.

5.4 Mixture quality for cohesive systems

As was seen in section 2.2, the value of the variance of a completely randomized mixture of particles has provided a datum on which the performance of particulate solid mixers can be based. These limiting values were based on the assumption that the constituent particles were capable of independent movement. In a free-flowing powder there is virtually no association between particles as interparticulate forces are negligible compared with particle weight. In a cohesive powder the independent movement of particles is restrained by a variety of interparticulate forces and the limiting values of S_0^2 and S_R^2, as previously defined, are no longer applicable.

A generalized solution for agglomerating systems has been derived for a two-component mixture of spheres in which there is a size differential between, but not within, components[2]. In this case

$$S_0^2 = p \left\{ \frac{\sum_i \sum_j p_{ij} \cdot j \cdot \Phi(i,j)}{\sum_i \sum_j j \cdot \Phi(i,j)} - p \right\} \tag{5.14}$$

and

$$S_R^2 = \frac{N_t}{W_B} \frac{w_f^2}{W} \cdot \sum_i \sum_j \{i \cdot p - j \cdot m^3 (1-p)\}^2 \Phi(i,j) \tag{5.15}$$

where

p is the mass fraction of the coarse component;
m is the ratio of the radii of the coarse and fine particles $(m \geqslant 1)$;
w_f is the weight of a fine particle;
W is the weight of the sample;
N_t is the total number of agglomerates in the batch;
W_B is the total weight of the batch;
i is the number of fine particles in the agglomerate;
j is the number of coarse particles in the agglomerate;
p_{ij} is the mass fraction of coarse particles in agglomerate (i,j);
$\Phi(i,j)$ is the probability of any agglomerate containing 'i' fine and 'j' coarse particles.

For a non-agglomerating mixture equations (5.14) and (5.15) reduce to the limiting equations of section 2.2. $p_{ij} < 1$ so that the value of S_0^2 for an agglomerating mixture is always smaller than that for a non-agglomerating mixture and complete segregation is not normally possible. Equations (5.14) and (5.15) have no practical value unless the composition distribution of the aggregates $\phi(i \cdot j)$ is known.

Solutions to the equations were obtained for the particular case of a coated mixture[2]. A model was postulated in which coarse spherical particles act as nuclei which are coated with a single layer of fine spherical particles. If fine particles were in excess it was assumed that they would agglomerate in a similar way around a nucleus of a fine particle. Using this model, particular solutions to the general equations (5.14) and (5.15) can be derived to give:

$$S_0^2 = p \left\{ \sum_{i=0}^{n} \frac{m^3}{i+m^3} f_i - p \right\} \tag{5.16}$$

$$S_R^2 = \frac{w_f(m^3 + nP) \cdot p \left\{ \frac{m^3}{(m^3 + nP)} - p \right\} + p^2 \{\bar{m} - (m^3 + nP)w_f\}}{W} \tag{5.17}$$

where

f_i is the relative frequency of occurrence of aggregates containing i fine particles;

$\bar{m}$ is the average aggregate mass;

P is the probability that a site on a coarse particle is occupied by a fine particle.

This expression takes into account the relative bonding strengths between coarse/fine particles and fine/fine particles by means of the probability term. With some re-arrangement equation (5.17) can be shown to be directly comparable in form with the Stange equation (2.5) for the random variance of independent particles. It is directly comparable with Buslik's[21] expression for random variance of a binary system.

A graphical solution is provided to the equations for a value of $m=3$ in *Figure 5.6*. When $P=0$ there is no attraction between particles and the plot reduces to the two coordinate solutions of equations (2.1) and (2.2) for independent particles. The coordinate P represents a degree of agglomeration, or cohesion, which will be at its greatest value when $P=1$. As the objective of the mixing operation is to reduce the variance of the mixture, it can be seen that any degree of agglomeration is helpful in reducing the value of the limiting variance values. An ordered mixture is obtained (i.e. $S_R^2 = 0$), when $P=0$ and $p=0$, $P=0$ and $p=1.0$, and when $P=1$ and $p=m^3/(m^3+n)$. In the last case the coarse particles will be completely covered in fines and the mixture effectively becomes a one-component system. It is seen that once the state of agglomeration has been defined in terms of P, the limiting values of p will not be 0 and 1 but 0 and $m^3/(m^3+nP)$.

Equations (5.16) and (5.17) describe idealized models of the 'coating' operation but are helpful in that they emphasize the importance of

(i) the strength of the interparticulate bond;
(ii) the absolute size and the size range of the agglomerating particles;

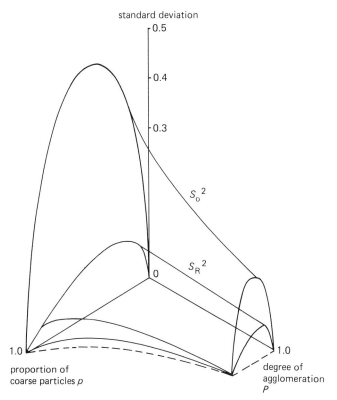

standard deviation

S_o^2

S_R^2

0

1.0

1.0

proportion of
coarse particles p

degree of
agglomeration
P

Figure 5.6   The variation of the limiting values of standard deviation with concentration of coarse particles, p, and the degree of agglomeration, P

(iii) the need to control the particle size distribution if a high-quality mixture is to be produced.

As was seen in section 2.2, expressions for limiting variance are very useful in providing a guide to the most important process variables in the mixing operation but provide no guarantee that such mixture quality will be obtained in practice.

References

1 COELHO, M. C. and HARNBY, N. (1978) *Powder Technology*, **20**, 197.
2 COELHO, M. C. (1976) Ph.D. Thesis, University of Bradford.
3 MASSIMILLA, L. and DONSI, G. (1976) *Powder Technology*, **15**, 253.
4 COELHO, M. C. and HARNBY, N. (1978) *Powder Technology*, **20**, 201.
5 ADAMSON, A. W. (1976) *Physical Chemistry of Surfaces*, Interscience, New York, 2nd Edn.
6 HOTTA, K., *et al.* (1974) *Powder Technology*, **10**, 231.
7 TUCKER, C. L. and SUH, N. P. (1976) *Polymer Engineering Science*, **16**, 657.
8 ENSTAD, G. G. (1981) 2nd European Symposium, 'Mixing of Particulate Solids', Birmingham.
9 SKINNER, S. M. (1955) *Journal of Applied Physics*, **26**, 498.

10 SMILGA, V. P. and DERYAGIN, B. V. (1963) *Research in Surface Forces*, Consultants Bureau, New York, p. 58.
11 LAPPLE, C. E. (1970) *Advances in Chemical Engineering*, **8**, 1.
12 RUMPF, H. (1958) *Chem. Ing. Tech.*, **30**, 14.
13 KAUZMANN, W. (1957) *Quantum Chemistry*, Academic Press, New York, p. 514.
14 JEHLE, H. W. C. (1969) *Ann. New York Academy of Science*, **153**, 240.
15 LONDON, F. (1930) *Z. Physik*, **63**, 245.
16 LIFSHITZ, E. M. (1956) *Soviet Physics, J.E.T.P.*, **2**, 73.
17 HAMAKER, H. C. (1937) *Physica*, **4**, 1058.
18 DERJAGUIN, B. V., ABRIKOSOVA, I. I. and LIFSHITZ, E. M. (1956) *Quarterly Reviews*, p. 245.
19 RUMPF, H. (1962) *Agglomeration* (Krepper, W. A., ed.), Wiley, New York, p. 379.
20 TURNER, G. A. and BALASUBRAMANIAN, M. (1974) *Powder Technology*, **10**, 121.
21 BUSLIK, D. (1950) *Bull. Amer. Soc. Test. Mat.*, **165**, 927.
22 COUGHLIN, R. W., ELBIRLI, B. and VERAGA-EDWARDS, L. (*1982*) *Journal of Colloid and Interface Science*, **1**, 18.
23 HARPER, W. R. (1967) *Contact and Frictional Electrification*, Oxford University Press.

Chapter 6

The dispersion of fine particles in liquid media

G D Parfitt

6.1 Introduction

One of the most important practical problems associated with powders concerns their dispersion in liquids, and this is relevant to a wide variety of industries. The term 'dispersion' does not mean the same to everyone—in the present context it refers to the complete process of incorporating a powder into a liquid medium such that the final product consists of fine particles distributed throughout the medium. Often 'dispersion' is used to refer to the whole or individual parts of this process, and misunderstandings readily arise.

In this review we shall consider the dispersion of particles having colloidal dimensions, i.e. at least one dimension of the particles lies between 1 nm and 1 μm. Hence the surface area of the powder is large and interfacial phenomena play a dominant role in the process.

The dispersion process involves a number of distinctly different stages, each of which can be reasonably well defined and related to established fundamental principles of colloid and interface science. But, because these stages overlap in practice and because most practical systems are multicomponent, it is often difficult to identify what is controlling the observed effects, and to correct for failures with anything but empirical means. Nevertheless there is a lot of knowledge available, and some guiding principles which could be put to good use; much of this is related to studies of the solid/liquid interface. In the early stages of the process, the solid/air interface is replaced by one between solid and liquid.

A powder consists of aggregates/agglomerates of small particles and mechanical work is performed in dispersing particles into a liquid. The forces that exist at the interface determine the ease with which the process can be brought about. Once dispersed, the particles are free to move in their new environment, and flocculation is prevented by various chemical and physical processes, all of which relate to the character of the solid/liquid interface. It is not surprising therefore that studies of this interface form an integral part of our understanding of the dispersion process.

6.2   Stages in the dispersion process

It is convenient to break down the overall process of dispersion into four stages, which are distinct and can be treated theoretically using principles that are well established. In practice these stages overlap, but it is quite likely that one of the stages is dominant with respect to a particular property of the dispersed system. The four stages concerned are:

(a) incorporation;
(b) wetting;
(c) breakdown of particle clusters (aggregates and agglomerates);
(d) flocculation of the dispersed particles.

A wide variety of milling equipment is used for dispersion, the choice depending on the nature of the components involved and the required characteristics of the end product. During milling each of these four processes occurs simultaneously in some part of the liquid phase, and the final result depends on a variety of factors such as shear rate, surface tension and viscosity, cohesiveness of the powder, and contact angle between liquid and solid, as well as the various interactions that occur between the solid surface and the molecules in the liquid phase.

We will now consider each stage separately, identifying the principles involved, and using illustrations from the literature some of which relate to the use of titanium dioxide (TiO_2) pigments in coating technology.

6.2.1   Incorporation

Powders contain three types of particles: primary particles, aggregates, and agglomerates. Aggregates are groups of primary particles joined at their faces and having a surface area significantly less than the sum of the areas of their constituent particles. An agglomerate is a collection of primary particles and aggregates which are joined at edges and corners, and the surface area of the whole is not markedly different from the sum of the areas of the individual components.

Every bag of TiO_2 pigment contains powder in an agglomerated condition; four volumes of 'pigment' contain three volumes of air, i.e. the pigment volume concentration (p.v.c.) is 25%, the continuous medium being air. In many cases the finished paint product in which the TiO_2 is to be incorporated also has a p.v.c. of approximately 25%, thus the process of dispersion consists of replacing air by liquid at the same time ensuring that the pigment particles remain separated, although the state of dispersion is normally quite different. Usually only very few aggregates exist in a commercial TiO_2 powder, so the agglomerate structure is a feature of the whole pigment mass.

Various interparticulate forces exist in a 'dry' powder. The most basic are the van der Waals' forces which are normally attractive. Electrostatic forces are also relevant where excess electric charges are retained by the particles, although this force is very much smaller than the van der Waals' interaction. There is also a force of attraction between particles carrying an adsorbed film of liquid that are in contact, but this is probably significant only at high relative humidity. When the amount of liquid between the particles is considerably larger then interparticulate liquid bridges are formed which increase the cohesion.

Finally, gravity is the force keeping the particles together in a container, and the energy required to raise the powder into a dust cloud is readily calculated. Simple expressions are available for assessing the magnitude of these forces between particles of well-defined geometry and composition, but as usual difficulties arise when these are applied to practical systems. However, one can estimate the magnitude of the major forces involved (van der Waals and gravity), to illustrate the problem of disagglomeration of the powder during the early stages of incorporation.

In general terms, the smaller the particles of a powder the more important become the interparticulate forces in controlling packing and flow behaviour. For particles to pack closely together and to flow freely, the force of gravity on each particle must be greater than the force holding the particles to each other. For example, it can be shown[1] for titanium dioxide-pigment particles that below a radius of about 100 μm the attractive force exceeds the separation force, and this accords with the common experience that powders tend to be free flowing when their particle size is above this radius, and cohesive when it is less. Pigment particles having a radius of 0.1 μm are therefore intrinsically cohesive.

Nevertheless, the pigment does flow as loose agglomerates, which under an optical microscope appear to have a diameter of 30 to 50 μm. If we assume that loose agglomerates are themselves loosely packed assemblies of particles with a density of not much more than unity, and that only one or two particles in their surfaces contribute to the attractive force, a similar calculation shows that agglomerates above 30 to 50 μm would be expected to flow fairly freely.

Titanium dioxide powder is very cohesive—it has a high tensile strength. The measurement of tensile strength of powders has been described[1,2], but relatively little has been published on its dependence on the surface chemistry of the particles. Surface treatment with other oxides and organic molecules greatly reduces the cohesiveness of TiO_2 pigments, and aids the breakdown of pigment/air agglomerates when the pigment and medium are brought into contact[1].

Surface moisture also has a profound effect—water bridges between particles increase agglomerate strength. *Figure 6.1* shows the effect of adsorbed moisture on the tensile strength of a TiO_2 pigment. So measurements of tensile strength and cohesiveness throw light on the ease of incorporation, i.e. on the initial submergence process when pigment and medium are brought into contact and pigment/air interfaces start to be replaced by pigment/liquid interfaces. Experience shows that pigments of high tensile strength are those that tend to be difficult to incorporate, particularly in poorly dispersing media.

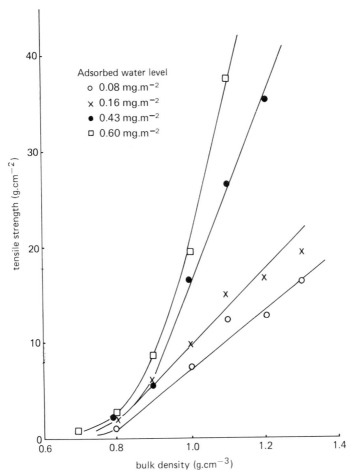

Figure 6.1   Tensile strength of a titanium dioxide pigment containing various levels of adsorbed water. Water level required for one statistical monolayer is 0.3 mg/m²

6.2.2   Wetting

We start with powder being added to the surface of the liquid medium. As the medium begins to penetrate the undersurface of the mass, small agglomerates and crystals will be engulfed by the advancing interface, and become detached from the bulk. Larger agglomerates will also tend to be engulfed, but medium can also be penetrating the agglomerate, at a slower rate than that of the engulfing process, as the interparticle spacing is smaller in the agglomerate.

At some degree of penetration, the bulk density of the agglomerate becomes greater than that of the medium, and the agglomerate detaches. Two things can now happen—the medium continues to penetrate, and the air content of the agglomerate escapes as bubbles, while the outer particles detach, until finally the whole agglomerate is dispersed. Alternatively, particularly if the binding in the agglomerate is slightly stronger than average, the medium penetrates from all sides, until the internal air pressure disrupts the agglomerate and the process starts again with the fragments.

The basic principles of wetting of powder surfaces for simple geometry are well established[3]. The initial stage of the wetting process involves both the external surface of the particles and the internal surfaces within the agglomerates. The characteristics of the wetting process are therefore dependent on the properties of the liquid phase, the character of the surface, the dimensions of the interstices in the agglomerates, and the nature of the mechanical process used to bring together the components of the system.

The simplest way to describe the wetting phenomena that occur when a solid particle is immersed in a liquid is to consider a cube of unit side length (*Figure 6.2*). There are three stages involved, namely adhesion, immersion, and

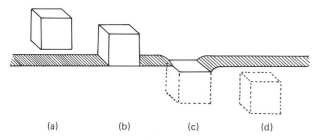

(a)   (b)   (c)   (d)

Figure 6.2   The three stages in the wetting of a cube by a liquid (a) to (b) adhesional wetting; (b) to (c) immersional wetting; (c) to (d) spreading wetting

spreading[4]. Using the Young–Dupré equation

$$\gamma_{S/V} = \gamma_{S/L} + \gamma_{L/V} \cos \theta \tag{6.1}$$

where $\gamma_{S/V}$, $\gamma_{S/L}$, and $\gamma_{L/V}$ are the interfacial free energies at the solid–vapour, solid–liquid, and liquid–vapour interfaces respectively, and θ is the contact angle; it can readily be shown that the appropriate work terms are related to the energy changes that take place by

$$W_a = \gamma_{S/L} - (\gamma_{L/V} + \gamma_{S/V}) = -\gamma_{L/V}(\cos \theta + 1) \tag{6.2}$$

$$W_i = 4\gamma_{S/L} - 4\gamma_{S/V} = -4\gamma_{L/V} \cos \theta \tag{6.3}$$

$$W_s = (\gamma_{S/L} + \gamma_{L/V}) - \gamma_{S/V} = -\gamma_{L/V}(\cos \theta - 1) \tag{6.4}$$

It is assumed that the solid surface before wetting is in equilibrium with the vapour of the liquid. From these relations we may predict whether or not any particular stage of the process is spontaneous, i.e. when the appropriate W is negative. When W is positive then work must be expended on the system for the process to take place. The conclusions are that

(i) adhesional wetting is positive when θ is less than 180°; the process is invariably spontaneous;

(ii) immersional wetting is positive and immersion is spontaneous only when θ is less than 90°; and

(iii) spreading wetting is positive only when $\theta = 0$, and work must be done to achieve spreading at all larger values of this angle.

The total work W_d for the overall dispersion process is given by the sum of those for the three separate stages

$$W_d = W_a + W_i + W_s = 6\gamma_{S/L} - 6\gamma_{S/V} = -6\gamma_{L/V} \cos \theta \tag{6.5}$$

a result that could have been predicted in terms of the overall energy expenditure. By splitting the operation down into the three stages it is possible to predict whether some intermediate state of the dispersion process may not be spontaneous and therefore retard the overall effect.

The penetration of liquid into the channels between and inside the agglomerates is more difficult to define precisely. Taking the simplest case, the pressure P required to force a liquid into a tube of radius r is

$$P = -2\gamma_{L/V} \cos \theta / r \tag{6.6}$$

and hence penetration is only spontaneous (P negative) when $\theta < 90°$. Using equation (6.1) we see that

$$P = -2(\gamma_{S/V} - \gamma_{S/L})/r \tag{6.7}$$

so $\gamma_{S/L}$ should be made as small as possible since $\gamma_{S/V}$ is virtually constant. Furthermore, to increase $\cos \theta$ we must reduce $\gamma_{L/V}$, but when $\theta = 0°$ a large value of $\gamma_{L/V}$ is desirable (equation (6.6)). Choosing the surface-active agent to achieve the best results may be difficult. The relevance of these equations to the practical situation is somewhat tenuous, since the channels between particles in agglomerates are not readily defined in geometric terms. Equation (6.6) indicates that in the absence of air a liquid will enter the channel spontaneously if $\theta < 90°$, but agglomerates are normally filled with air and once the liquid has penetrated the channel the air pressure will increase and complete wetting would therefore appear to be impossible. Heertjes and Witvoet[5] have extended the simple treatment to the more complex situation found in practice, and, besides demonstrating the relevance to the wetting behaviour of the structure of the agglomerate, have shown that when filled with air complete wetting can only occur when $\theta = 0$.

Although the energetics of wetting must be such that the process will take place, of more concern is the rate at which the pigment is wetted by the medium, and the rate of penetration of the liquid into the interstices of the agglomerates is particularly important, since this may decide if the particle clusters are incorporated with all surfaces wetted or just the external area. Again for the simplest case of horizontal capillaries, or in general where gravity may be neglected, the rate of penetration of a liquid into the tube is given by the Washburn equation[6]:

$$l = \sqrt{\frac{rt\gamma_{L/V} \cos \theta}{2\eta}} \tag{6.8}$$

where l is the depth of penetration in time t and η is the viscosity of the liquid.

For a packed bed of solid particles the radius should be replaced by a factor K, which contains an effective radius for the bed and a tortuosity factor which takes into account the complex path formed by the channels between the particles and aggregates. Thus

$$l^2 = \frac{Kt\gamma_{L/V}\cos\theta}{2\eta} \tag{6.9}$$

and for any particular (stable) powder bed a plot of l^2 versus t should be linear; this is often found in practice. To achieve rapid penetration high $\gamma_{L/V}$, low θ, and low η are desirable, with K as large as possible, i.e. a loosely packed powder. High $\gamma_{L/V}$ and low θ are normally incompatible; a low θ is an important requisite, but when $\theta = 0$ further lowering of $\gamma_{L/V}$ will reduce the penetration rate.

Hence the wetting phenomena are directly related to $\gamma_{L/V}$ and θ. In general, the overall process is likely to be more spontaneous the lower θ and the higher $\gamma_{L/V}$, although these factors usually operate in opposite senses. Significant reduction in $\gamma_{S/L}$ is desirable, i.e. the agent is active (strongly adsorbed) at the solid/liquid interface, and it would help if the surface-active agent did not cause appreciable lowering of the tension at the liquid/vapour interface.

It is difficult to find in the literature details of experiments that relate contact angle to the efficiency of the dispersion process. A correlation between contact angle and flotation rate for quartz in aqueous solutions of dodecylammonium acetate, with adsorption from solution and electrokinetic (zeta potential) data, was reported by Fuerstenau[7]. In a later review Fuerstenau discussed the relevance of the adsorption process to flotation technology illustrating the direct relationship between the amount of surfactant adsorbed and the contact angle[8]. Numerous values of θ are published in the literature. Most have been measured on flat surfaces and even with these problems arise due to surface roughness[9]. For powders the measurement is more difficult, although a number of methods have been reported[10,11] and reviewed[12].

6.2.3 Breakdown of agglomerates and aggregates

When the powder has been partially wetted the next stage is to break down the remaining agglomerates by some mechanical force. The ways in which this disruption occurs in various mills are not completely understood and much remains to be elucidated. The milling action disrupts the agglomerates except those that are strongly bound (aggregates) which form the 'grit' content. The more intense the milling action the fewer grits remain. Highly dispersible powders contain few aggregates and so have a low grit content. Agitation during wetting increases the extent of breakdown of agglomerates and so speeds up the whole process.

It is generally accepted that disagglomeration is brought about by shearing and/or impact. Disruption by shearing relies on viscous drag, whereas the comminution process takes place most easily when unhindered by viscous resistance. The two mechanisms therefore operate under opposing conditions and a particular mill works best within fairly close viscosity limits. Comparison of various types of equipment used, for example, in paint making, shows that mills that work mainly by an impact process require a low millbase viscosity while those relying on shearing need a high one[1].

Clearly the addition of surface-active agents is important to the wetting and disagglomeration processes both in terms of reducing the work involved and increasing the rate. Surface-active agents also play an important role in maintaining the stability of the dispersion, but in the milling stage it is more important to disrupt agglomerates and aggregates than to secure a stable dispersion of primary particles. The liquid medium should therefore be formulated to obtain the fastest breakdown of over-size particles, the final stability being of only secondary consideration. This means maximizing the speed of wetting and the speed of disagglomeration, provided that the rate of flocculation of primary particles is not so rapid that an unfavourable equilibrium results.

Surface-active agents may also contribute to the mechanical breakdown of the aggregates during the milling stage. Rehbinder[13] has demonstrated that the adsorption of surface-active agents at structural defects in the surface considerably facilitates the fine grinding of solids to create new interfaces, and therefore may contribute to the breaking of aggregates during milling.

6.2.4   Stability to flocculation

We now assume that the wetting and disagglomeration stages are complete, and the particles are uniformly distributed throughout the medium. The aim is to maintain this distribution and prevent flocculation, the process that reduces the number of particles due to collisions between particles, both under static (Brownian motion) and dynamic (under shear) conditions. In dilute dispersions it is sufficient to consider only the interaction between pairs of particles, but in the very concentrated dispersions that one meets in practical systems it is necessary to take multiparticle interactions into account. Most of the published theoretical work has been concerned with pair interaction; relatively little has been written on multiparticle interactions.

At least three major types of interaction are involved in the approach of colloidal particles, namely

 (i) the London–van der Waals' force of attraction;
(ii) the Coulombic force (repulsive or attractive) associated with charged particles;
(iii) the repulsive force arising from interaction between adsorbed layers of polymer on the particle surfaces.

These forces originate from entirely different sources and therefore may be evaluated separately. The interplay of (i) and (ii) forms the basis of the classical theory of flocculation of lyophobic dispersions, first proposed by Derjaguin and Landau in Russia[14] and independently by Verwey and Overbeek in the Netherlands[15] and now known as the DLVO theory. The interplay of (i) and

(iii) is commonly termed 'steric stabilization', and much has been written on this protective mechanism, although a workable understanding has developed only during the last decade.

6.2.4.1   DLVO theory and its application

The classical DLVO theory of lyophobic dispersions considers the interaction between two charged particles in terms of the overlap of their electric double layers leading to a repulsive force which is combined with the attractive term to give the total potential energy as a function of distance for the system. To calculate the potential energy of attraction V_A between solid spherical particles we may use the Hamaker expression[16]

$$V_A = -\frac{A}{6}\left[\frac{2}{s^2-4}+\frac{2}{s^2}+\ln\left(\frac{s^2-4}{s^2}\right)\right] \tag{6.10}$$

where A is the Hamaker constant for the system and $s = R/a$; R is the distance between the centres of the particles of radius a.

For repulsion, Verwey and Overbeek[15] show that to a good approximation the following equations may be used to calculate the potential energy:

FOR AQUEOUS SYSTEMS

$$V_R = \tfrac{1}{2}\varepsilon a\psi_0^2 \ln\left[1+\exp\left(-\kappa H\right)\right] \tag{6.11}$$

FOR MEDIA OF LOW DIELECTRIC CONSTANT

$$V_R = \frac{\varepsilon a\psi_0^2}{H+2a}\exp\left(-\kappa H\right) \tag{6.12}$$

where ε is the dielectric constant of the medium between the particles, H is the shortest distance between particles' surfaces, ψ_0 the surface potential, and κ the reciprocal of the 'thickness of the double layer'.

There is considerable evidence indicating that the zeta potential obtained from an electrokinetic experiment is an adequate substitute for ψ_0, particularly for dispersions in hydrocarbon media. Electrophoresis is therefore a useful technique for studying colloidal systems.

The total potential energy of interaction is given by the sum of the two terms $V_{tot} = V_R + V_A$, and with the increased availability of computers a large amount of theoretical data has accumulated, all concerned with the interaction between two particles. In a typical potential energy diagram for a dispersion in aqueous media (*Figure 6.3*), V_R decreases exponentially with distance, the range depending on the thickness of the electric double layer since the repulsive force arises from the overlap of the double layers of the approaching particles.

V_A shows an approximate inverse relationship with the square of the distance, and the form of the total potential energy curve is largely controlled by the V_R term and hence depends on the electrolyte concentration. The existence of two minima in the V_{tot} curve is important—the primary minimum

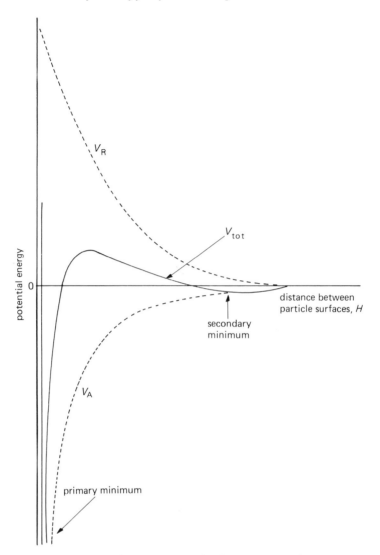

Figure 6.3   Potential energy curves for the interaction of two charged surfaces in aqueous media

relates to particles in close contact and therefore difficult to separate; the secondary minimum represents a more loosely bound state which may be broken down under shear (and may reform when the shear is removed, as in thixotropy).

We must now relate the energetics to the kinetics. The theory of rapid flocculation was first proposed by Smoluchowski[17], who treated the problem as one of diffusion (Brownian motion) of the spherical particles of an initially monodisperse dispersion, with every collision, in the absence of a repulsive force, leading to a permanent contact. In general, we may express the rate of flocculation, i.e. the rate of decrease of the total number of particles, as

$$-dN/dt = k'N^2 \tag{6.13}$$

where k' is the rate constant, and this becomes, when integrated with $N = N^0$ at $t = 0$

$$1/N = 1/N^0 + k't \tag{6.14}$$

For the most rapid rate of disappearance of particles of all types, Smoluchowski's theory gives

$$k'_0 = 8\pi Da \tag{6.15}$$

where D is the diffusion coefficient for the particles.

A theoretical value for k'_0 may be evaluated using the Einstein expression $D = kT/6\pi\eta a$. Hence the reciprocal of the total number of particles should increase linearly with time during flocculation and the slope of the $1/N$ against t plot for rapid flocculation might be compared with that predicted by the Smoluchowski theory. Such linear plots have been recorded and the k'_0 values are of the expected order of magnitude. However, Smoluchowski's theory does not take into account the long-range force of attraction between the particles (only attraction on contact is considered). Taking account of this force leads to

$$-\frac{dN}{dt} = \frac{4\pi DN^2}{\displaystyle\int_{2a}^{\infty} \frac{\exp{(V_A/kT)}}{R^2}\,dR} \tag{6.16}$$

This expression becomes that of Smoluchowski when $V_A = 0$. Smoluchowski's theory predicts that for other than the most dilute dispersion, flocculation would be expected to be fast on any practical time scale, e.g. for aqueous dispersions at 25°C with initially N^0 particles per cm^3, $t_{1/2} \approx 2 \times 10^{11}/N^0$ sec, and therefore instant separation of the disperse phase of a practical system (containing ~ 1–10% solids) would occur provided no other factors oppose the collision process.

We now consider the kinetics of the flocculation of a system for which there is an energy barrier. The rate is a function of the probability of particles having sufficient energy to overcome this barrier. The problem is again one of diffusion and was first solved by Fuchs[18]. His theory leads to a factor W by which the rapid rate (Smoluchowski) is reduced by the presence of a repulsive force; W is called the stability ratio and is related to the height of the potential energy barrier by

$$W = 2\int_{2}^{\infty} \frac{\exp{(V_{\text{tot}}/kT)}}{s^2}\,ds \tag{6.17}$$

However, Smoluchowski's value for the rapid rate does not assume the existence of long-range attractive forces. Introduction of the appropriate term leads to the following equation[19]

$$W = \frac{\displaystyle\int_2^\infty \frac{\exp\,(V_{tot}/kT)}{s^2}\,ds}{\displaystyle\int_2^\infty \frac{\exp\,(V_A/kT)}{s^2}\,ds} \tag{6.18}$$

so that $W = 1$ when $V_{tot} = V_A$ at all distances.

Given the appropriate potential energy diagrams from the DLVO theory, the stability ratio may be calculated by graphical or numerical integration and then compared with experimental values of $W = k'_0/k'$, the ratio of the experimental rate constants for rapid and slow flocculation. Such a comparison is a severe test of the applicability of theory to experiment, and the observed deviations, although often not appreciable, reflect the assumptions and approximations which are necessary in the calculation of the potential energy terms.

Flocculation kinetics can be readily measured for dilute dispersions, e.g. by particle counting (equation (6.7)), or from light-scattering data[20], and numerous experimental data for aqueous systems confirm that the DLVO theory is essentially correct although there are deviations in some of the fine detail.

Extension of the DLVO theory to non-aqueous systems has attracted relatively little attention. For media of intermediate dielectric constant, e.g. alcohols, ketones, etc., we might expect the behaviour to be similar to that already established with aqueous systems, but the actual ionic concentrations being one or two orders of magnitude lower.

Fundamental studies on such systems are scarce, but some attention has been paid to dispersion in hydrocarbon solutions of surface-active agents for which the ionic concentrations are extremely small, e.g. $\sim 10^{-10}$ mol·dm^{-3}, corresponding to $1/\kappa \sim 10$ μm. Since $1/\kappa$ is large the capacity of the double layer is small and only a small surface charge density is necessary to obtain an appreciable surface potential. Furthermore, the slow decay in potential from the surface means that the zeta potential, readily obtained from electrophoresis experiments, may be equated with considerable accuracy to the surface potential.

Another simplification compared with aqueous systems is that an absolute value of A is not required since the V_R term dominates the V_{tot} against distance relationship. A secondary minimum is theoretically not feasible for colloidal particles in media of low dielectric constant, unless factors other than charge, e.g. force (iii) are relevant.

It is therefore possible to predict, with some confidence, the stability of a dispersion from readily obtainable parameters, and this has been done for hydrocarbon media for a range of particle size and surface potential[21]. The validity of the DLVO theory has been demonstrated both by qualitative[22,23] and by rigorous experimental tests[24,25] on systems in which there is no significant influence of the interaction of adsorbed layers of surface-active material, i.e. only the charge mechanism is operative. The origin of the charge on the particles is still subject to debate.

Fowkes[26], and Tamarabuchi and Smith[27] have suggested an acid-base mechanism, involving transfer of charge between the adsorbed molecules and the surface, followed by desorption of the charged species. The presence of trace amounts of water have a profound effect on the stability of hydrophilic powders in non-polar media, presumably through changes in surface energy and/or charge but the details have not yet been worked out. The subject of charge stabilization in non-polar media has been reviewed[28].

For practical systems, when the particle concentration is high, the validity of the DLVO theory is open to question. Particle–particle interaction might not begin at 'infinite' distance of separation, i.e. one particle is, on average, sufficiently close to another to reduce the effective energy barrier which must be overcome for the particles to come into contact. This is particularly relevant to dispersions in hydrocarbon media for which the double-layer thickness is large compared with the average distance between the particles.

The most useful attempt to describe the stability of concentrated systems has been made by Levine[29,30], who concluded that electric double-layer effects are unable to provide the necessary energy barrier for dispersions in hydrocarbon media. A similar conclusion was reached from experimental observations of the stability/opacity of dispersions of titanium-dioxide pigments in alkyd resin solutions[31].

So far we have considered only the interaction between particles which are identical in nature, size and surface potential. When more than one type of particle is present, or if there is a large disparity in size, the situation becomes more complex. Interaction between dissimilar particles is termed *hetero-flocculation*. Approximate equations based on DLVO theory have been derived[32] for spheres with $\kappa a > 10$ and values of ψ_0 less than 50 mV. The potential energy of repulsion for two particles of radius a_1 and a_2 and potentials ψ_{01} and ψ_{02} is given by

$$V_R = \frac{\varepsilon a_1 a_2 (\psi_{01}^2 + \psi_{02}^2)}{4(a_1 + a_2)} \left[\frac{2\psi_{01}\psi_{02}}{(\psi_{01}^2 + \psi_{02}^2)} \ln \left(\frac{1 + \exp(-\kappa H)}{1 - \exp(-\kappa H)} \right) + \ln(1 - \exp(-2\kappa H)) \right]$$

(6.19)

Combining this with the appropriate expression of Hamaker[16] for the attraction between two spheres gives the total potential for the interaction from which energy barriers, stability ratios, etc., may be evaluated.

Two facts emerge from the theoretical calculations[3]. Firstly, it is the particle with the smaller radius and/or potential that determines the height of the potential energy barrier. Secondly, although the different particles may have the same sign of surface charge and potential (but different magnitude), they may nevertheless attract each other when the double layers overlap; this is a consequence of the relative changes which occur in the charge and potential as the particles approach[33]. Similar conclusions have been reached for non-polar media using Maxwell's equations for charged conducting spheres[34].

The interaction of oppositely charged particles has been recently analysed by Prieve and Ruckenstein[35]. Constant charge is assumed. One type has strongly acidic (negative) surface sites and the other strongly basic (positive) sites. It is shown that, when the ratio of the absolute surface charge densities is significantly different from unity ($\sim 10:1$), stability ratios $> 10^4$ are obtained over a significant range of ionic strength (*Figure 6.4*). At large ionic strength ($\sim 10^{-1}$ mol dm^{-3}), rapid heteroflocculation occurs; similarly at very low ionic strength ($\sim 10^{-4}$ mol dm^{-3}), while in the intermediate range the system is stable despite the fact that the particles are oppositely charged. This is due to the osmotic pressure that arises from the excess electrolyte between the particle surfaces as they approach. Experimental data supporting these conclusions have not yet been identified.

In recent years there has been an increasing interest in heteroflocculation in aqueous media, and at least qualitative agreement found between experiment and theory, although not in every case. Systems studied include chromium

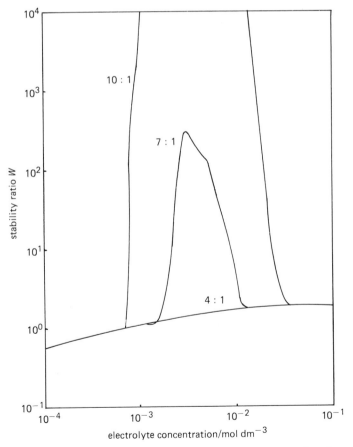

Figure 6.4   Stability ratio v. electrolyte concentration as a function of the ratio of the absolute surface charge densities of two interacting particles[35]

hydroxide with polyvinyl chloride latex of similar size[36], silica with larger-size polyvinyl chloride latex[37], aluminium oxide and haematite with a smaller-size carboxylated polymer latex[38], and two polystyrene latexes of the same size but different surface groups[39].

6.2.4.2   Steric stabilization

The term 'steric stabilization' is now used to embrace all aspects of the stabilization of colloidal particles by adsorbed non-ionic macromolecules, and is equally effective in aqueous and non-aqueous media.

Polymers adsorbed at particle/medium interfaces may take up any configuration from lying flat (strong interactions between the surface and many points along the polymer chain) to higher extended into the medium (interaction at few points). In the context of steric stabilization we are particularly interested in the thickness of the adsorbed layer, the configuration of the polymer chain, the fraction of segments adsorbed and the segment density distribution, both normal and parallel to the surface.

For the extreme case of parallel orientation the only effects on flocculation are (a) the modification of the attractive force between particles which is likely to be small in most cases, and (b) the effect on the zeta potential which could be significant depending on the chemical character of the surface and polymer. When the polymer chains are extended into the liquid phase there is a major effect which arises when particles approach sufficiently close for the adsorbed layers to interact.

Two extreme cases can be envisaged, one in which the polymer chains mix, while the other involves no mixing but the layers become compressed on close approach of the surfaces. Both involve a reduction in the configurational freedom of the adsorbed polymer molecules, which results in a repulsive force. One or both types of interaction could be present in any particular system, depending on the nature of the polymer and solvent, and the structure of the adsorbed layer. There has been considerable theoretical analysis of both cases, and the relative merits of the two approaches is still the subject of much debate in the literature[40,41].

Experience shows that the most effective stabilizers are amphipathic block or graft copolymers, which contain 'anchor' groups that are strongly held to the surface and are insoluble in the medium, and stabilizing chains that are soluble in the medium and extended from the surface. High surface coverage is essential.

The solvency of the stabilizing chain is the important factor, and instability is readily induced if the solvency is reduced by temperature and pressure charges, or by adding a non-solvent for the chain to the dispersion medium. When the solvency is reduced, the dispersions often show sharp transitions from long-term stability to rapid flocculation, and the 'critical' flocculation values correlate strongly with the corresponding θ point for the stabilizing moiety in free solution[41]. This provides strong evidence of mixing effects as two adsorbed polymer layers interact. Examples of the close correlation

Table 6.1 Comparison of critical flocculation temperature and θ temperature

Polymer stabilizer	Molecular weight	Medium	CFT/°K	θ/°K
Poly(ethylene oxide)	96 000	0.39 mol dm^{-3} MgSO$_4$	316 ± 2	315 ± 2
Poly(acrylic acid)	51 900	0.2 mol dm^{-3} HCl	283 ± 2	287 ± 2
Poly(vinyl alcohol)	57 000	2.0 mol dm^{-3} NaCl	301 ± 3	300 ± 3
Polyisobutylene	23 000	2-methylbutane	325 ± 1	325 ± 2

between θ temperature and critical flocculation temperature are given in *Table 6.1* (taken from Napper[41]).

The free energy of the interaction ΔG_R is related to the enthalpy (ΔH_R) and entropy (ΔS_R) changes by

$$\Delta G_R = \Delta H_R - T\,\Delta S_R$$

and the sign and magnitude of ΔG_R are dependent on the relative magnitudes of the enthalpy and entropy terms. If ΔG_R is negative, flocculation is promoted; if zero, flocculation proceeds as if no adsorbed layers were involved and, if positive, the particles are protected from flocculation. For ΔG_R to be positive, we require that either

(i) both the entropy ($T\,\Delta S_R$) and enthalpy terms are negative with the former being the larger—termed *entropic* stabilization indicating a dominating effect due to reduction in the number of configurations of the polymer molecules in the adsorbed layer,

(ii) both terms are positive with the enthalpy change predominating—termed *enthalpic* stabilization,

(iii) ΔH_R is positive and ΔS_R negative, and therefore both terms contribute to stability—termed *enthalpic–entropic* or 'combined' stabilization.

A characteristic feature of sterically stabilized systems is their different responses to temperature change. Those in (i) flocculate on cooling, while for (ii) flocculation occurs on heating, and for (iii) there is no accessible temperature for flocculation. Entropic stabilization seems to be more common in non-aqueous media, whereas enthalpic stabilization is more frequently encountered in aqueous media. Examples of the three types of stabilization were given by Napper[42].

The mixing effects described above are readily interpreted in terms of polymer solution theory, and apply when the separation between the approaching surfaces is one to two adsorbed layer thicknesses. At closer approach it is likely that both mixing and elastic compression occurs. The following equation was derived by Napper[41] for the free energy of the interaction involving both effects:

$$\Delta G_R = \frac{2\pi akTn^2\Gamma^2 V_s^2}{V_1}(\tfrac{1}{2}-\chi)S_{mix} + 2\pi akTTS_{el} \tag{6.20}$$

The first term refers to the mixing of the adsorbed layers, and the second to the elastic compression effect at distances less than one-layer thickness. V_s and V_1 are the volumes of polymer segment and solvent molecule respectively, n is the number of segments per polymer chain, Γ the number of chains per unit area of surface, χ the Flory interaction parameter, and S_{mix} and S_{el} are geometrical terms.

When the stabilizing chain is in a good solvent environment ($\chi < \tfrac{1}{2}$) the mixing term is positive, but when θ conditions are exceeded ($\chi > \tfrac{1}{2}$) it changes sign and the force is attractive. Attraction between polymer segments in worse than θ solvents is well known[43]. However, the elastic term remains repulsive, and operates at short distances. Hence the flocculation behaviour depends on the structure of the adsorbed layer (solvency, molecular weight, thickness, segment distribution), and the magnitude of the repulsive and/or attractive interactions arising from adsorbed layer overlap.

Significant advances have been made over the last two decades in our understanding of steric stabilization, and recently measurements have been reported on the force between two surfaces containing adsorbed polymer[44,45]. Such data will help to clarify the mechanism.

An interesting example of the application of the concept of steric stabilization is the interpretation of the optical performance of alkyd paints pigmented with TiO_2 by Franklin *et al.*[31]. Having established from electrophoresis and opacity measurements that the electric charge on the particles is not the controlling factor in flocculation, a study was made of the adsorption characteristics of the resin using pigments coated with different levels of silica/alumina such that the surfaces created varied from predominantly silica to mostly alumina.

Interpretation of the adsorption isotherms indicated that for the acidic silica surface the basic resin molecules interacted strongly with the surface and adopted a parallel orientation, thus making little contribution to preventing flocculation. On the other hand, for the predominantly alumina-coated surface, the resin only made contact with the surface with its limited number of acid groups, the rest of the molecule being extended into the medium and providing a steric barrier to flocculation.

When the surface coverage of polymer is low then 'bridging' flocculation is possible. This is brought out by the polymer becoming simultaneously adsorbed on two or more particles—a 'bridging' mechanism leading to a rather open structure in the flocculates. The mechanism has been summarized by Kitchener[46]. Since polymer adsorption is usually irreversible, the method of mixing of the components can have a profound effect on the flocculation process. Tadros has reported[47] a good demonstration of the effect, using silica and aqueous solutions of polyvinyl alcohol. The effect depends on the pretreatment of silica, hence the surface hydroxyl population, and on the pH of the solution which determines the particle charge, both factors influencing the adsorption sites for PVA adsorption.

References

1 PARFITT, G. D. *FATIPEC Congress Book*, 1978, 107
2 SUTTON, H. M., in *Characterization of Powder Surfaces* (Parfitt, G. D. and Sing, K. S. W., eds.), Academic Press, London, Chapter 3.
3 PARFITT, G. D. (1973) in *Dispersion of Powders in Liquids* (Parfitt, G. D., ed.), Applied Science, London, Chapter 1.
4 OSTERHOFF, H. J. and BARTELL, F. E. (1930), *J. Phys. Chem.*, **34**, 1399.
5 HEERTJES, P. M. and WITVOET, W. C. (1970) *Powder Technol.*, **3**, 339.
6 WASHBURN, E. D. (1921) *Phys. Rev.*, **17**, 374.
7 FUERSTENAU, D. W. (1957) *Trans. AIME, Mining Eng.*, 1365.
8 FUERSTENAU, D. W. (1970) *Pure and Appl. Chem.*, **24**, 135.
9 HUK, C. and MASON, S. G. (1977) *J. Colloid Interface Sci.*, **60**, 11.
10 KOSSEN, N. W. F. and HEERTJES, P. M. (1965) *Chem. Eng. Sci.*, **20**, 593.
11 BIKERMAN, J. J. (1941) *Ind. Eng. Chem.*, **13**, 443.
12 HEERTJES, P. M. and KOSSEN, N. W. F. (1967) *Powder Technol.*, **1**, 33.
13 REHBINDER, P. A. (1958) *Colloid J. USSR*, **20**, 493.
14 DERJAGUIN, B. V. and LANDAU, L. D. (1941) *Acta Physicochim.*, **14**, 633.
15 VERWEY, E. J. W. and OVERBEEK, J. Th. G. (1948) *Theory of the Stability of Lyophobic Colloids*, Elsevier, Amsterdam.
16 HAMAKER, H. C. (1937) *Physica*, **4**, 1058.
17 VON SMOLUCHOWSKI, M. (1917) *Zeits. Physik. Chem.*, **92**, 129.
18 FUCHS, N. (1934) *Zeits. Physik.*, **89**, 736.
19 McGOWN, D. N. L. and PARFITT, G. D. (1967) *J. Phys. Chem.*, **71**, 449.
20 OSTER, G. (1949) *J. Colloid Sci.*, **2**, 291.
21 McGOWN, D. N. L. and PARFITT, G. D. (1967) *Kolloid Zeits.*, **219**, 48.
22 KOELMANS, H. and OVERBEEK, J. Th. G. (1954) *Disc. Faraday Soc.*, **18**, 52.
23 McGOWN, D. N. L., PARFITT, G. D. and WILLIS, E. (1965) *J. Colloid Sci.*, **20**, 650.
24 LEWIS, K. E. and PARFITT, G. D. (1966) *Trans. Faraday Soc.*, **62**, 1652.
25 McGOWN, D. N. L. and PARFITT, G. D. (1966) *Disc. Faraday Soc.*, **42**, 225.
26 FOWKES, F. M. (1966) *Disc. Faraday Soc.*, **42**, 246.
27 TAMARABUCHI, K. and SMITH, M. L. (1966) *J. Colloid Sci.*, **22**, 204.
28 PARFITT, G. D. and PEACOCK, J. (1978) *Surface and Colloid Science* (Matijevic, E., ed.), Plenum Press, New York, **10**, 163.
29 CHEN, C. S. and LEVINE, S. (1973) *J. Colloid Interface Sci.*, **43**, 599.
30 FEAT, G. R. and LEVINE, S. (1976) *J. Colloid Interface Sci.*, **54**, 34.
31 FRANKLIN, M. J. B., GOLDSBROUGH, K., PARFITT, G. D. and PEACOCK, J. (1970) *J. Paint Technol.*, **42**, 740.
32 HOGG, R., HEALY, T. W. and FUERSTENAU, D. W. (1966) *Trans. Faraday Soc.*, **62**, 1638.
33 BIERMAN, A. (1955) *J. Colloid Sci.*, **10**, 231.
34 PARFITT, G. D., WOOD, J. A. and BALL, R. T. (1973) *J. Chem. Soc. Faraday Trans. I.*, **69**, 1908.
35 PRIEVE, D. C. and RUCKENSTEIN, E. (1980) *J. Colloid Interface Sci.*, **73**, 539.
36 BLEIER, A. and MATIJEVIC, E. (1976) *J. Colloid Interface Sci.*, **55**, 510.
37 BLEIER, A. and MATIJEVIC, E. (1978) *J. Chem. Soc. Faraday Trans. I.*, **74**, 1346.
38 HANSEN, F. K. and MATIJEVIC, E. (1980) *J. Chem. Soc. Faraday Trans. I.*, **76**, 1240.
39 JAMES, R. O., HOMOLA, A. and HEALY, T. W. (1977) *J. Chem. Soc. Faraday Trans. I.*, **73**, 1436.
40 VINCENT, B. (1974) *Adv. Colloid Interface Sci.*, **4**, 193.
41 NAPPER, D. H. (1977) *J. Colloid Interface Sci.*, **58**, 390.
42 EVANS, R. and NAPPER, D. H. (1975) *J. Colloid Interface Sci.*, **52**, 260.
43 FLORY, P. J. (1971) *Principles of Polymer Chemistry*, Cornell University Press, Ithaca, New York.
44 KLEIN, J. (1980) *Nature*, **288**, 248.
45 ISRAELACHVILI, J. N., TANDON, R. K. and WHITE, L. R. (1980) *J. Colloid Interface Sci.*, **78**, 430.
46 KITCHENER, J. A. (1972) *Br. Polym. J.*, **4**, 217.
47 TADROS, Th. F. (1978) *J. Colloid Interface Sci.*, **64**, 36.

Chapter 7

A review of liquid mixing equipment

M F Edwards

Schools of Chemical Engineering, University of Bradford

7.1 Introduction

The wide range of mixing equipment available commercially reflects the enormous variety of mixing duties required in the chemical, paint, food, and pharmaceutical industries. Some of these duties are listed below:

Blending of miscible liquids;
Contacting of immiscible liquids, e.g. in solvent extraction processes;
Emulsification processes to produce stable products;
Suspending coarse solids in low-viscosity liquids;
Dispersing fine solids in high-viscosity liquids;
Dispersing gas in liquids, e.g. fermentation processes;
Contacting gas/solid/liquid in catalytic chemical reactions.

It is clear that no single item of mixing machinery will be able to carry out such a range of duties efficiently, i.e. with low capital and operating cost. Thus a number of distinct types of mixer have been developed over the years, e.g.

Mechanically agitated vessels;
Jet mixers;
In-line static mixers;
In-line dynamic mixers;
Dispersion mills;
Valve homogenizers;
Ultrasonic homogenizers;
Extruders.

The above list is not exhaustive and within each of the above types there is still a wide range of possible designs. Very little has been done in the way of standardization of equipment and no design codes are available.

In the following sections the main mechanical features of each type of equipment are described and the range of operating duty is discussed. This is done in a largely qualitative way and it is hoped that this will set the scene for the detailed quantitative treatments of liquid mixing processes which are presented in the subsequent chapters.

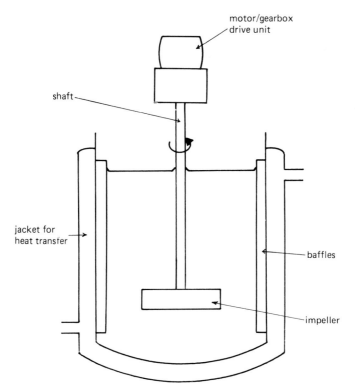

Figure 7.1   Typical arrangement for a mechanically agitated vessel

7.2   Mechanically-agitated vessels

A typical arrangement for a mechanically-agitated vessel is shown in *Figure 7.1*. This diagram serves to illustrate the overall configuration and the main features of the component items are discussed below.

7.2.1   Vessels

These are often vertically mounted cylindrical tanks which typically will be filled to a depth equal to about one tank diameter. However, in some gas–liquid contacting systems a liquid depth up to about three tank diameters is used with multiple impellers on the shaft. Vessel diameters can range from 0.1 m for small bench units up to 10 m or more in large industrial installations.

The base of the tanks may be flat, dished or conical, depending upon factors such as ease of emptying or solids suspension. In some cases, to prevent deposition of solids on the tank bottom, a specially contoured base has been proposed[1], see *Figure 7.2*. In the design of mixer/settler units for solvent-extraction purposes it has been common practice to use square tanks because of their low cost for large-capacity applications and because of the ease of combination with the settler.

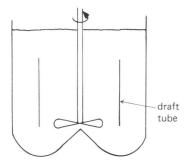
draft
tube

*Figure 7.2   Contoured base of vessel for
solids suspension*[1]

Some tanks are mounted horizontally, particularly for batch handling of viscous pastes and doughs using ribbon impellers and Z-blade mixers, etc. In such units the working volume is often relatively small and the mixing blades are massive in construction.

7.2.2   Baffles

To prevent gross vortexing behaviour when low-viscosity liquids are agitated in a vertical cylindrical tank with a centrally mounted impeller, baffles are often fitted to the walls of the vessel. Typically four baffles will be used, each having a width equal to about one-tenth of the tank diameter. In some cases the baffles are mounted flush with the wall, although occasionally a small clearance is left between the baffle and the wall.

Baffles are generally not required with high-viscosity liquids where gross vortexing is not a problem.

7.2.3   Impellers

Some of the impellers in common use are shown in *Figure 7.3*.

Propellers, turbines, paddles, anchors, helical ribbons, and helical screws are usually mounted on a central vertical shaft inside a vertical cylindrical tank and their range of application depends to a great extent upon liquid viscosity, see below.

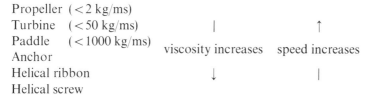

Propeller (<2 kg/ms)
Turbine   (<50 kg/ms) | ↑
Paddle   (<1000 kg/ms)   viscosity increases   speed increases
Anchor
Helical ribbon   ↓ |
Helical screw

Thus propellers, turbines, and paddles are generally used with relatively low-viscosity systems and operate at high rotational speeds. A typical tip speed for a turbine is in the region of 3 m/s with a propeller being somewhat faster and the paddle slower. These impellers are classed as remote clearance impellers having diameters in the range $\frac{1}{4}-\frac{2}{3}$ of the tank diameter.

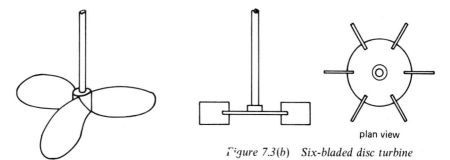

Figure 7.3(a)   Three-bladed propeller

plan view

Figure 7.3(b)   Six-bladed disc turbine

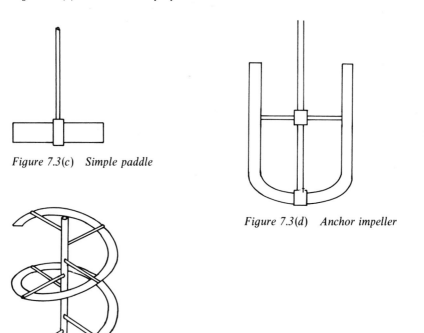

Figure 7.3(c)   Simple paddle

Figure 7.3(d)   Anchor impeller

Figure 7.3(e)   Helical ribbon

The turbine illustrated in *Figure 7.3b* is a six, flat-bladed Rushton turbine and this has a wide range of application, particularly in the chemical industry. It is possible to obtain retreating blade turbines, angled-blade turbines, four-to twenty-bladed turbines, and several patented varieties are commercially available. However, the reasons for changes in the basic design are in some cases rather obscure. Turbines are often used for gas-dispersing duties with some form of sparging device to introduce the gas under the impeller. Gas liquid contacting is considered in detail in Chapter 17.

Propellers are usually of the three-bladed marine type. They are sometimes used for in-tank blending operations with low-viscosity liquids and often for this duty angled side-entry units are used, see *Figure 7.4*.

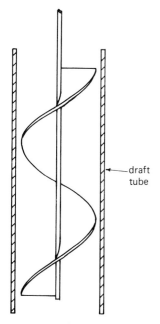

Figure 7.3(f) Helical screw with draft tube

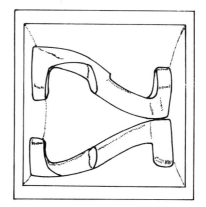

Figure 7.3(g) Z-blade mixer

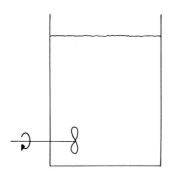

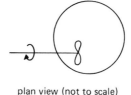

plan view (not to scale)

Figure 7.4 Side-entering propeller

In some applications, e.g. fermentation vessels, the liquid depth is large compared with the vessel diameter, and in such cases it is common practice to use more than one impeller on a single shaft to provide adequate liquid mixing throughout the vessel volume. For large installations there is some advantage to be gained by using side- or bottom-entering impellers to avoid the large length of unsupported shaft necessary when the motor/gearbox unit is mounted directly above the vessel. In this latter case the use of a bottom bearing to position the impeller may be necessary. Mechanical design aspects are further discussed in Chapter 14.

Anchors, helical ribbons, and screws are generally used for high-viscosity liquids (see Chapter 11). The anchor and ribbon have only a close clearance with the vessel wall whilst the helical screw has a smaller diameter and is often used with a draft tube to promote circulation throughout the vessel (see *Figure 7.3f*). Helical ribbons or interrupted ribbons are often used in horizontally mounted cylindrical vessels.

Kneaders, Z- and sigma-blade units are generally used for the mixing of high-viscosity liquids, pastes, rubbers, and doughs, etc., which cannot be handled by helical ribbons and anchors. The tanks are usually mounted horizontally and two impellers are often used. The impellers are massive and the clearances between blades and between blade and vessel wall are very small The rotating blades cut through most of the vessel volume ensuring that no dead spaces are possible (see *Figure 7.3g*).

From the above it is clear that a considerable amount of overlap exists between the various impellers. However, the following table may be of use in equipment selection.

Propellers ⎫ low-viscosity blending
Turbines ⎪ dispersing gases in liquids of low viscosity
Paddles ⎬ liquid/liquid contacting
⎭ suspending solids in low-viscosity liquids

Anchors ⎫
Helical ribbons ⎬ high-viscosity blending
Helical screw ⎭

Kneaders ⎫ blending pastes, rubbers, doughs
⎭ dispersing fine solids in viscous liquids, etc.

Typically power units will be in the range 0.2–4 kW/m³, the lower figure being appropriate for gentle agitation of low-viscosity fluids and the larger figure applying to high-viscosity blending duties[2]. More details of these particular operations can be found in subsequent chapters.

7.3   Jet mixers

Mixing in a vessel can be achieved for low-viscosity applications by the use of a submerged nozzle from which a high-velocity jet of liquid emerges. A pump is used to withdraw part of the liquid from the vessel and recycle it via the nozzle to the vessel. The momentum transferred from the high-velocity jet to the liquid in the vessel causes the mixing action and circulation within the tank (see *Figure 7.5*).

For blending in large tanks, several nozzles may be used. The operating characteristics of such systems are treated in detail in Chapter 9 together with jet mixing in tubular devices.

Similarly, mixing in vessels and lagoons can be achieved by gas injection with no mechanical agitation. In the simplest case, bubble columns using a perforated plate distribution cause a swarm of bubbles to rise through the

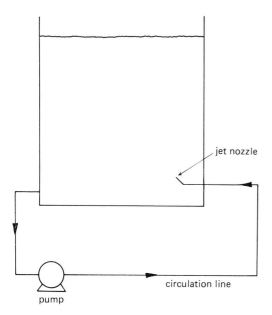

Figure 7.5   Jet-mixing arrangement

liquid and this gives an agitation effect to the liquid phase. If liquid circulation is important then air-lift devices, often coupled with a draft tube, can be used, see *Figure 7.6*. Again, many geometrical variations are possible and it is not feasible to cover all these in this present chapter.

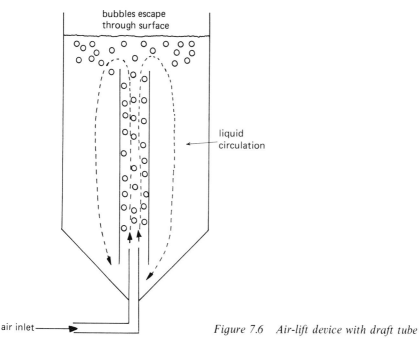

Figure 7.6   Air-lift device with draft tube

7.4   In-line static mixers

In recent years the in-line static mixer has been adopted for a large number of blending and dispersing operations. Many in-line mixers are now commercially available. The geometrical details of these devices differ greatly but the principles of operation are basically the same in each case.

The stationary units are housed in a pipe and the mixer is fed by a pumping system, see *Figure 7.7*. For the case of blending of viscous liquids in the laminar regime, mixing is achieved by a slicing and folding mechanism, see *Figure 7.8*. This mixing process clearly gives an improvement in 'product mixedness' as the number of repeated mixing elements increases. To a first approximation the mixing is independent of fluid flow rate and fluid properties in the laminar region.

In the case of liquid/liquid and gas/liquid dispersion, the above mixing mechanism does not hold and the operation is usually carried out in the turbulent region. The break-up of gas bubbles or liquid droplets to give a

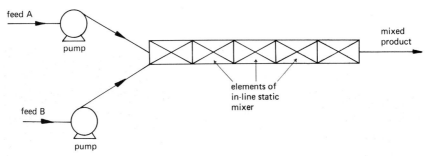

Figure 7.7   Mixing with an in-line static unit

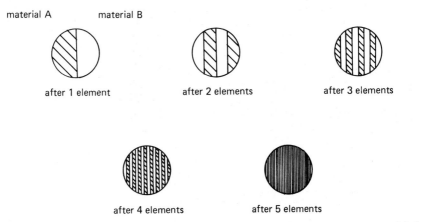

Figure 7.8   Laminar blending in a Chemineer-Kenics in-line static mixer: simplified mixing mechanism

system of high interfacial area for mass transfer is brought about by the shear stresses in the system. These stresses are related to the pressure drop and hence flow rate through the mixer. Thus to get a smaller droplet size the fluid flow rate must be increased. It will not be effective to merely increase the number of elements (as was the case in laminar blending).

For blending and dispersing duties, in-line static mixers offer the advantage of continuous operation and small working volume. Units are available to fit pipes from about 1 cm diameter up to 0.5 m or more. Further consideration is given to these devices in Chapter 13.

7.5 In-line dynamic mixers

For mixing operations which require the continuous production of finely dispersed solids, emulsions, stable foams, etc., the in-line dynamic mixer in one of its several forms could be used. These usually consist of a rotor which spins at high speed inside a casing and the feed materials are pumped continuously to the unit. Inside the casing the fluid is subjected to extremely high shear forces which are required for the dispersing operation, see *Figure 7.9*.

Figure 7.9   In-line dynamic mixer (Courtesy of E. T. Oakes, Macclesfield, UK)

Often the dispersing unit itself produces little or no pressure differential and the unit has to be fed with its input materials by separate pumping units. However, some in-line dynamic mixers do incorporate a pumping action and in such cases additional pumping may not be necessary.

7.6   Mills

Several duties involving the dispersion of fine solids and emulsification cannot be carried out in the conventional mechanically agitated vessel because it is not possible to generate sufficiently high stresses to break-down the aggregated particles to achieve the required dispersion quality or to create a stable emulsion.

In such cases various types of mill may be considered and some of these are illustrated in *Figure 7.10* (ref. 3). The range of application of the various mills and their mode of operation, i.e. whether particle dispersion is achieved by crushing or shearing, is shown in *Figure 7.11* (ref. 3).

More detailed consideration is given to the dispersion of fine solids in liquids in Chapter 6 and also in Chapter 12.

7.7   High-speed dispersing units

This type of equipment is similar to the in-line dynamic mixer but in this case it is used in a vessel. The mixing head consists of a high-speed rotor inside a housing where the fluid is subjected to an intense shearing action. The fluid is

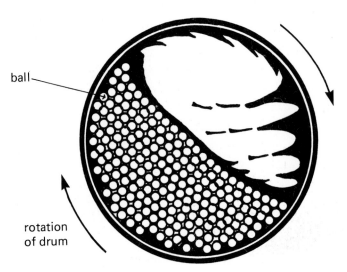

Figure 7.10(a)   Ball mill

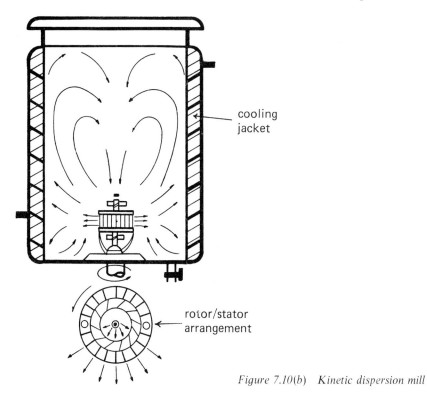

cooling
jacket

rotor/stator
arrangement

Figure 7.10(b)   Kinetic dispersion mill

drawn into the unit and expelled at high velocity thus setting up a circulation pattern in the vessel, see *Figure 7.12*.

Various geometrical designs of rotor/stator and vessel are possible and such units find application in emulsification, solids dispersion and size reduction, dissolving duties, etc., for which an intense shearing action is essential.

7.8   Valve homogenizers

These units employ a pumping section to supply the materials to be dispersed through a small (often adjustable) orifice. The high pressures generated (34–550 bar) to force the fluid through the orifice create high shear forces which are capable of producing emulsions and colloidal suspensions, etc., in a continuous manner, see *Figure 7.13*. In addition to the shearing action the effects of cavitation and particle impact are important.

Single- and multiple-stage units are available. Typically a high-pressure unit giving 300 bars at a throughput of 7000 l/hr may require a 60 kW drive.

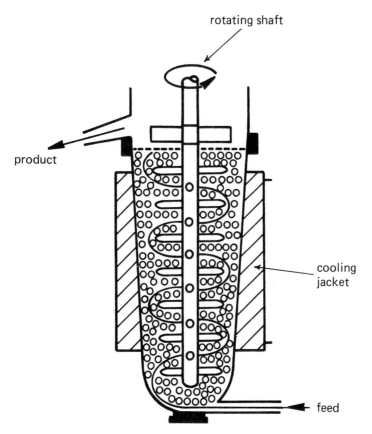

rotating shaft

product

cooling
jacket

feed

Figure 7.10(c)   Attrition mill

7.9   Ultrasonic homogenizers

The principle of operation of an ultrasonic homogenizer is illustrated in *Figure 7.14*. The process material is pumped at high pressure (up to about 150 bar) through a specially designed orifice to produce a high-velocity stream over a blade which vibrates at ultrasonic frequencies. This produces a high level of cavitation, ideal for the production of high-quality emulsions and dispersions.

Typical throughputs for such units are 2–800 l/min and a typical power requirement for the pumping unit is 1.5 kW per 1000 kg/hr of product.

7.10   Extruders

Mixing duties in the plastics industry are often carried out in either single- or twin-screw extruders. The feed to such units will usually contain the base polymer in either granular or powder form, together with additives such as stabilizers, plasticizers, colouring pigments, etc. During processing in the

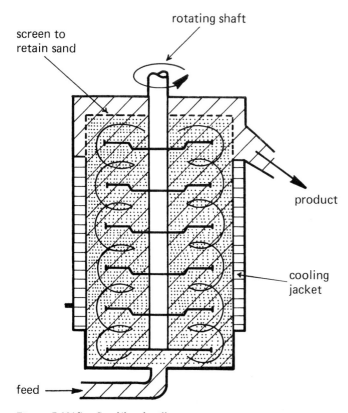

Figure 7.10(d)   Sand/bead mill

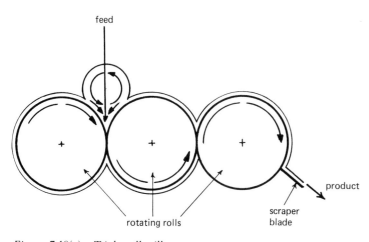

Figure 7.10(e)   Triple roll mill

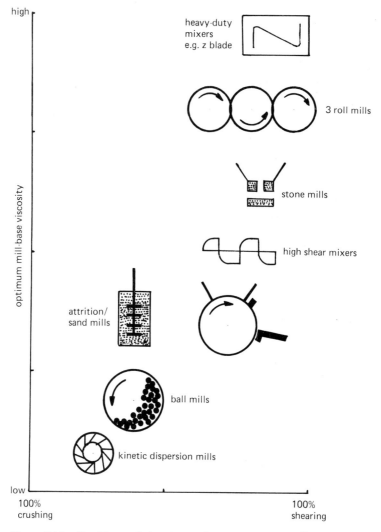

Figure 7.11   Crushing and shearing in dispersion equipment[3]

extruder the polymer is melted and the additives mixed. The extrudate is delivered at high pressure and at a controlled rate from the extruder for shaping either by means of a die or a mould.

For the case of single-screw units, see *Figure 7.15*, the shearing which occurs in the helical channel between the screw and the barrel is not intense. Thus the mixing is not as effective as in the case of co- and counter-rotating twin-screw machines. Here there are regions where the rotors are in close proximity and the shear stresses are extremely high, see *Figure 7.16*. Clearly, twin-screw units can yield a product of better mixture quality than a single-screw machine. However, the capital cost and power requirement will be greater.

It is possible to improve the mixing performance of single-screw machines

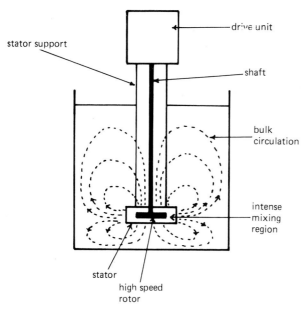

Figure 7.12 Sketch of a high-speed dispersing unit

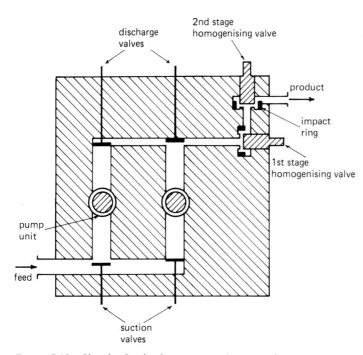

Figure 7.13 Sketch of valve homogenizer (two-stage)

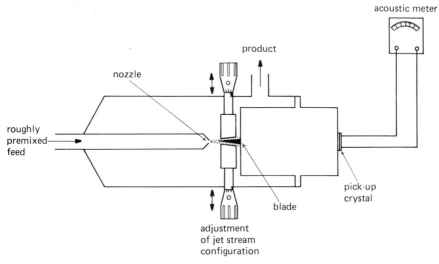

Figure 7.14   Ultrasonic homogenizer

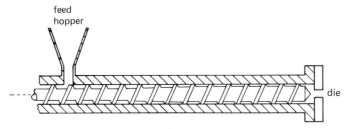

Figure 7.15   Single-screw extruder

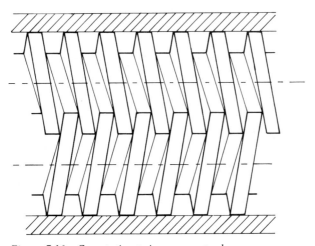

Figure 7.16   Co-rotating twin-screw extruder

Figure 7.17   Cavity transfer mixer (Courtesy RAPRA, Shrewsbury, UK)

by the use of additional mixing heads or by the use of in-line static mixers described earlier. Several such devices are available commercially and mixing will be improved in comparison with the single screw alone. However, additional capital cost and power requirement will be incurred. *Figure 7.17* shows the Cavity Transfer Mixer[4] which is just one of the dervices available to improve the performance of single-screw units.

Similar considerations apply to mixing in extruders in the food industry. Some relevant theoretical developments are presented in Chapter 12 for the mixing of high-viscosity materials in single- and twin-screw units.

7.11   Equipment selection

The preceding sections serve to illustrate the remarkably wide range of mixing equipment that is available. Many of the different mixing units overlap in their range of application and equipment selection is a major problem confronting the industrial engineer.

It is hoped that the chapters which follow will give more detailed and quantitative information on the various types of mixers and that this will be helpful in design and selection of equipment. However, much will depend upon the particular duty, e.g. batch or continuous, whether the materials are shear-sensitive, whether there are corrosion or special materials of construction requirements, whether heat removal or addition is a controlling factor.

Many of these considerations are not directly related to the science of mixing but can be critical in deciding the appropriate mixer for a given duty.

References

1   BOURNE, J. F. and SHARMA, R. N. (September 1974) *Paper No. B3, Proc. First European Conference on Mixing and Centrifugal Separation*, BHRA Fluid Engineering, Cranfield, UK.
2   *Agitator Selection and Design*, Engineering Equipment Users' Association, Constable (1963).
3   *Dispersion of Tioxide Pigments in Non-Aqueous Media*, Tioxide International, Kynoch Press, Birmingham, UK.
4   GALE, M. (May 1982) *Paper No. 18*, '*Polymer Extrusion II*', Plastics and Rubber Institute, London.

Chapter 8

Mixing of low-viscosity liquids in stirred tanks

M F Edwards

Schools of Chemical Engineering, University of Bradford

8.1 Introduction

In this chapter the behaviour of single-phase liquids in mechanically agitated vessels is considered. Particular emphasis is given to 'low-viscosity systems' although some of the basic concepts set down will be equally applicable to 'high-viscosity systems'. A more detailed coverage of the mixing of highly viscous and non-Newtonian liquids is presented in Chapters 11 and 12.

Initially the power requirement of a rotating impeller will be examined because this not only enables the design engineer to size the motor/gearbox unit for a stirred tank but also clearly demonstrates the distinction between 'low' and 'high' viscosity liquid mixing duties. Although work on single-phase liquids is of limited application industrially, an attempt is made to show how single-phase work on flow patterns, flow-rate-head concepts, turbulence and mixing time for low-viscosity liquids can be used to lay the foundation for subsequent chapters on solid/liquid and gas/liquid systems.

The published literature on the topics covered below is vast and it is not feasible to refer to all relevant studies. Instead only a few references have been cited in order to support the concepts which are presented.

8.2 Power input

Consider the stirred vessel illustrated in *Figure 8.1* where a Newtonian liquid of density, ρ, and viscosity, μ, is agitated by an impeller of diameter, D, rotating at rotational speed, N. Let the tank diameter be T, the impeller width W, and the liquid depth H.

The power requirement of the impeller, P, under these conditions will depend upon the independent variables as follows:

$$P = fn(\rho, \mu, N, g, D, T, W, H, \text{other dimensions}) \tag{8.1}$$

P is the impeller power requirement, i.e it represents the rate of energy dissipation within the liquid. The electrical power required to drive the motor will exceed P by the motor, gearbox, and bearing losses.

It is not possible to obtain the functional relationship in equation (8.1) by

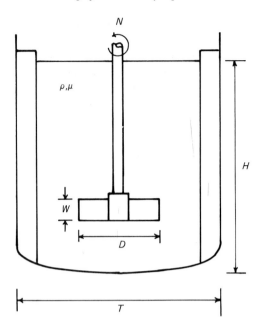

Figure 8.1   Some parameters influencing power input

analysing the fluid mechanics of the flow in the vessel because of the complex geometry of the vessel, impeller, and any other inserts, e.g. heating coils. However, using dimensional analysis the number of variables describing the problem can be minimized and equation (8.1) reduces to:

$$\frac{P}{\rho N^3 D^5} = fn \left\{ \frac{\rho N D^2}{\mu}, \frac{N^2 D}{g}, \frac{T}{D}, \frac{W}{D}, \frac{H}{D}, \text{etc.} \right\} \tag{8.2}$$

T = tank diam.
W = impeller
H = height of liquid

where

$P/\rho N^3 D^5$ is the Power number, Po;
$\rho N D^2 / \mu$ is the Reynolds number, Re; and
$N^2 D / g$ is the Froude number, Fr.

The Froude number is usually important only in situations where gross vortexing exists and this can be neglected if the Reynolds number is less than about 300 (ref. 1). For higher Reynolds numbers the Froude number effects are eliminated by the use of baffles or off-centre stirring.

Thus in cases where the Froude number can be neglected we have:

$$Po = fn(Re, \text{geometrical ratios}) \checkmark \tag{8.3}$$

and if we consider *geometrically similar* systems:

$$Po = fn(Re) \tag{8.4}$$

The functional relationship for a given system geometry has now to be found by conducting experiments in which the power requirement is measured with the impeller rotating at various speeds in fluids of different density and viscosity. Several vessels which are geometrically similar may be used. The data are then plotted as *Po* vs. *Re*, usually on log–log paper to give the power curve for the system. All the single-phase experimental data should fall on one unique curve for a given geometrical design of vessel and impeller.

A typical 'power curve' is shown in *Figure 8.2*, from which it can be seen that at low values of the Reynolds number, less than about 10, a *laminar* region

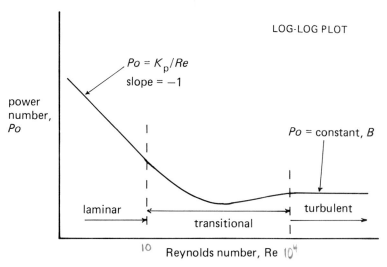

Figure 8.2   General characteristics of power curve

exists and here the flow is dominated by viscous forces. The slope of the power curve (on log–log paper) is −1 indicating:

$$Po = K_p/Re \tag{8.5}$$

where K_p is a constant depending only upon the system geometry.

In the laminar region the mixing is slow and initially blending is achieved by the velocity distribution in the vessel because turbulent dispersion is absent and molecular diffusion effects in liquid systems are very slow. Eventually the velocity distribution will create a large interfactial area between the components which are being blended and only then will molecular diffusion play a significant role. This low Reynolds number regime is the area in which viscous liquids are processed and more attention will be devoted to this in Chapters 11 and 12.

At high Reynolds numbers, greater than about 10^4, the flow is *turbulent* and mixing is rapid due to the motion of the turbulent eddies. In this region the power number is essentially constant, i.e.

$$Po = B \tag{8.6}$$

Often gas/liquid, solid/liquid, and liquid/liquid contacting operations are carried out in this region and such processes may be slowed down by mass transfer considerations.

In between the laminar region at low Reynolds numbers and the turbulent region at high Reynolds numbers there exists a gradual transition zone in which no simple mathematical relationship exists between the power number and Reynolds number.

Power curves for many different impeller geometries can be found in the literature[1,2,3,4,5] but it must be remembered that, whilst a power curve is applicable to any single-phase Newtonian liquid with any impeller speed, the curve will only be valid for one system geometry, i.e. any vessel size provided the impeller, vessel, inserts, etc., are all geometrically similar. Thus, these power curves (e.g. see *Figure 8.3*) can easily be used to calculate the impeller power requirement if the vessel/impeller geometry, fluid properties and

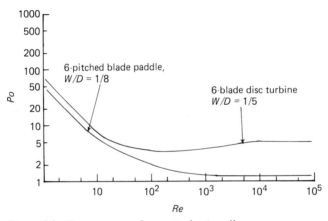

Figure 8.3   Power curves for particular impellers

impeller speed are known. If no curve is available for the vessel/impeller geometry of interest, experimental work must be conducted to establish the power curve for this geometry. This, of course, can be done on a small, bench-top scale if necessary.

Some typical energy consumptions for stirred tanks are given below[6] which provide a rough guide to the magnitude of the power requirement for certain duties. These values are only intended to be rule-of-thumb values and do not remove the need to carry out correct sizing calculations using appropriate power curves.

Low power	suspending light solids, blending low-viscosity liquids	$0.2\,\text{kW/m}^3$
Moderate power	some heat transfer, gas dispersion, liquid/liquid contacting, suspending moderate density solids	$0.6\,\text{kW/m}^3$
High power	suspending heavy solids, emulsification, gas dispersion	$2\,\text{kW/m}^3$
Very high power	blending pastes, doughs	$4\,\text{kW/m}^3$

It must be borne in mind that the power requirement which is calculated from power curves gives a measure of the energy dissipated within the liquid. It does not give any indication of whether or not this power is dissipated effectively. Furthermore, it does not contain the losses which occur in the motor/gearbox/bearings which must be considered in sizing the drive unit.

It is worth noting that the calculation of the power requirement demands a knowledge of the impeller speed and it is this value which is difficult to determine, e.g.:

1. What impeller speed is necessary to blend the contents of a tank in a given time?
2. What impeller speed will be required to bring about a desired mass transfer rate in gas/liquid contacting?

Answers to questions such as these require a detailed consideration of the mixing/mass transfer mechanisms and these are covered in later chapters. In the present chapter we have established a method of calculating power input to a single-phase system once the impeller speed is known.

It is possible to extend the procedure developed above the certain multiphase applications in which the power requirement has to be estimated. In the case of low-viscosity liquid/liquid systems, as encountered in solvent extraction, and for coarse solids suspended in low-viscosity liquids at low concentrations, the operation is likely to be carried out in the turbulent region.

The single-phase power curves can be used in such instances with the mean density being used in both the power number and Reynolds number. However, such an approach must not be used for gas/liquid systems where predictions based on average density values can lead to gross over-estimates of the power requirement. This is considered in detail in Chapter 17.

Finally, for liquid/liquid systems which form stable emulsions and for solid/liquid systems of high concentrations, it is likely that non-Newtonian properties will be encountered. In such cases an alternative approach is required for the prediction of the power input. This is considered in detail in Chapter 11 which covers the mixing of viscous and non-Newtonian liquids.

8.3   Flow patterns

The flow patterns of single-phase liquids in tanks agitated by various types of impeller have been reported in the literature[2-5]. The experimental techniques used include the use of coloured tracer liquid, mutually-buoyant particles, hydrogen bubble generation and mean velocity measurements using pitot probes, hot-film devices and lasers.

A qualitative picture of the flow field created by an impeller in a single-phase liquid is useful to establish whether there are 'dead zones' in the vessel, or whether particles are likely to be suspended in the liquid, etc. Some typical primary flow patterns are illustrated in *Figures 8.4* and *8.5* for a propeller and a disc turbine respectively. The propeller creates a mainly axial flow through the

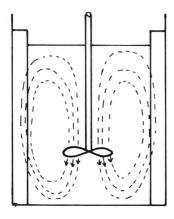

Figure 8.4   Axial flow pattern for propeller

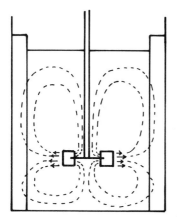

Figure 8.5   Radial flow pattern for disc turbine

impeller and this central axial flow may be upwards or downwards depending upon the direction of rotation.

The predominant circulation pattern for a downward pumping propeller is shown in *Figure 8.4*. Of course, the velocities at any point will be three dimensional and unsteady but circulation patterns such as those in *Figure 8.4* are useful in the avoidance of dead zones and in the selection of appropriate impellers for a given duty.

The flat-bladed turbine produces a strong radial flow outwards from the impeller, see *Figure 8.5*, creating circulation zones in the top and bottom of the tank. The type of flow can be altered by changes in the impeller geometry. For example, if the turbine blades are angled to the vertical a stronger axial flow component is produced and this could be advantageous for duties involving solids suspension. The flat paddle produces a flow field with large tangential components of velocity and this is not effective in many mixing operations. However, axial velocities can easily be induced using an angled paddle.

Although the propeller, turbine and paddle are the principal impellers for use with low-viscosity systems, i.e. in the turbulent and transitional regions, a brief mention will be made of flow visualization studies for anchors, helical ribbons, and helical screws. The flow in an anchor agitated vessel has been studied in detail by Peters and Smith[7]. The impeller promotes fluid motion close to the vessel wall but the region near the shaft may be relatively stagnant. In addition, there is little top-to-bottom turnover. In order to promote a top-to-bottom motion a helical ribbon may be used and often a screw is added to the shaft to obtain motion in the central regions of the vessel. This combined impeller system would have a ribbon pumping upward near the wall of the vessel with the screw, twisted in the opposite sense, pumping downwards near the shaft.

Clearly, the fluid flow pattern depends critically upon the impeller/vessel/baffle geometry, and in selecting the appropriate combination of equipment the designer must ensure that the resulting flow pattern is suitable for his particular duty. For example, for suspending coarse solid particles it is desirable to have high axial velocity components.

8.4 Flow-rate-head concepts

The action of a turbine or propeller in a liquid of low viscosity can be likened to that of the impeller in a radial or axial flow pump respectively. That is, we can regard the mixer as being a caseless pump. The use of the flow rate (or discharge), Q, from a pump and the fluid head, H', is well known. Thus for a pump of unit efficiency these variables can be related to the power input to the runner

$$P = \rho Q g H' \qquad (8.7)$$

For a mixing vessel, a given input of power, P, to the impeller can create a 'flow rate' Q (and thus a circulation throughout the vessel) and a 'head' H' which is dissipated on circulation through the vessel. For low-viscosity liquids the head can be thought of in terms of the turbulence which is generated. This is most intense in the region of the impeller and decays in regions away from the impeller.

These regions of high intensity of turbulence are suitable for 'dispersive' processes such as liquid/liquid, gas/liquid contacting and for generating good conditions for mass transfer. It is desirable to 'circulate' the liquid through the regions of high intensity of turbulence as frequently as possible. Thus mixing in low-viscosity systems is seen to be influenced by:

1. The intensity of turbulence (head);
2. The rate of circulation (flow).

Good impeller selection will ensure that the power input to the agitator provides the correct balance between flow and head.

The argument can be extended to the entire Reynolds number range and impellers can be classified as the 'shear type', generating large turbulent shear stresses for dispersion or the 'circulation' type which sets up a strong circulation throughout the vessel. Thus, as a very broad classification for operations in the turbulent region, marine propellers produce a large circulation, whereas turbines generate higher shear stresses in the impeller discharge region and are well suited to dispersive mixing processes. Anchors, helical ribbons and screws are circulation-type impellers which set up a gross circulation pattern and velocity profile within the vessel upon which the mixing process depends. This latter class are good for laminar blending applications. Paddles hold a somewhat intermediate position.

It is possible to express equation (8.7) in dimensionless form to give:

$$Po = (Q/ND^3)(gH'/N^2D^2) \tag{8.8}$$

where Q/ND^3 is a flow coefficient, flow number or pumping number, and gH'/N^2D^2 is the head coefficient.

There are data available in the literature on the flow generated by different impellers. However, care must be taken in using such information since the circulation rate through the vessel will not necessarily be the same as the discharge flow from the impeller. This is because the discharge flow leaving the impeller entrains liquid from elsewhere in the vessel. For example, Revill[8] reviewed the data available in the literature on the pumping capacity of disc turbine agitators and recommended for design purposes:

$$Q/ND^3 = 0.75 \quad \text{for } 0.2 < D/T < 0.5 \tag{8.9}$$

where Q is the radial discharge rate from the turbine.

The total circulation rate was found to be of the order of twice the discharge rate. However, it should be mentioned that the value of discharge rate Q is of itself not a useful value unless it can be related to measures of process performance such as blend times, etc. Nevertheless the general concepts of 'flow' and 'head' are useful in understanding certain mixing processes.

8.5   Turbulence measurements

Mean and fluctuating velocities in stirred tanks have been studied to determine the mean velocity fields and the nature of the turbulence. This work has generally been carried out in the turbulent or transitional regions using hot-film, hot wire, or laser anemometers.

These measurements of turbulence require careful and detailed experimentation to study just one single-phase liquid in one vessel/impeller geometry. The data collected are useful in interpreting the nature of mixing when two miscible reactants are agitated at high Reynolds number, in establishing droplet and bubble sizes in two phase systems and in understanding mass-transfer processes. Prior to a consideration of some of the experimental findings it is worthwhile to introduce a few of the basic concepts which can be applied to turbulence.

The principal feature of turbulent flows is the presence of eddies which are large compared with the molecular scale and which aid the mixing process (turbulent dispersion). This mixing mechanism in liquids is often rapid compared with the other processes of bulk flow and molecular diffusion. The eddies vary in size, having a maximum scale, L, which is of the order of the scale of the equipment (i.e. impeller or baffles) down to a minimum value, l, which, according to Kolmogoroff[2,3,9], for isotropic turbulence depends only upon the power input per unit mass to the system and the kinematic viscosity of the liquid:

$$l = (\mu^3 V/P\rho^2)^{1/4} \tag{8.10}$$

Energy is transferred from the largest eddies, set in motion by the impeller, etc., down through the size range of eddies until it is eventually dissipated from the smallest eddies by the action of viscosity. Thus if two miscible liquids, initially separated, are contacted in a mixing vessel in the transition or turbulent zone, complete mixing down to the molecular level is not possible by turbulence alone. Molecular diffusion must be present to complete the mixing process. For details of the influence of this type of turbulent mixing upon reacting liquids the reader is referred to Chapter 10.

Further examination of isotropic turbulence shows[2,3] that an appropriate mean square velocity over distance, d, is:

$$\overline{u^2} = C_1 (Pd/\rho V)^{2/3} \tag{8.11}$$

and that the turbulent shear stress for a scale size, d, is ($l \ll d \ll L$):

$$\pi = C_2 (P\rho/dV)^{2/3} \tag{8.12}$$

These relationships are important in examining and interpreting the rate of mass transfer in solid/liquid, gas/liquid, and liquid/liquid systems and in examining droplet and bubble deformation and break-up in liquid/liquid and gas/liquid systems respectively. More details on these concepts can be found in subsequent chapters.

Some experimental measurements have been made in mixing vessels of the mean and fluctuating velocities but, as mentioned earlier, it is time consuming to collect data throughout the vessel for just one single-phase liquid and one impeller speed. It is not yet possible to generalize the results for design purposes including the full effects of fluid properties and impeller/baffle/vessel geometries.

Turbine agitation in baffled vessels has received the most attention. For a six-bladed disc turbine in a 19.2 cm-diameter vessel fitted with four baffles with the impeller Reed *et al.*[10] measured mean and fluctuating velocities at various positions in the vessel. In the jet region from the impeller a significant periodic effect was observed due to the passage of the turbine blades. Correcting for this periodic effect, turbulence intensities of about 30–60% have been observed[11,12]. In the region away from the impeller the periodic component due to the turbine blades disappears. However, a strong influence of the baffles is observed in their close proximity[10].

In view of our incomplete knowledge at present on the details of the turbulence in mixing vessels in the impeller region and in the bulk of the liquid in the vessel, the developments based on equations (8.10) to (8.12) should be interpreted with caution. More work is required to complete our knowledge even for a modest range of impeller designs with single-phase liquids.

8.6   Mixing time

Consider a single-phase liquid in a stirred tank to which a volume of tracer material is added. The 'mixing time' is the time measured from the instant of addition until the vessel contents have reached a specified degree of uniformity when the system is said to be 'mixed'.

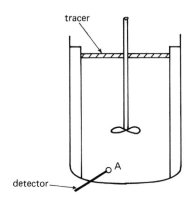

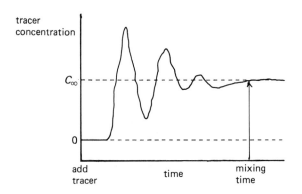

Figure 8.6   Mixing time measurement

Let us examine a 'mixing time' experiment represented in *Figure 8.6* where a tracer material is to be mixed into the liquid in a vessel. The tracer has the same viscosity and density as the main liquid with which it is miscible. That is, we are observing a simple blending operation. By means of a suitable detector the tracer concentration is measured as a function of time at a particular point in a vessel.

If the volume of tracer added is known it is a simple matter to calculate the equilibrium concentration, C_∞. The tracer concentration at point A will approach this equilibrium concentration as shown in *Figure 8.6*. The mixing time may then be defined as the time required from the instant of tracer addition for the concentration at A to reach the equilibrium value. However, this time depends upon the way in which tracer is added and the location of point A. Furthermore, the concentration approaches C_∞ in an asymptotic manner and the end point of the experiment is difficult to detect with precision.

In order to eliminate the influence that the position of the detector has upon the measured mixing time, another experiment could be based on recording the tracer concentrations at several, say, n, positions. At any time the concentration variance σ^2 about the equilibrium value can be calculated as:

$$\sigma^2 = \frac{1}{n-1} \sum_{i=0}^{n} (C_i - C_\infty)^2 \tag{8.13}$$

where C_i is the concentration at time t recorded by the ith detector.

The change of this variance with time would take the form shown in *Figure 8.7*. Again the 'mixing time' from the start of the experiment to the predetermined cut-off point depends upon the experimental technique adopted.

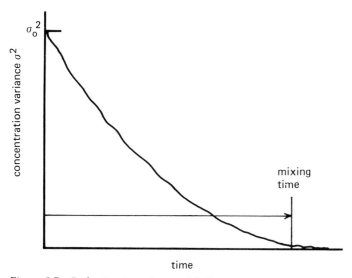

Figure 8.7 Reduction in variance with time

Practical mixing times can be measured by a variety of experimental techniques[13], e.g.

Acid/base/indicator reactions
Electrical conductivity variations
Temperature variations
Refractive index variations
Light-absorption techniques.

In each case it is necessary to specify the manner of tracer addition, the position and number of recording points, the sample volume of the detection system, and the criterion adopted for deciding the cut-off point of the end of the experiment. Each of these factors will influence the measured mixing time and therefore care must be exercised in comparing results from studies employing different measuring techniques.

The mixing time t_M measured by a given experimental technique will depend upon the process and operating variables as follows:

$$t_M = fn(\rho, \mu, N, g, D, T, \text{other geometrical parameters}) \tag{8.14}$$

Using dimensional analysis a dimensionless mixing time Nt_M can be introduced such that:

$$Nt_M = fn(\rho N D^2/\mu, N^2 D/g, T/D, \text{other geometrical ratios}) \tag{8.15}$$

i.e.

$$Nt_M = fn(Re, Fr, \text{geometrical ratios}) \tag{8.16}$$

For *geometrically similar systems*, neglecting the effect of the Froude number, the dimensionless mixing time becomes a function of Reynolds number only:

$$Nt_M = fn(Re) \tag{8.17}$$

In presenting experimental data on mixing times workers have generally used relationships of the form of equations (8.16) and (8.17). For studies carried out in the laminar, low Reynolds number region it has often been found that the dimensionless mixing time is essentially independent of Reynolds number. A similar trend is observed at high Reynolds number and this typical behaviour is sketched in *Figure 8.8*.

In addition to the sensitivity of mixing time to the experimental technique employed, results in the literature also show the marked dependence upon impeller/vessel geometry. This geometrical influence is due to the effect that geometry has upon gross flow patterns and intensity of turbulence.

Some typical mixing-time data obtained by Norwood and Metzner[14] for turbine impellers in baffled vessels using an acid/base/indicator technique are presented in *Figure 8.9*. Much of the mixing-time data presented in the literature is concerned with blending liquids of equal density and viscosity. However, these studies will often underestimate the blend times required for components of unequal density and/or viscosity.

In many multiphase systems, e.g. gas/liquid or liquid/liquid mass-transfer operations the process is not controlled by mixing time considerations. Often this mixing will be rapid compared with the mass transfer which is the rate controlling step. However, this may not be the case for all operations and particularly with large vessels and non-Newtonian liquids the mixing time can be a critical factor.

Mixing times are considered in further chapters of this book. In Chapter 9, the mixing times in jet mixers (both tank and tubular devices) are discussed, whilst in Chapter 10 the significance of mixing times for the study of reactive

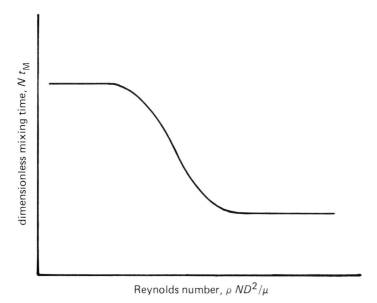

Figure 8.8 Mixing time behaviour

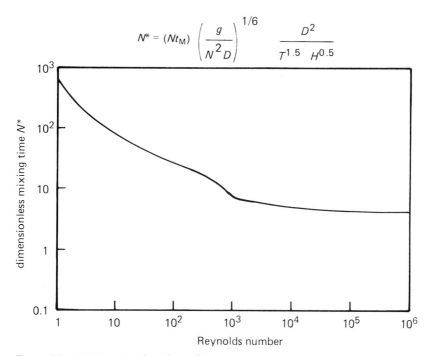

$$N^* = (Nt_M) \left(\frac{g}{N^2 D} \right)^{1/6} \frac{D^2}{T^{1.5} H^{0.5}}$$

Figure 8.9 Mixing time data for turbines

systems is considered. Data on mixing times for viscous and non-Newtonian liquids receive attention in Chapter 11.

Notation

C	concentration
d	particle, drop, bubble diameter
D	impeller diameter
Fr	Froude number
g	gravitational acceleration
H	liquid height
H'	head generated by impeller
l	minimum eddy scale
L	maximum eddy scale
N	impeller speed
P	power input
Po	power number
Q	impeller pumping rate
Re	Reynolds number
T	vessel diameter
t_M	mixing time
W	impeller width

Greek letters

ρ	liquid density
μ	liquid viscosity
τ	shear stress
σ^2	concentration variance

References

1   SKELLAND, A. H. P. (1967) *Non-Newtonian Flow and Heat Transfer*, Wiley.
2   NAGATA, S. (1975) *Mixing: Principles and Applications*, Wiley.
3   UHL, V. W. and GRAY, J. B. (1966) *Mixing: Theory and Practice, Vol. 1*, Academic Press.
4   HOLLAND, F. A. and CHAPMAN, F. S. (1966) *Liquid Mixing and Processing in Stirred Tanks*, Reinhold.
5   STERBACEK, Z. and TAUSK, P. (1965) *Mixing in the Chemical Industry*, Pergamon.
6   ENGINEERING EQUIPMENT USERS' ASSOCIATION (1965) *Agitator Selection and Design*, Constable.
7   PETERS, D. C. and SMITH, J. M. (1966) *Trans. Instn. Chem. Engrs.*, **44**, 224.
8   REVILL, B. K. (1982) *Paper R1, Fourth European Conference on Mixing*, BHRA Fluid Engineering, Cranfield, UK.
9   BRODKEY, R. S. (ed.) (1975) *Turbulence in Mixing Operations*, Academic Press.
10 REED, X. B., PRINCZ, M. and HARTLAND, S. (1977) *Paper B1, Second European Conference on Mixing*, BHRA Fluid Engineering, Cranfield, UK.
11 MUJUMDAR, A. S., HUANG, B., WOLF, D., WEBER, M. E. and DOUGLAS, W. J. (1970) *Can. J. Chem. Engng.*, **18**, 475.
12 GUNKEL, A. A. and WEBER, M. E. (1975) *A.I.Ch.E. Jnl.*, **21**, 931.
13 FORD, D. E., MASHELKAR, R. A. and ULBRECHT, J. (1972) *Proc. Technol. International*, **17**, 803.
14 NORWOOD, K. W. and METZNER, A. B. (1960) *A.I.Ch.E. Jnl.*, **6**, 432.

Chapter 9

Jet mixing

B K Revill

Heavy Chemicals New Science Group, Imperial Chemical Industries p.l.c.

9.1 Introduction

The use of turbulent jets for mixing miscible liquids is very common in the chemical process industries. In jet mixing, a fast-moving stream of liquid, the jet or primary liquid, is injected into a slow-moving or stationary liquid, the bulk or secondary liquid. The velocity difference between the jet and bulk liquids creates a mixing layer at the jet boundary. This mixing layer grows in the direction of the jet flow, entraining and mixing bulk liquid into the jet.

Jet mixing in tanks is usually used either to blend a fresh feed in with the contents of a tank or to obtain an homogeneous mixture within the tank. Batch or continuous operation is possible and a side entry (i.e. through the tank wall) or an axial jet (i.e. along the axis of the tank) may be used. The jet is positioned either near the tank floor pointing towards the liquid surface or near the liquid surface pointing towards the tank floor. Industrial applications of tank jet mixing have ranged from the homogenization of hydrocarbon and LNG storage tanks[1,2] to acid mixing[3].

Some typical jet/tank geometries are shown in *Figure 9.1*. A batch operation where a liquid B is added directly to a liquid A contained in the mix tank and one in which B is added to the recycle line are shown in *Figures 9.1(a)* and *(b)*. A continuous operation, with A and B added directly to the mix tank, is shown in *Figure 9.1(c)*. Note that the product can be taken directly from the mix tank by placing the off-take line in an area of the tank remote from the direct action of the jet.

Tubular jet mixers are used in turbulent flow either to blend two streams or to mix reactants prior to reaction. They are high-throughput continuous-flow devices and accurate flow control is, therefore, essential. Mixing times as low as a few milliseconds on the small scale or a few seconds on the large scale are possible.

Two common types of jet mixer, the coaxial jet mixer and the side-entry jet mixer, are shown in *Figure 9.2*. In the coaxial jet mixer the jet stream is introduced through a small-diameter pipe running concentrically inside a large-diameter pipe. In the side-entry jet mixer, the jet stream is injected radially into the main pipe flow.

Multiple jet mixers, or a series arrangement of single jet mixers can be used

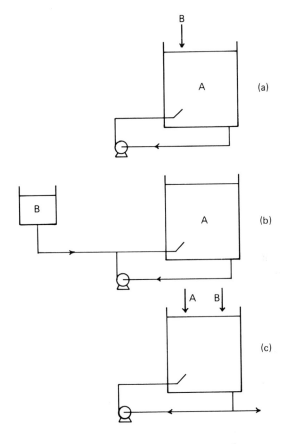

Figure 9.1   Jet mixed tanks

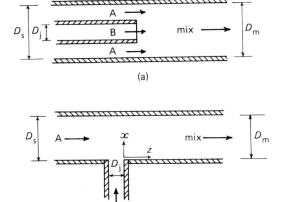

Figure 9.2   Tubular jet mixers.
(a) Coaxial jet mixer; (b) side-entry jet mixer

to mix more than two streams. The series arrangement can also be used either when the flow-rate ratio of the feedstreams is very high or when the temperature rise due to mixing would be unacceptably high if the mixing were to be carried out in one stage. Batch jet mixing can also be carried out if storage and recycle are exploited. This can be particularly attractive if precise control of mixing is required at some stage of a batch operation.

Unfortunately, the design information available in the literature for the various types of tank and tubular jet mixing systems is limited and, in some cases, contradictory. Therefore, to set the scene, section 9.2 of this chapter summarizes the results of the many experimental studies of the fluid dynamics of turbulent jets. These studies provide a valuable insight into the whole phenomenon of jet mixing. The criteria important to the design of tank and tubular jet mixing systems are then discussed in sections 9.3 and 9.4.

9.2   Fluid dynamics of turbulent jets

There have been a large number of studies of the behaviour of turbulent jets (see, for example, Abramovich[4] or Rajaratnam[5]). A jet is fully turbulent at a jet Reynolds number, Re_j, above about 1000–2000 and laminar for Re_j below about 100. Here

$$Re_j = \frac{\rho_j V_j D_j}{\mu_j}$$

The majority of the experimental work on jets has been carried out on free jets, i.e. jets issuing into unbounded bulk liquids. The jets used in tank mixing can be considered to be free. In turbulent flow, in tubular jet mixers such as the coaxial jet mixer, the jet behaves like a free jet up to the point where the jet boundaries hit the mix pipe walls if[6]

$$Ct > 0.75$$

where Ct is the Craya-Curtet number defined by

$$Ct = \frac{(V-1)D^2 + 1}{D[V^2 - V - 0.5D^2(V-1)^2]^{0.5}}$$

where

$$V = \frac{\text{jet velocity, } V_j}{\text{secondary fluid velocity, } V_s}$$

and

$$D = \frac{\text{jet diameter, } D_j}{\text{secondary pipe diameter, } D_s}$$

If

$$Ct < 0.75$$

then recirculation zones occur near the mix pipe walls, i.e. the secondary liquid flow rate is insufficient to 'feed' the entrainment capacity of the jet.

As the jet penetrates the bulk liquid it entrains bulk liquid and 'expands' at the jet angle, α. This jet angle is difficult to measure because of the ill-defined nature of the jet boundary. However, it is reported as varying between about 15° and 25° (ref. 7) for $Re_j > 100$.

Similarly, the volumetric flow rate of bulk liquid entrained by the jet, Q_e, is difficult to measure and the available experimental data are widely scattered. However, it appears that

$$\frac{Q_e}{Q_j} = \frac{fz}{D_j} \tag{9.1}$$

where

$$f = fn(Re_j)$$

f is a strong function of Re_j for $100 < Re_j < 2000$, but only a weak function for $Re_j > 2000$.

Finally, a turbulent jet flow can be divided into two distinct regions, the core region and the fully developed region (see *Figure 9.3*). In the core region, which

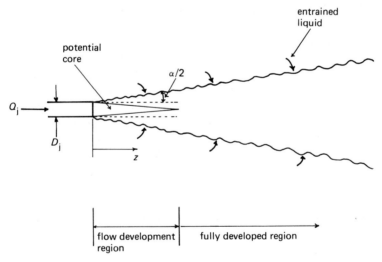

Figure 9.3   Jet-flow behaviour

extends up to about $6D_j$ from the nozzle, the mixing layer penetrates inwards towards the jet centreline or axis and there is a cone-like volume of jet liquid of velocity V_j. This cone of liquid is known as the potential core. In the fully developed flow region which starts at about $10D_j$ from the nozzle, the mixing layer has penetrated to the jet axis.

In the fully developed region of a turbulent jet, the centreline jet velocity V_m is given by

$$\frac{V_m}{V_j} \approx \frac{6D_j}{z} \tag{9.2}$$

and the centreline jet concentration C_m by

$$\frac{C_m}{C_j} \approx \frac{4.5D_j}{z}$$

(9.3)

The 'constants' here are probably weak functions of Re_j.

From equations (9.2) and (9.3) it can be seen that the centreline jet velocity V_m and concentration C_m fall to about 5% of the initial values after an axial distance of 100 jet diameters. It is generally considered that after about 400 jet diameters the velocities have become so low that the mixing effect of the jet is insignificant at more remote positions.

9.3   Jet mixing in tanks

9.3.1   Measurement of mixing time

Mixing times in tanks are usually measured either by tracer techniques or by visual observation. These techniques measure macromixing, i.e. they can be used to measure the time taken to achieve a given degree of homogeneity throughout the whole vessel. They are of no value if local concentrations close to a feed point are required.

In the tracer technique, the tracer (usually a pulse of electrolyte solution) is injected into the tank. The tracer concentration is then measured with respect to time at a point, or at several points, within the tank. The mixing time is taken as the time at which the tracer concentration, C, at the measurement location has reached, or has nearly reached, the expected final mean tracer concentration $\bar{C}$. If there is no tracer initially present in the tank then a mixing time, t_m, can be defined as the time from tracer addition to the time when

$$\frac{|C - \bar{C}|}{\bar{C}} = m$$

where m is the maximum acceptable absolute value of the relative deviation of the mix[10]. At the start of the mixing process, $m = 1.0$, and when complete homogeneity has been achieved, $m = 0$.

In most experimental studies t_{95} is measured, i.e. the time from tracer addition to the time when $m = 0.05$. This is probably the lowest value of m which can be both easily and accurately measured. Experimental evidence suggests that towards the end of the mixing process the rate of mixing is a first-order process[19]. We can thus relate t_{95} to any arbitrary mixing time t_m by the relationship

$$\frac{t_m}{t_{95}} = -\frac{1}{3}\log_e\left(\frac{100 - m}{m}\right)$$

thus, for example,

$$t_{99} \approx 1.5t_{95}$$

The major disadvantages of the tracer technique are:

(a) The tracer concentration is often measured at only one, or sometimes a few, places in the tank. Thus, unless flow visualization is used to first identify poorly mixed regions, the measured mixing time will be unrepresentative of the whole tank.
(b) The measurement probes can interfere with the flow in the tank.
(c) The size of the probe, i.e. the scale of observation, determines the measured time. Thus, the bigger the probe, the shorter the recorded time.

In the visual observation technique, the tank liquid is first made weakly acidic and an indicator added. Strong base, in a quantity just sufficient to neutralize the acid, is then added. The mixing time is taken as the time from the moment of base addition to the time at which the last whisp of indicator colour disappears. Poorly mixed regions of the tank are the last areas to change colour.

The major disadvantages of the visual method are:

(a) The method provides a qualitative rather than quantitative measure of mixing.
(b) Optical resolution and scale of observation are highly subjective.

It is not possible to relate quantitatively the mixing times obtained using tracer techniques with the mixing times obtained using visual observation. Indeed, it is usually not possible to compare even the mixing times obtained in different experimental studies where the same technique has been used, because the necessary background information, e.g. size of concentration probe, is often not available.

This incomparability of techniques and differences in the way in which a given method has been applied to obtain measurements using a given technique are the major reasons for the widespread discrepancies between the mixing times reported in the literature.

9.3.2   Fluid dynamics of jet-mixed tanks

Mixing occurs in a jet-mixed tank because of:

(a) Bulk transport of the jet liquid from the jet nozzle to remote areas of the tank.
(b) Bulk transport, induced by the jet flow in remote areas of the tank.
(c) Bulk transport, induced by entrainment of secondary liquid into the jet.
(d) Mixing of the jet and secondary liquids within the jet flow.

The flow pattern created by a side-entry jet is shown in *Figure 9.4(a)* and by an axial jet in *Figure 9.4(b)* (ref. 8). The overall flow can be described as:

(a) Lateral expansion of the jet due to entrainment as the jet penetrates the secondary liquid. The velocity and turbulence of the jet flow decrease because the jet flow momentum is spread over a steadily increasing flow area.

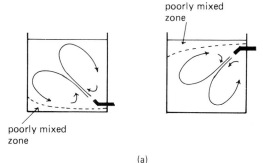

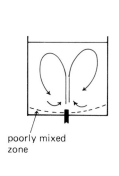

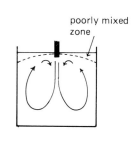

Figure 9.4   Flow patterns in jet mixed tanks. (a) Side-entry jets; (b) axial jets

(b) Rollover of the jet flow when it hits the tank wall or base or liquid surface.
(c) After rollover, a very weak liquid motion driven by the jet flow along the tank wall or base or liquid surface.
(d) Liquid flow induced by jet entrainment from remote regions towards the jet.

The poorly mixed regions shown in *Figure 9.4* are the last regions of the tank in which a given degree of mixing will be achieved. The size of these zones (and, therefore, the mixing time) is affected by:

(a) The relative positions of the jet nozzle and the recycle suction.
(b) The relative sizes of the tank and the jet.
(c) Jet protrusion.
(d) Liquid-level variation.
(e) Shape of tank base.

The objective of any design must be to minimize the size of the poorly mixed regions by producing liquid motion throughout the whole tank.

9.3.3   Theoretical prediction of batch mixing time

Following Coldrey[9] (see also Revill[10]), it is possible to derive a simple expression for the mixing time based on the results of the studies of the behaviour of turbulent jets. The basic premise of the derivation is that the mixing time is directly related to the amount of liquid entrained by the jet[9].

Suppose that mixing is essentially complete when a volume of liquid equal to R times the volume of liquid in the vessel has been entrained by the jet. Thus

$$R = \frac{tQ_e}{V_T}$$

If we substitute for Q_e from equation (9.1), then at large distances from the jet nozzle

$$t = \frac{R}{fz} \frac{V_T D_j}{Q_j}$$

with

$$V_T = \frac{\pi}{4} T^2 H$$

$$Q_j = \frac{\pi}{4} D_j^2 V_j$$

we thus have

$$t = \frac{R}{fz} \frac{T^2 H}{V_j D_j} \tag{9.4}$$

The derivation of equation (9.4) forms the basis of a publication by Lehrer[11]. z can be thought of as an 'effective' jet length. As the jet travels extensively away from the nozzle it slows down and becomes less and less turbulent. Thus, although the jet will still be entraining liquid, it will become less efficient as a mixer.

The distance through which the jet travels before its mixing efficiency becomes negligible is z. Obviously z will depend on the geometry of the jet and the tank and on the jet velocity, V_j.

The form of mixing time correlation developed in equation (9.4) may be compared with empirically based correlations in the following section.

9.3.4   Experimental correlations for jet mixing time

The available experimental studies of jet mixing in tanks are summarized briefly below. Van de Vusse[12] 'layered' liquids of different density in a 36.0 m-diameter storage tank and measured the mixing time from the start of the recirculation pump to the time when the densities of samples drawn from the tank were within a few percent of the expected mean tank density. He correlated his results at an Re_j of about 1.5×10^6 by

$$t = 3.7 \frac{T^2}{V_j D_j}$$

This correlation is of very limited value because it is based on only three experimental runs in which T, D_j and μ_j/ρ_j were kept constant and only H and V_j varied. The degree of mixing measured in these runs cannot be quantified, although it is likely to be high, at least t_{99}, because a very small sample size was used in comparison with the tank size.

Fossett and Prosser[1], using a 1.52 m-diameter tank, injected a small amount of tracer solution (usually as a very long pulse) into the recirculation loop. They defined the mixing time as the time from the start of injection to the time when there was 'no' difference between the concentration measured by a probe in the tank and the concentration measured by a reference probe in a solution made up to the final mean bulk concentration. For single-nozzle jets they correlated their data by

$$t = 9.0 \frac{T^2}{V_j D_j}$$

The injection of the tracer occupied a considerable proportion of this measured 'total' mixing time and the tank contents were usually 50–90% homogeneous before injection was complete. This point was emphasized in their paper and Fossett[13] later proposed that for a short pulse addition of the tracer a more appropriate correlation would be

$$t = 4.5 \frac{T^2}{V_j D_j}$$

Because of the obvious difficulty in interpreting their measured mixing time and because they kept T and μ_j/ρ_i constant and only varied H, D_j, and V_j, this work is of limited quantitative use.

Okita and Oyama[14], using 0.4 m- and 1.0 m-diameter tanks, injected a tracer pulse at the centre of the tank near the liquid surface. They defined the mixing time as the time between tracer addition and the moment at which there was 'no' difference between the concentration measured by two probes, one located near the tank floor at the wall and the other near the liquid surface at the wall. They correlated their data, with an accuracy of $\pm 30\%$ for both axial jet and side-entry jet configurations by

$$t = \frac{2.8 \times 10^4}{Re_j} \frac{T^{1.5} H^{0.5}}{V_j D_j} \quad \text{for } 1000 < Re_j < 5000$$

$$t = \frac{5.5 T^{1.5} H^{0.5}}{V_j D_j} \quad \text{for } 5000 < Re_j < 80000$$

They did not vary μ_j/ρ_j.

Fox and Gex[15], for side-entry jet configurations, used the visual technique to obtain mixing times for tanks of 0.29 m and 1.52 m diameter. They then used another technique, iodine value determination, to obtain mixing times for a 4.27 m-diameter tank. They correlated the data for all three tank scales to within $\pm 50\%$ by an expression of the form

$$t = \frac{ATH}{V_j D_j} \left(\frac{D_j}{H} \right)^{0.50}$$

where

$$A = \frac{7.8 \times 10^5 Fr_j^{0.17}}{Re_j^{1.33}} \quad \text{for} \quad 250 < Re_j < 2000$$

or

$$A = \frac{120 Fr_j^{0.17}}{Re_j^{0.17}} \quad \text{for} \quad 2000 < Re_j < 1.6 \times 10^5$$

Here Fr_j, the Froude number, is $V_j^2 / g D_j$.

Because of the large data scatter Okita and Oyama[14] felt justified in recorrelating this data in the form

$$t = \frac{1.8 \times 10^4}{Re_j} \frac{T^{1.5} H^{0.5}}{V_j D_j} \quad \text{for} \quad 1000 < Re_j < 5000$$

$$t = \frac{2.6 T^{1.5} H^{0.5}}{V_j D_j} \quad \text{for} \quad 5000 < Re_j < 80000$$

However, the data scatter did not improve significantly. Note also that these equations have the same form as Okita and Oyama's own correlations presented above but the constants are less by a factor of about 2.

Hiby and Modigell[16] and Racz and Wassink[17] used the tracer technique to measure t_{95} (i.e. t when $m \leqslant 0.05$) for axial jet systems. In both studies H, T, D_j, and μ_j / ρ_j were kept constant with only V_j being varied. Hiby and Modigell[16] correlated their data by

$$t_{99} = 3.2 \frac{T^2}{V_j D_j} \quad \text{for} \quad 6000 < Re_j < 36000$$

and Racz and Wassink[17] proposed

$$t_{99} = 3.9 \frac{T^2}{V_j D_j}$$

Lane and Rice[18-20] used the tracer technique to measure t_{95} and changed their probe (sample) size in proportion to changes in the tank size. For both side-entry and axial jets they were able to correlate their data in the same form as the Fox and Gex[15] correlation. However, they obtained, for side-entry jets

$$A = \frac{1.13 \times 10^6 Fr_j^{0.17}}{Re_j^{1.34}} \quad \text{for} \quad 200 < Re_j < 2000$$

or

$$A = \frac{154 Fr_j^{0.17}}{Re_j^{0.17}} \quad \text{for} \quad 2000 < Re_j < 60000$$

for

$$0.31 \text{ m} \leqslant T \leqslant 0.57 \text{ m}$$

$$0.90 \leqslant H/T \leqslant 1.10$$

For axial jets, they found

$$A = \frac{7.08 \times 10^5 Fr_j^{0.17}}{Re_j^{1.30}} \quad \text{for} \quad 100 < Re_j < 2000$$

or

$$A = \frac{145 Fr_j^{0.17}}{Re_j^{0.17}} \quad \text{for} \quad 2000 < Re_j < 100\,000$$

or

$$A = 22.8 Fr_j^{0.17} \quad \text{for } 100\,000 < Re_j < 140\,000$$

within the ranges

$$0.31 \text{ m} \leqslant T \leqslant 0.57 \text{ m}$$

$$0.50 \leqslant H/T \leqslant 3.0$$

These values of A are for upward-pointing jets, the side-entry jet nozzle being $5D_j$ from the tank wall and the tank base, and the axial jet nozzle flush with the tank base. Lane[8] found that for downward-pointing jets the mixing times are longer than those for upward-pointing jets by about 90% at $Re_j < 2000$ and by about 50% at $Re_j > 2000$. In addition, the mixing times depend upon the amount by which the jet protrudes into the tank. For upward axial jets when $T = 0.31$ m and $320 < Re_j < 55000$ the mixing times increase with increased protrusion. For protrusions of 3.3, 5.0, and 10.0 jet diameters the mixing times are increased by factors of 2.2, 2.9, and 3.0, respectively, compared with the flush jet.

Finally, note that close inspection of the Fox and Gex data[15] reveals that the data are, in fact, much better represented (to within $\pm 20\%$) by three separate but similar correlations, one for each tank size[21].

Thus for $250 < Re_j < 2000$

$$A = \frac{7.82 \times 10^5 Fr_j^{0.17}}{Re_j^{1.33}} \quad \text{for } T = 0.29 \text{ m}$$

$$A = \frac{5.27 \times 10^5 Fr_j^{0.17}}{Re_j^{1.33}} \quad \text{for } T = 1.52 \text{ m}$$

and for $2000 < Re_j < 1.6 \times 10^5$

$$A = \frac{120 Fr_j^{0.17}}{Re_j^{0.17}} \quad \text{for } T = 0.29 \text{ m}$$

$$A = \frac{85.5 Fr_j^{0.17}}{Re_j^{0.17}} \quad \text{for } T = 1.52 \text{ m}$$

$$A = \frac{182 Fr_j^{0.17}}{Re_j^{0.17}} \quad \text{for } T = 4.27 \text{ m}$$

Different correlations for the different scales are reasonable because they used a different measure of mixing on each of the three tank scales.

The use in a design procedure of mixing time equations such as those presented above is discussed in section 9.3.9.

9.3.5   Recommended jet/tank geometry

Single axial jets should only be used when the ratio of tank height to diameter lies in the range

$$0.75 \leqslant \frac{H}{T} \leqslant 3.0$$

similarly single side-entry jets should only be used when

$$0.25 \leqslant \frac{H}{T} \leqslant 1.5$$

and multiple side-entry jets should be used if

$$\frac{H}{T} \leqslant 0.25$$

or if

$$\frac{H}{T} > 3.0$$

Thus, more than one jet need only be used when designing a large-diameter, shallow tank (small H/T) or a tall, small-diameter tank (large H/T). In both of these cases the tank should be considered as a number of smaller 'tanks', stacked either vertically or horizontally. Each of these small tanks should then be designed as a separate vessel.

Since a jet will lose momentum when it hits the tank wall or base, the jet should be positioned and directed along the longest tank dimension, X (see *Figure 9.5*). X should be no more than $400D_j$ (see section 9.2). For side-entry jets,

$$X \approx (T^2 + H^2)^{1/2}$$

and for axial jets

$$X \approx H$$

This ensures that most of the jet momentum is spread by entrainment and mixing the jet and bulk liquids, and that a reasonable secondary circulation is set up within the bulk liquid—the object of any design must be to produce liquid motion throughout the whole tank.

Side-entry jets should be installed along a radius to the tank wall and should protrude no more than $5D_j$ from the tank wall. The nozzle should be no more than $5D_j$ from the tank floor or liquid surface. Axial jets should be installed perpendicular to, and no more than $5D_j$ from, the tank floor or liquid surface. The jet nozzle should always be submerged during the actual mixing operation.

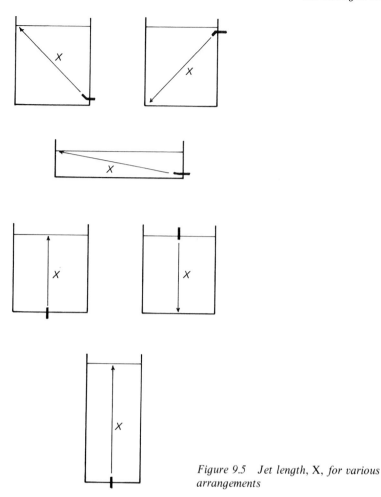

Figure 9.5 Jet length, X, for various arrangements

If the relative density difference between the two liquids to be mixed is in the range

$$(\rho_2 - \rho_1)/\rho_2 \leqslant 0.05$$

then stratification (see section 9.3.6) will not be a problem. Here, the jet can be positioned either near the tank floor pointing towards the liquid surface or near the liquid surface pointing towards the tank floor. The recycle suction pipe should be placed in any area away from the direct action of the jet. If, however,

$$(\rho_2 - \rho_1)/\rho_2 > 0.05$$

then stratification (see section 9.3.6) could be a problem.
    If

$$x_1 > 0.5$$

where

$$x_1 = V_1/V_T$$

and

V_1 = volume of light liquid

V_T = total batch volume

use an upward-pointing jet with the recycle suction just beneath the liquid surface.

Alternatively, if

$$x_1 < 0.5$$

use a downward-pointing jet with the recycle suction as near to the tank floor as possible.

Note that the recycle line need be reduced to the actual jet diameter only about $2D_j$ before the jet nozzle to minimize the pumping power in the external loop.

9.3.6 Stratification

When there is a relative density difference $(\rho_2 - \rho_1)/\rho_2 > 0.05$ between the jet liquid and the bulk liquid then there is a critical jet velocity, V_c, below which layers of high- and low-density liquids form and no mixing occurs. Fossett and Prosser[1] studied this phenomenon which they attributed to gravitational effects. They correlated their data by

$$V_c = \left[\frac{2gGH\left(\frac{\rho_2 - \rho_1}{\rho_2} \right)}{\sin^2 \theta} \right]^{0.50}$$

Here

$\theta = (\beta + 5)°$;
β = angle of inclination of jet nozzle to horizontal;
ρ_2 = density of heavy liquid;
ρ_1 = density of light liquid.

For $H/D_j > 80$, G is constant for a given $(\rho_2 - \rho_1)/\rho_2$ (see *Figure 9.6*). For $H/D_j < 80$, G decreases as H/D_j decreases for a given $(\rho_2 - \rho_1)/\rho_2$. Thus, for a given $(\rho_2 - \rho_1)/\rho_2$ the value of G obtained from *Figure 9.6* will overestimate V_c for $H/D_j < 80$.

The above equation for V_c is based only on data for the injection of a heavy liquid into a light liquid using an upwards-pointing jet. On the basis of intuition, Fossett[13] has suggested that if the jet liquid is lighter than the bulk liquid then a downwards-pointing jet should be used with a V_c of 1.5 times the V_c predicted by their equation. This argument appears reasonable.

With an upwards-pointing jet, if the jet is heavier than the bulk liquid, then the jet possesses 'negative' buoyancy. If V_c is adequate to mix such a system,

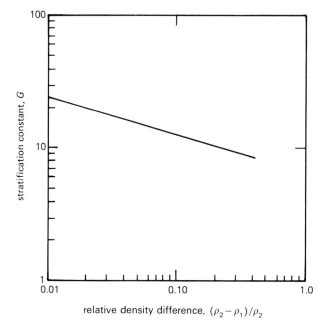

Figure 9.6 Stratification data[1]

then one would expect that V_c would also be adequate to mix a light jet injected upwards into a heavy bulk liquid—the jet here possesses 'positive' buoyancy. Similarly, if $1.5V_c$ is adequate to mix a light jet injected downwards into a heavy bulk liquid, then it must also be adequate to mix a heavy jet injected downwards into a light bulk liquid.

Coldrey[9] and Revill[10] have questioned several aspects of the Fossett and Prosser work but state that, since no other data on the phenomenon are available, V_c should always be exceeded for complete design safety.

It is usually accepted that a jet entrains liquid over a length of up to about $400D_j$. In the design procedure (see section 9.3.9) the nozzle is sized such that

$$X \leqslant 400D_j$$

where X is the jet-path length. If $X > 400D_j$, then the jet will not 'stir' the liquid in the farthest reaches of the tank. Coldrey[9] and Revill[10] reason that since Fossett and Prosser used tanks with $X/D_j > 600$ the jet would have directly influenced only a small part of the tank. Furthermore, with a large X/D_j, the velocity of the jet stream in the later stages of jet dissipation would have been relatively small.

The ability of a jet to entrain liquid is a function of the jet velocity, whereas the tendency of a liquid not to be entrained is a function of $(\rho_2 - \rho_1)g$ and the distance between it and the jet. Thus, it is easy to see how gravitational effects, with subsequent stratification, could become dominant in Fossett and Prosser's work. It must also be remembered that the interface between the two different density liquids is remarkably resilient and that the energy contained in a slow-moving jet approaching such an interface could easily be dissipated simply as surface waves.

If the design method suggested in section 9.3.9 is adopted and the jet diameter sized such that the jet is effective over the entire length of the tank, then the risk of stratification should be much less than in Fossett and Prosser's work. The critical velocity V_c is correspondingly less than that predicted by their equation.

9.3.7   Liquid level variation

The mixing time correlations in section 9.3.4 should only be used if the depth, H, of the liquid in the tank is reasonably constant. If the level varies by less than 20% of the usual operating level, then base the size of the jet on the distance from the nozzle to the point on the far wall just below the normal operating level. However, use the maximum operating liquid height to size the jet recycle rate from the mixing time equations.

Note that with this design the jet will break the surface when the tank is operating below its normal liquid level. If this mode of operation is not possible or if the variation in liquid is more than about 20% of the usual operating level, then a multiple jet system should be considered.

9.3.8   Continuous mixing

If the tank is being run at a constant level, H, and if the input is through one jet nozzle, then the tank can be safely assumed to be well mixed if

$$t_R > 5t_{99}$$

where t_R is the mean tank residence time.

9.3.9   Design procedure

The basic design steps are summarized below:

(a) Given the batch volume, decide on T and H;
(b) Calculate H/T—this ratio governs whether a side-entry or axial jet should be used;
(c) Calculate the jet path length, X (see *Figure 9.5*) for axial jets

$$X \approx H$$

and for side-entry jets placed as recommended in section 9.3.5

$$X \approx (T^2 + H^2)^{0.5}$$

(d) For side-entry jets, calculate the angle of inclination of the jet nozzle to the horizontal, β. For upward-pointing jets, placed as recommended

$$\beta \approx \tan^{-1} \frac{H}{T}$$

and for downward jets

$$\beta \approx \tan^{-1}\frac{T}{H}$$

Calculate the jet velocity, V_j. For an upward-pointing jet

$$V_j \geqslant V_c$$

and for a downward jet

$$V_j \geqslant 1.5V_c$$

where the critical velocity, V_c, is given in section 9.3.6.

(f) Choose the jet diameter, D_j, such that

$$50 \leqslant \frac{X}{D_j} \leqslant 400$$

Initially choose a small value of D_j so that the flow rate in the external loop will not be excessive.

(g) Calculate the mixing time, t, using the most appropriate correlation given in section 9.3.4.

(h) Calculate the jet recycle flow rate, Q_j, from

$$Q_j = \frac{\pi}{4} D_j^2 V_\gamma$$

(i) Estimate the required pump discharge pressure from

$$P = \Delta P_1 + \Delta P_2 + P_3$$

where

$\Delta P_1 = $ pressure drop in pipework between the tank take-off and jet nozzle;
$\Delta P_2 = $ pressure drop through jet nozzle;
$P_3 = $ static pressure of liquid in the tank.

9.3.10 When to use jet mixed tanks

The process of jet mixing is similar to the mixing action of a propeller agitator. With jet mixing the jet motion is induced by an external pump. With a propeller the agitator itself induces a jet-like flow in the tank. We can get a rough idea of the relative efficiency of two types of mixer from the rather simple analysis which follows.

For an axial jet, for example, the Hiby and Modigell[16] correlation for the mixing time, t, is (see section 9.3.4)

$$t_j = 3.2 \frac{T^2}{V_j D_j}$$

and for a propeller agitator in a baffled tank[22] the mixing time t_p is

$$t_p = 5.6 \frac{T^2}{ND^2}$$

Thus for equal mixing times

$$t_j = t_p$$

and we have

$$\frac{D_j V_j}{ND^2} = 0.57$$

Consider only the power expended at the jet nozzle P_j given by

$$P_j = \frac{\pi}{8} \rho D_j^2 V_j^3$$

The power input P_P to a three-bladed propeller, with $D = T/3$ and $Re_T = 100\,000$, is given by[23]

$$P_P = 0.35 \rho D^5 N^3$$

Thus the ratio of power requirements (jet/pump) becomes

$$\frac{P_j}{P_P} = 1.12 \frac{D_j^2 V_j^3}{D^5 N^3}$$

After some manipulation, we obtain, for $t_j = t_p$,

$$\frac{P_j}{P_P} = 0.21 \frac{D}{D_j}$$

Now, for an axial jet,

$$X \approx H$$

and if

$$T \approx H$$

then we have, typically,

$$D_j = \frac{T}{100}$$

Also for a propeller

$$D = \frac{T}{3}$$

Thus typically

$$\frac{P_j}{P_P} = 7.0$$

Thus, for equal mixing times, the power input to the jet is about seven times the power input to a propeller. This factor of seven depends strongly on which correlations are used for mixing times with propellers and jets and upon the other assumptions made. However, all the available (reliable) evidence does suggest that the energy costs of a jet mixing system are usually higher than of an equivalent mechanically agitated system.

Note that the above power values are for power dissipated by the jet or propeller and do not include motor efficiency and bearing losses for the agitator, or pump efficiency and external pipe friction losses for the jet. The head loss by the recirculating fluid in a jet mixer can be significant proportion of the total fluid head required for mixing.

The capital cost of a jet mixing system is usually lower than the capital cost of a mechanically agitated system for the same duty and the pump unit is usually cheaper than the agitator assembly. Furthermore, for jet mixing there is no need for the tank to be reinforced to support the agitator and drive unit as is sometimes required for mechanical agitators. If an existing pump can be used, then the capital cost of a jet mixing system will be very low.

The maintenance cost of a jet mixing system is also usually lower than that of a mechanically agitated system—the jet mixing system is very simple, and the pump is remote from the tank. Pump maintenance will not require the tank to be drained and purged.

Jet mixers seem to have gained wide acceptance for blending in large tanks but their use need not be restricted to such systems. There is no reason why they should not be used for all tank sizes. The mixing-time equations presented in this chapter have been proved to apply to mixing times of several minutes to several hours.

9.4   Jet mixing in tubes

9.4.1   Design basis

The first step in the design of any jet mixer is to establish the following design objectives:

(a) The required quality of mix;
(b) The required mixing time;
(c) The available pressure drop;
(d) The required turndown.

The quality of an 'homogeneous mixture' is usually defined by the specification $M = 0.01$ (see, for example, Gross-Roll[24]) where M is the variation coefficient. M is the ratio of the standard deviation of the concentration to the mean expected concentration. This definition of homogeneity will be adopted in this chapter, and the mixing time, t, defined as the time taken by the liquid to travel from the mixing point to the point at which $M = 0.01$.

The required mixing time and the available pressure drop are obviously functions of the particular duty being considered. For example, site considerations may limit the length of the unit and this, together with the throughput, will define the mixing time.

If turndown is required, then mixing-time calculations should be based on the lowest flow rates (i.e. lowest velocities) to be processed and pressure-drop calculations on the highest flow rates (i.e. highest velocities).

9.4.2   Coaxial jet mixer design

In the coaxial jet mixer the jet flow is introduced through a small pipe running coaxially with a large pipe (see *Figure 9.2*). The jet expands and entrains and mixes with the secondary liquid. For such a mixer, good mixing will be achieved provided

$$Re_j > 2000 \tag{9.5a}$$

$$Re_s > 5000 \tag{9.5b}$$

$$Re_m > 5000 \tag{9.5c}$$

and
$$\rho_j V_j^2 > 2\rho_s V_s^2 \tag{9.5d}$$

If recirculation does not occur in the mixer, i.e. if

$$Ct > 0.75$$

(see section 9.2), then the jet can be considered to be a free jet.

For a coaxial jet mixer, satisfying equations (9.5) and the Ct criterion, the experimental mixing-time data can be correlated by[21]

$$t \approx \left(6.6 + 1.7 \log\left[\frac{Q_s}{Q_j}\right]\right)\frac{L}{V_m} \tag{9.6}$$

where L is the axial distance the jet moves before its boundary hits the mix pipe wall. Thus, with reference to *Figure 9.2*,

$$L = \frac{\left(\dfrac{D_m}{2} - \dfrac{D_j}{2}\right)}{\tan \alpha/2}$$

where α is the jet angle (see section 9.2).

For a coaxial jet mixer satisfying equations (9.5) but with

$$Ct < 0.75$$

the mixing time predicted by equation (9.6) will be conservative.

The pressure drop for the jet and secondary flows can be calculated from

$$\Delta P_j = \left(1 - \frac{A_j}{A_m}\right)^2 \frac{\rho_j V_j^2}{2} \tag{9.8a}$$

$$\Delta P_s = \left(1 - \frac{A_s}{A_m}\right)^2 \frac{\rho_s V_s^2}{2} \tag{9.8b}$$

whilst the pressure drop along the mix pipe (of length $V_m t$) is given by

$$\Delta P_m = \left(\frac{\rho_m V_m^2}{2}\right)\left(\frac{V_m t}{D_m}\right)4f \qquad (9.9)$$

where f is the Fanning friction factor, see, for example, Coulson and Richardson[25].

The pressure drop in the lines leading up to the mixing point can be considerable and should be checked before a design is completed. Finally, the design should be checked for mechanical integrity, i.e. support may be needed for the jet pipe.

Note that the above design equations can be applied to multi-jet coaxial mixers of the types shown in *Figure 9.7* if L in equation (9.7) is redefined as the axial distance the jet moves before its boundary intersects the boundary of another jet. Thus in *Figures 9.7(a)* and (b)

$$L = \frac{s/2}{\tan \alpha/2} \qquad (9.10)$$

α, the jet angle, is defined as before, see section 9.2.

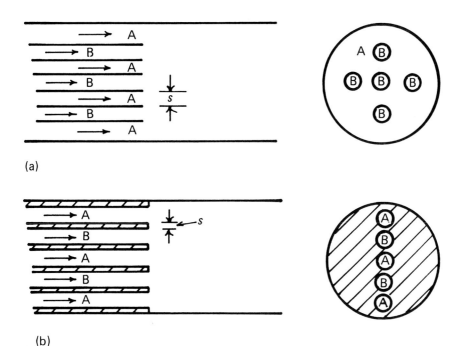

(a)

(b)

Figure 9.7 *Multiple jets in pipes. (a) Multiple jets of B; (b) multiple jets of A and B*

9.4.3   Side-entry jet mixer design

This type of mixer is more suitable for handling flows of unequal size than the coaxial jet mixer. It is also easier to fabricate and maintain.

The jet stream is injected radially into the secondary stream and the jet expands and entrains and mixes with the secondary flow. It is rapidly deflected towards the mainstream flow. The behaviour of a side-entry jet injected into a cross flow has been extensively studied, both theoretically and experimentally. Reilly[26] has reviewed the theoretical work, which is not of much use for practical design purposes. Experimentally, the general form of the expression for the jet penetration, x, is (see *Figure 9.2*)

$$\frac{x}{D_j} = 1.0 \left\{ \frac{\rho_j V_j^2}{\rho_s V_s^2} \right\}^{0.4} \left\{ \frac{z}{D_j} \right\}^{0.3} \tag{9.11}$$

Obviously, the momentum flux ratio, M_f, where

$$M_f = \frac{\rho_j V_j^2}{\rho_s V_s^2}$$

is the single most important parameter influencing the behaviour of the jet. However, if the flows into and out of the mixer satisfy equation (9.5) then the effect of the absolute values of $\rho_j V_j^2$ and $\rho_s V_s^2$ is negligible[27]. For a given value of M_f, increasing D_j will increase x at a given axial distance z.

At low values of $M_f < 1.5$, the jet momentum may not be sufficient to allow the jet flow to escape from the boundary layer of the exit plane[28]. The jet will then simply be deflected along the pipe wall. The mixing time in this situation will then approach the mixing time in open pipe flow. For mixing in an open pipe the experimental data are well correlated by[21]

$$t = 150 \frac{D_m}{V_m}$$

When the jet does penetrate the secondary flow, a reasonable estimate of t is given by equation (9.6). However, L is now defined as the axial distance from the mixing point where either

(a) The jet axis hits the far wall, or
(b) The jet axis becomes angled at less than $10°$ to the main stream flow—this angle may be found by differentiating equation (9.11).

L is the smaller of these two distances.

9.4.4   Use of tubular jet mixers

In tubular jet mixers, as in all tubular mixing devices, the energy required for mixing is provided by the pressure drop. *Figure 9.8* shows the pressure drop, in terms of $\rho V_m^2/2$ (see equation (9.9)), required by the various types of mixer to achieve $M = 0.01$ as a function of dimensionless mixer length, $V_m t/D_m$.

The empty pipe has the lowest pressure drop but the longest length (mixing

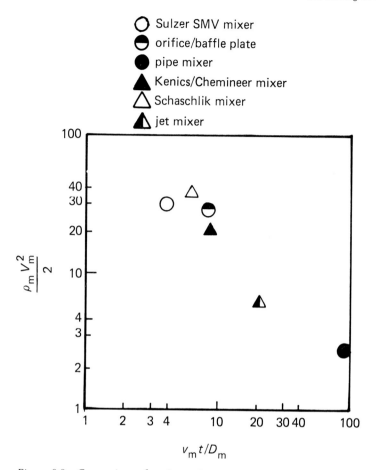

Figure 9.8 Comparison of various mixers

time). The motionless mixer, the orifice/baffle plate mixer and the baffle plate mixer have the highest pressure drop but the shortest length (mixing time). Thus, if the mixing time is not critical, and space is available, an ordinary pipe run is the most energy efficient mixer. If mixing time is critical, or if space is limited, then another type of mixer may have to be used.

The capital cost of a motionless mixer is very high compared with the capital cost of, say, an orifice/baffle plate mixer or jet mixer.

Notation

A constant in mixing-time correlation
Ct Craya-Curtet number
D propeller diameter
D_j jet nozzle diameter
D_m mix pipe diameter

D_s secondary flow pipe diameter

Fr_j jet Froude number

g gravitational acceleration

H liquid height in tank

L axial distance from mixing point

m_e mass flow rate of entrained liquid

m_j jet mass flow rate

N propeller rotation speed

P_j jet power consumption

P_P propeller power consumption

Q_e flow rate across jet cross-section at axial distance z

Q_j flow rate at jet nozzle

R jet entrainment factor

Re_j jet Reynolds number

Re_m mix Reynolds number

Re_s secondary flow Reynolds number

Re_T tank Reynolds number

t mixing time

T tank diameter

x penetration of a side-entry jet

X effective jet length

V_c critical jet velocity

V_j jet velocity

V_m mix velocity

V_T tank volume

z axial distance, effective jet length

Greek symbols

α jet angle

β jet nozzle inclination to the floor of a jet mixed tank

ρ_j jet fluid density

ρ_s secondary fluid density

μ_j jet fluid viscosity

References

1 FOSSETT, H. and PROSSER, L. E. (1949) *Proc. I. Mech. E.*, **160**, 224–232.
2 SARSTEN, J. A. (1972) *Pipeline and Gas J.*, Sept., 37–39.
3 SUDHINDRA, N. and SINHA, S. N. (1982) *Paper C4, 4th Europ. Conf. on Mixing*, Leeuwenhorst, Netherlands (27–29 April).
4 ABRAMOVICH, G. N. (1963) *The Theory of Turbulent Jets*, MIT Press.
5 RAJARATNAM, N. (1976) *Turbulent Jets*, Elsevier Scientific Publ. Co.
6 BECKER, H. A., HOTTEL, H. C. and WILLIAMS, G. C. (1962) Paper presented at 9th Int. Symp. on Combustion.
7 LANE, A. G. C. and RICE, P. (1982) *Trans. I. Chem. E.*, **60**, 245–248.
8 LANE, A. G. C. (1982) *Ph.D. Thesis*, Loughborough Univ. of Tech.

9 COLDREY, P. W. (1978) Paper to I. Chem. E., 'Mixing in the Process Industries', course, Bradford University, UK (July).

10 REVILL, B. K. (1981) Paper to I. Chem. E., 'Mixing in the Process Industries', course, Bradford University, UK (July).

11 LEHRER, I. H. (1981) *Trans. I. Chem. E.*, **59**, 247–252.

12 VAN DE VUSSE, J. G. (1959) *Chem.-Ing.-Tech.*, **31**, 583–587.

13 FOSSETT, H. (1951) *Trans. I. Chem. E.*, **29**, 322–332.

14 OKITA, N. and OYAMA, Y. (1963) *Kagaku Kogaku*, **27**, 252–259.

15 FOX, E. A. and GEX, V. E. (1956) *A.I.Ch.E.J.*, **2**, 539–544.

16 HIBY, J. W. and MODIGELL, M. (1978) Paper presented at 6th CHISA Conf., Prague.

17 RACZ, I. and WASSINK, J. G. (1974) *Chem.-Ing.-Tech.*, **46**, 261.

18 LANE, A. G. C. and RICE, P. (1982) *Trans. I. Chem. E.*, **60**, 171–176.

19 LANE, A. G. C. and RICE, P. (1981) *Paper K1*, I. Chem. E. Symp. Series No. 64.

20 LANE, A. G. C. and RICE, P. (1982) *Poster v, 4th Europ. Conf. on Mixing*, Leeuwenhorst, Netherlands (27–29 April).

21 REVILL, B. K. (1981) unpublished work.

22 LANDAU, J. and PROCHAZKA, J. (1961) *Coll. Czech. Chem. Communs.*, **26**, 1976–1990.

23 UHL, V. and GRAY, J. B. (1966) *Mixing, Theory and Practice*, Academic Press.

24 GROSS-ROLL, F. (1980) *Int. Chem. Engng.*, **20**, 542–549.

25 COULSON, J. M. and RICHARDSON, J. F. (1970) *Chemical Engineering*, Pergamon Press.

26 REILLY, R. S. (1968) *Ph.D. Thesis*, Univ. of Maryland.

27 WALKER, R. E. and KORS, D. L. (1973) *NASA CR 121217*.

28 KEFFER, J. F. and BAINES, W. D. (1963) *J. Fluid Mech.*, **15**, 481–497.

Chapter 10

Mixing in single-phase chemical reactors

J R Bourne

Swiss Federal Institute of Technology, Zurich

10.1 Introduction

This chapter discusses the mixing of reagents with each other in order to produce chemical reaction within a single phase. With turbulent flow and well-designed reactors, the time needed for this mixing is often of order 0.1–10 s. Thus, if the chemical reaction is slower (e.g. needing more than 100 s), mixing is practically complete before reaction has proceeded to any significant extent; mixing does not therefore influence the reaction. When, however, the half-life times of these two processes are of similar magnitude or when mixing is slower than reaction, mixing and reaction proceed simultaneously, not consecutively. Reaction becomes localized into zones, whose width shrinks to a plane as the inherent reaction rate rises. (This seems to happen in practice as the chemical half-life falls to times of order ms or less.) In such cases only a minute fraction of the total volume available is usefully employed in achieving reaction.

Put a little differently, the reaction is throttled by insufficient mixing and is not proceeding at its inherent rate. Concentration gradients are then present and when multiple reactions occur, their *relative* rates can be influenced by inhomogeneity, i.e. the product distribution obtained can depend upon the mixing intensity. This often means that reagents are being wasted on manufacturing unwanted products and, further, that energy is being wasted separating the desired product, which has been manufactured at low selectivity, from unwanted materials. Thus, although only one fluid phase is present, attention should be paid to mixing if reaction is 'fast'.

10.2 Mechanisms of mixing

In simple laminar flows (e.g. one-dimensional shear or elongation) the time evolution of some mixing parameters (e.g. striation thickness or intensity of segregation) has been calculated in a number of cases (Chapter 12 for general information). In turbulent, single-phase fluids it is very difficult to model the transport phenomena in full physical detail. Qualitatively, however, the following sequence[1] may be visualized after the feed streams have met:

(a) *Distributive mixing.* Relatively large eddies exchange positions and convect material so that, at a scale of observation which is coarser than the eddy size, macroscopic uniformity of concentration results. At a scale much smaller than the eddy size no significant mixing occurs.

(b) *Dispersive mixing.* The larger eddies of (a) decay in size through the effect of turbulent shear and a finer-grained mixture is formed. At molecular scale the mixture remains, however, highly segregated.

(c) *Diffusive mixing.* Diffusion within the finely dispersed structure operates over short distances and proceeds to randomize the mixture at the molecular scale. The result is termed a homogeneous mixture.

Figure 10.1 indicates these three stages, which although mainly consecutive are also to some extent simultaneous.

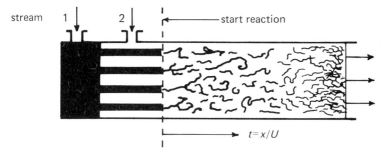

Figure 10.1   Distributive and dispersive mixing in a pipeline[2]

It will be convenient here to use the terms diffusive mixing for stage (c) and convective mixing for stages (a) and (b). It has not yet proved possible to develop a complete physical description of mixing in turbulent fluids[2,3]. The present review will confine itself to simple models, which are supported by experimental work and which are (or could be) used industrially.

10.2.1   Convective, distributive mixing of a single-feed stream

When a feed stream enters a mixing tank, one might wish to know how long the process of mixing this feed with its surroundings lasts. Moreover the concentration profile in the region where the mixing takes place is also of interest. Frequently the time should be short and the region small, compatible with a reasonable power input for the mixing.

This type of mixing is sometimes termed blending. Examples include distributing a supersaturated stream within a crystallizer, a feed stream containing catalyst throughout a reactor or a fresh monomer stream (also possibly containing initiator) throughout a polymerizer.

By injecting a pulse of tracer (electrolyte, dyestuff, miscible liquid having a different refractive index, radioactive or photochromic substances) into the tank at the position of interest, the change of tracer concentration (electrical

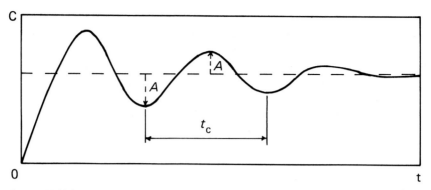

Figure 10.2(a)   Tracer concentration as function of time in batch-operated tank[4]

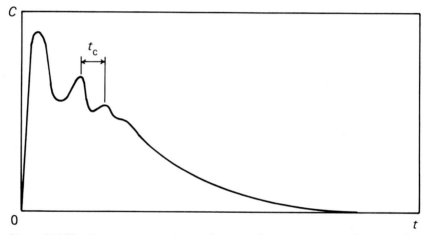

Figure 10.2(b)   Tracer concentration as function of time in continuously operated tank[10]

conductivity, pH, optical density, refractive index, radioactivity) at one or more other positions is followed as a function of time[4]. For batch (closed) and continuous (open) tanks the signal at one position changes with time as shown in *Figure 10.2*.

Frequently distinct peaks, corresponding to circulations of the whole fluid, appear, their amplitude being damped within some three to five circulations[5]. Using a disc turbine having six flat blades, located half the liquid depth, and three sizes of baffled tanks with up to four turbine sizes for each tank, the following correlation for the circulation time, t_c, was established in the turbulent flow regime:

$$Nt_c = 0.85(T/D)^2 \tag{10.1}$$

The damping of the concentration wave at the measuring point has been employed to deduce the turbulent diffusivity[6,7]. This damping was subjectively judged to be complete after five circulations, i.e. the mixing time,

$t_M \approx 5t_c$. The mixing time can be shortened somewhat by setting the turbine at one-third the liquid depth, in which case the trough of a concentration fluctuation transported back to the impeller by the lower (upper) circulation loop arrives with the peak from the upper (lower) loop[4]. Depending somewhat on the method used to assess macroscopic homogeneity, t_M will then be roughly $3t_c$.

Clearly the magnitude of the mixing time must depend upon the criterion of mixedness. Adding tracer gradually changes the tracer concentration in the tank from C_i to C_f. One definition of mixing time is then the time needed to bring the concentration at the measuring point $C(t)$ to the value $C_f - X(t)(C_f - C_i)$. $X(t)$ is then a fractional unmixedness, equal to:

$$X(t) = \frac{C_f - C(t)}{C_f - C_i} \tag{10.2}$$

Experiments on tracer distribution in baffled tanks operated in the turbulent regime were correlated by the expressions[8]:

Propeller	$Nt = 3.480(T/D)^{2.05} \log (2/X)$	(10.3)
Pitched-blade turbine (45°)	$Nt = 2.020(T/D)^{2.20} \log (2/X)$	(10.4)
Disc turbine (6 blades)	$Nt = 0.905(T/D)^{2.57} \log (2/X)$	(10.5)

A draft tube may be used with the axially pumping impellers[9].

The amplitude of the concentration oscillations A shown in *Figure 10.2* is damped exponentially by the circulations and simultaneous turbulent diffusion, according to equation (10.6):

$$A = 2 \exp (-Kt) \tag{10.6}$$

The rate constant depends on the agitator speed and diameter according to the following empirical correlations, valid in the turbulent regime[10]:

Propeller (pitch = 1.5)	$\dfrac{N}{K} (D/T)^{2.0} = 0.9$	(10.7)

Disc turbine (6 blades)	$\dfrac{N}{K} (D/T)^{2.3} = 0.5$	(10.8)

These correlations were established from the damping of concentration fluctuations in two sizes of tanks, each operated with two impellers of each type having different diameters.

Homogenization with Pfaudler, anchor, gate, and other impellers has also been studied[11,12].

The correlating equations (10.3), (10.4), (10.5), (10.7), and (10.8) were established using probes which measured the concentration in volumes of order 1 cm³. At this scale distributive and, to some extent also, dispersive mixing can be followed. The larger the sample size becomes, the sooner will a given degree of mixing be attained[13].

The form of the correlations already presented has been derived, as outlined above, from models of the circulating flow. Further support comes from casting the basic transport equations—Navier–Stokes equation and convective—diffusive mass balance—into dimensionless form[14]. The 'degree of mixing' is defined as the concentration distribution developed throughout the mass by mixing. Equal degrees of mixing are then attained when the dimensionless concentrations at positions having the same dimensionless co-ordinate values (corresponding points) are equal. Thus the following functional form can be derived:

$$X = X_A \text{ at corresponding points}$$
$$X_A = f(Nt, ND^2\rho/\mu, \mu/\rho\mathscr{D}, N^2D/g) \tag{10.9}$$

for a given feed point, an isothermal Newtonian fluid and an absence of free convection as well as for an apparatus of given shape.

It seems that in well-baffled tanks $Fr = N^2D/g$ exerts no influence on the blending; similarly $Re = ND^2\rho/\mu$ has no effect in the fully turbulent regime. Since the probes used to measure concentration are large compared with molecular dimensions, diffusion contributes little compared with turbulent mass transfer, and so the influence of Schmidt number $Sc = \mu/\rho\mathscr{D}$ is insignificant. Thus the concentration distribution, which is the result of convective mixing, depends only on the number of impeller revolutions Nt. For a given degree of mixing Nt_m should be a constant and this result evidently agrees well with the correlations given above.

These correlations, together with equation (10.9), have an important implication for scale-up, namely that in geometrically similar tanks the same distributive mixing time will be produced at different scales if N remains constant. This implies, however, an enormous increase in the specific power consumption upon scale-up, namely $P/V \sim T^2$ in the turbulent region. It is thus unlikely that t could be kept constant, the energy consumption and resultant viscous heating being excessive. Using the common scale-up rules of either (a) constant power per unit volume, or (b) constant tip speed ($ND =$ constant), the blending time would increase as (a) $t \sim T^{2/3}$, and (b) $t \sim T$. Distributive mixing thus normally slows down when scaling-up.

Blending of highly viscous materials in the laminar flow regime (e.g. using helical ribbon impellers) has been investigated and correlated using analogous methods. The laminar regime falls, however, outside the scope of this chapter.

10.2.2   Diffusive mixing

Diffusion refers to the randomizing effect of molecular motions. It is the only mechanism enabling contact between individual molecules and is therefore the precursor to a chemical reaction. Thus whereas diffusive mixing is of less importance than convective mixing in some duties (e.g. blending of lubricating oils to achieve a required viscosity), it is always important in chemical reactions.

10.2.3   Approximate method

A simple method will be explained allowing an order of magnitude estimate of the diffusive mixing time, t_{MD}. It is supposed that convective mixing forms small fluid elements of size L and having a shape and size which do not further change with time. Moreover, these are initially completely unmixed (fully segregated) and they undergo diffusion until a prescribed level of mixing (i.e. a given concentration profile) is attained; the time needed is t_{MD}.

The diffusion time is highly dependent upon L, solutions of Fick's second law of diffusion taking the form $t \sim L^2/\mathscr{D}$, and depends also on the criterion of mixedness. Consider, for example, a sphere of radius R. The values of β in the relationship:

$$t_{MD} = \beta R^2 / \mathscr{D} \tag{10.10}$$

have the following values[15]:

(a) The concentration change at the centre of the sphere is required to be one-half of its value at infinite time, when diffusion is complete. $\beta = 0.15$.
(b) As in (a), but the change shall be 63.2% or $(1 - e^{-1})$ of its final value. $\beta = 0.18$.
(c) The average concentration change within the whole sphere shall be half its final value. $\beta = 0.036$.
(d) As (c), but the change shall be 63.2% of its final value. $\beta = 0.06$.
(e) All molecules within the sphere shall have been exchanged with molecules from the surroundings. $\beta = 0.5$.

Different values of β would be obtained for another geometry, e.g. for a slab. The scale of eddies present in a turbulent fluid covers a wide range, but is bounded below (i.e. for eddies of small scale) by viscous drag. Turbulent velocity fluctuations are damped out rapidly at a scale smaller than the Kolmogoroff velocity microscale, given by equation (10.11):

$$\lambda_k = (v^3/\varepsilon)^{1/4} \tag{10.11}$$

The simplest approximation for L is $L = \lambda_k$ (ref. 16), so that for a sphere $(R = \lambda_k/2)$ and definition (e):

$$t_{MD} = 0.5\lambda_k^2 / 4\mathscr{D} \tag{10.12}$$

Similar calculations may be made for other shapes and other criteria of mixedness. In turbulent aqueous solutions λ_K is of order 10–30 μm, and t_{MD} will be of order ms.

The simplest way of roughly estimating the time needed to mix freshly entering material into material already present in a stirred tank reactor is to sum the times for distributive (convective) and diffusive mixing. This total time refers to a final state of molecular homogeneity. Evidence for the success of this method is available[17].

10.2.4 More accurate methods

In calculating β in equation (10.10), solutions to Fick's second law have been used[18]. This formulation completely neglects reaction and is valid only if the half-life of any reaction greatly exceeds that for molecular diffusion. Faster reactions steepen the concentration gradients within eddies and accelerate diffusion.

Instantaneous reactions:

A good example of a reaction, which for practical purposes is infinitely fast, is neutralization (e.g. HCl and NaOH). Such reactions have rate constants of order $10^8 \, \text{m}^3 \, \text{mol}^{-1} \, \text{s}^{-1}$.

Supposing that it were possible to premix the ions completely and yet have no reaction, the rates at which these ions would diffuse to meet each other can be calculated. Since the ions would have zero concentrations where they met (their reaction is instantaneous), the rate of reaction would be equal to the rate of diffusion. The expression given by this so-called encounter-controlled model is[19]:

$$k = 8RT/3\mu \qquad (10.13)$$

which is the diffusion-controlled limiting value for a second-order rate constant. In dilute aqueous solutions at 298 K equation (10.13) gives $k \approx 7 \times 10^6 \, \text{m}^3 \, \text{mol}^{-1} \, \text{s}^{-1}$, which by comparison with the value $10^8 \, \text{m}^3 \, \text{mol}^{-1} \, \text{s}^{-1}$ given above shows that neutralization is encounter-controlled. It is important to recognize that second-order reactions having smaller values of k can be mixing-controlled and thus slowed down because the initially separated reagents have not completely mixed before reaction starts.

Solutions of Fick's second law when instantaneous reactions occur are available in closed form for a few simple cases[20,21]. There is no reaction term in the mass balance, which is simply Fick's second law, whereas the boundary conditions express the fact that reaction is instantaneous.

Analogous to the discussion of β in equation (10.10), the diffusive mixing time depends strongly upon the size of the region within which diffusion and reaction occur (L), upon the shape of this region (e.g. sphere, cylinder, slab), and upon the criterion of mixedness. It will not depend upon the rate constant (k), since this is presumed to be so large that a reaction plane is formed where reaction is instantaneous. However, the stoichiometric ratio will be important, since it helps to determine the steepness of the concentration gradients and hence the rate of diffusion. Considering again a sphere of radius R containing acid (base) of a particular concentration immersed in a large volume of base (acid) of the same concentration, the time needed for complete neutralization of the sphere is[20]:

$$t_N = 0.21 R^2 / \mathscr{D} \qquad (10.14)$$

This result may be compared with the slower exchange of molecules occurring by diffusion without instantaneous reaction, namely $t_{MD} = 0.5R^2/\mathscr{D}$ [case (e) above]. The reaction has clearly accelerated diffusion; diffusion has, however, slowed down an inherently instantaneous reaction. Acid-base reactions in stirred tanks have been interpreted using such diffusion equations[22].

Solution of Fick's second law, where the instantaneous reaction determines the boundary conditions, cannot always be written in closed analytical forms. Examples include a simultaneous depletion of acid (base) in the well-mixed environment as this material diffuses into the sphere, slab, cylinder, etc., as well as situations where the reaction plane moves with time. Numerical solution of the partial differential equation(s) representing the mass balance(s) is then necessary[21,23].

Fast reactions:

In this case the reaction plane formed when instantaneous reaction occurs spreads out to a reaction zone, within which both reagents are present. When the reaction is fast enough, this zone remains thin and free from turbulence, i.e. the mechanism of mass transport within the reaction zone is predominantly diffusive (*Figure 10.3*). For the single, second-order reaction[21,23]:

$$A + B \xrightarrow{k} P \tag{10.15}$$

the unsteady-state mass balance within the reaction zone is:

$$\frac{\partial C_A}{\partial t} = \mathscr{D}\frac{\partial^2 C_A}{\partial x^2} - kC_A C_B \tag{10.16}$$

The corresponding equation for B is obtained by exchanging suffices, provided

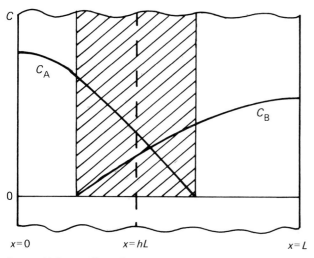

Figure 10.3 *Profiles of reagent concentrations at an arbitrary time during diffusion and reaction within a non-turbulent slab of fluid (reaction zone is shaded)*[23]

that A and B have the same diffusivity in the solvent. These balances contain no convective diffusion terms and are written for a reaction zone, which retains its shape and size during the processes of diffusion and reaction. These are significant simplifications, which are not always applicable as current research is revealing, and which tend to overestimate the role of diffusion. Nevertheless, they allow at least order-of-magnitude estimates of the effect of diffusive mixing on fast reactions and tend to be conservative.

Equation (10.16) together with the corresponding equation for B are coupled and no analytical solution exists. By casting the equations into dimensionless form and considering the boundary and initial conditions:

$$t=0, \quad 0<x<hL: C_A=C_{A_0}, \quad C_B=0$$

$$hL<x<L: C_A=0, \quad C_B=C_{B_0}$$

$$x=0 \quad \text{and} \quad L: \partial C_A/\partial x = \partial C_B/\partial x = 0$$

and general form of the solution must be:

$$\frac{C_A}{C_{A_0}} = f(x/L, kC_{A_0}t, C_{A_0}/C_{B_0}, V_A/V_B, kL^2C_{A_0}/\mathscr{D}) \tag{10.17}$$

The first two dimensionless groups define the position and time at which C_A is being evaluated. The next two define the concentration and volumetric ratios of the reactants and hence also their stoichiometric ratio. The last group is proportional to the ratio of the half-lives for diffusion and reaction, and is analogous to the Thiele modulus in heterogeneous catalysis and to the Hatta number in absorption with reaction:

$$t_{1/2\mathscr{D}} \sim L^2/\mathscr{D} \tag{10.18}$$

$$t_{1/2R} = (kC_{A_0})^{-1} \quad \text{provided } C_{A_0} = C_{B_0} \tag{10.19}$$

If reaction is slow compared with diffusive mixing, then $t_{1/2\mathscr{D}} \ll t_{1/2R}$ and diffusion precedes reaction; the reaction term in equation (10.16) drops out and equation (10.10) is valid. If, however, reaction is instantaneous, then $t_{1/2\mathscr{D}} \gg t_{1/2R}$ and reaction rate is limited by diffusion; again the reaction term vanishes in equation (10.16), but a reaction plane is formed. When, however, the half-lives are of the same order, diffusion and reaction are coupled and the reaction term remains in equation (10.16). Application of its numerical solution to fast reactions is available[21,23].

In a batch or plug flow reactor, the time t should be counted from the moment that convective mixing is complete (i.e. has produced a random distribution of fluid elements of size L) and will thereafter have the same value for all elements. If, however, a distribution of the residence times of such fluid elements arises, the following procedure may be adopted[24]. For any value of t, numerical solution of equation (10.16) is $C_i(x,t)$. The spatially averaged concentration is thus:

$$C_i(t) = \frac{1}{L} \int_0^L C_i(x,t) \, dx \tag{10.20}$$

The fraction of fluid elements leaving the reactor in the time interval t to $t+dt$ is $f(t)\,dt$, where $f(t)$ is the residence time frequency function. Thus the concentration of i leaving the reactor is:

$$C_{i,\text{out}}(t_R)=\frac{1}{L}\int_0^\infty \int_0^L C_i(x,t)\,dx\,f(t)\,dt \tag{10.21}$$

An application is available[25]. *Figures 10.4(a)* and *10.4(b)* show, for given values of V_A/V_B and C_{A_0}/C_{B_0}, the influences of time (t in *Figure 10.4(a)*, t_R in *Figure 10.4(b)*) as well as of diffusion ($L^2/\mathscr{D}$) on the conversion obtained in a batch and continuously operated stirred tank reactor respectively. The solid upper curve represents the chemical regime; an increasing effect of diffusion (rising value of $kC_{A_0}L^2/\mathscr{D}$) decreases the conversion obtainable after a given reaction time (t or t_R).

Conversion (%)

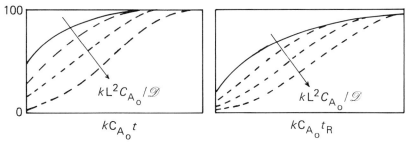

Figure 10.4(a)   Results of solving diffusion reaction equations for a plug flow reactor[25]. (b) Results of solving diffusion reaction equations for a perfectly macromixed stirred tank reactor[25]

The shortest time for diffusive mixing, which is attainable under normal operating conditions in a stirred tank reactor, will now be estimated. Typical solution viscosity and density are 1 mPa s and 1000 kg m^{-3}, respectively. The specific power consumption is unlikely to exceed 5 kW m^{-3}. Equation (10.11) gives then $\lambda_k \approx 20\,\mu$m and thus $R \approx 10\,\mu$m. Defining $t_{1/2\mathscr{D}}$ according to cases (a) and (c) given above and using $\mathscr{D}=10^{-9}$ m^2 s^{-1} gives $t_{1/2\mathscr{D}}$ values between 3.6 and 15 ms, i.e. of order 10 ms (values of order 1 ms have been attained in stopped-flow equipment, but at the expense of a high power input[26,27]). Thus we might expect reactions having half-lives of less than 0.1 s to be influenced by inhomogeneities at the molecular scale.

With a reagent concentration of 10^3 mol m^{-3}, this influence would exist if k were higher than $(t_{1/2R}C_{A_0})^{-1}$, i.e. $k>10^{-2}$ m^3 mol s^{-1}. Under these conditions the mixing-controlled regime extends from $k \approx 10^{-2}$ to $k \approx 10^6$ m^3 mol^{-1} s^{-1}, and the encounter-controlled regime occurs with still-higher k values.

The speed with which two reagent streams are mixed influences the product distribution obtained from certain multiple reactions. *Table 10.1* gives some examples, several of which can be described at least approximately as

Table 10.1   Reactions showing mixing effects on selectivity

1.   [structure of durene: benzene ring with H_3C, CH_3, H_3C, CH_3] + $NO_2^{\oplus}$ [28]

(Durene)

2.   HO—⟨ ⟩—CH_2—CH—COOH + I_2 [29]   3. H_3C—⟨ ⟩—OH + I_2 [30]

  |
  NH_2

(L—Tyrosine)

4. H_3C—COO—CH_2—CH_2—OOC—CH_3 + $OH^{\ominus}$ [30] [36]

(Glycol diacetate)

5.   m— and p—xylene, mesitylene, prehnitene + NO_2^+ [31]

6.   [structure: 1-Naphthol-6-sulfonic acid with OH, HO_3S] + [structure: phenyldiazonium ion with $N_2^{(+)}$]   [32]

$\left(\begin{array}{l}\text{1—Naphthol—6—}\\ \text{sulfonic acid}\end{array}\right)$ $\left(\begin{array}{l}\text{Phenyldiazonium}\\ \text{ion}\end{array}\right)$

7.   [structure: resorcinol with OH, OH] + Br_2 [33]   8.   [structure: 1,3,5-trimethoxybenzene with OCH_3, H_3CO, OCH_3] + Br_2 [34]

(Resorcinol) $\left(\begin{array}{l}\text{1,3,5—Trimethoxy—}\\ \text{benzene}\end{array}\right)$

9.   [structure: 1-Naphthol with OH] + [structure: 4-Sulfophenyl-diazonium ion with $N_2^{(+)}$, SO_3H]   [35] [37] [39]

(1—Naphthol) $\left(\begin{array}{l}\text{4—Sulfophenyl-}\\ \text{diazonium ion}\end{array}\right)$

competitive-consecutive (or parallel-series) reactions, having the following stoichiometry:

$$A + B \xrightarrow{k_1} R \tag{10.22}$$

$$R + B \xrightarrow{k_2} S \tag{10.23}$$

In the fully diffusion-controlled regime, the intermediate R does not survive and the only product formed is S. Defining the product distribution with the quantity X_S, which is the fraction of reagent B present in the secondary product S relative to the quantity of B in both products R and S:

$$X_S = 2C_S/(C_R + 2C_S) \tag{10.24}$$

then $X_S = 1$ in the diffusion-controlled regime. If, on the other hand, reaction is sufficiently slow that no concentration gradients arise at the molecular scale (i.e. for reaction in the chemical regime), X_S will have its minimum value.

In the mixed regime, where both chemical (e.g. rate constant, concentration level) as well as physical (e.g. stirring rate, viscosity of reaction medium) factors determine rate and selectivity, modelling using diffusion-reaction equations, analogous to equation (10.16), has been carried out[36,38]. The product distribution depends upon the following variables[36], written analogous to equation (10.17), where $X_S = X_S(x, t)$:

$$X_S = f(x/L, k_1 C_{A_0} t, C_{A_0}/C_{B_0}, V_A/V_B, k_1/k_2, k_2 L^2 C_{B_0}/\mathscr{D}) \tag{10.25}$$

The dependence on position (x) drops out if X_S is interpreted as the spatially averaged product distribution (refer equation (10.20)). t is again measured from the start of diffusive, molecular scale mixing (i.e. after coarse scale homogenization has been completed). In a stirred tank reactor, for example, averaging overall possible values of t using the residence time distribution is in principle necessary (refer equation (10.21)). However, when the reaction is fast enough, the stream leaving any reactor will no longer be reacting and the concentration of the limiting component B will be practically zero. (In this case the group $k_1 C_{A_0} t$—the first Damkoehler number—will be effectively infinite and exert no further influence on product distribution.)

The product of the next two groups defines the stoichiometric ratio N_{A_0}/N_{B_0}. This should exceed 0.5, otherwise the intermediate R will be fully converted to S as the reactions run to completion. The ratio of the rate constants has an obvious influence on product distribution in the chemical regime, where, all other things being equal, an increase in k_1/k_2 favours the formation of R and decreases X_S. The last group—also termed the second Damkoehler number—expresses the relative importance of diffusional and chemical rate-determining steps (refer equations (10.18) and (10.19)).

When both chemical and diffusion factors operate, the function in equation (10.25) must be evaluated by numerical integration of the diffusion-reaction equations. Results are available for the average product distribution X_S from spherical[38] and slab-like[36] reaction zones of constant size and can be represented as in *Figure 10.5*. The curves are generally S-shaped, whereby:

(a) $X_S \rightarrow 1$ in the diffusion-controlled regime (right-hand side), and
(b) $X_S \rightarrow$ constant minimum value in the chemical regime (left-hand side).

Figure 10.5(a) shows that as diffusion increases in importance, X_S depends less on k_1/k_2. *Figure 10.5(b)* reveals, at constant stoichiometric ratio, an influence on product distribution of the relative volumes of fluid which have to be mixed with each other. This is not the case in the chemical regime. Experiments have

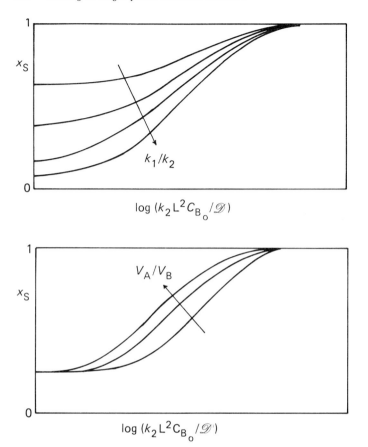

Figure 10.5(a)   Influence of ratio of rate constants on final product distribution
(t → ∞, V_A/V_B and N_{A_0}/N_{B_0} have given values)[38]. (b) Influence of volume ratio of
feed solutions on final product distribution (t → ∞, k_1/k_2 and N_{A_0}/N_{B_0} have given
values)[38]

confirmed the influence of V_A/V_B as well as that of the operating mode (semi-continuous and continuous tank reactors), stirrer type, scale, impeller speed, viscosity, etc.[31,32,35,37,39]. The agreement between experiments and the modelling using diffusion-reaction equations is in many cases good, allowing better than order of magnitude description of the observations. Several aspects are, however, topics for further research.

10.3   Conclusion

The literature also contains alternative ways of modelling the influence of mixing on rate and to a smaller extent on selectivity. In the space available and in order to give a consistent description, only diffusion-reaction models have been treated here. This is also dictated to some extent by personal experience;

to some extent, however, by the experimental confirmation of such models available today.

Notation

A	Amplitude of concentration oscillation
C	Concentration
D	Impeller diameter
$\mathscr{D}$	Diffusion coefficient
f	Residence time frequency function
h	Defined in *Figure 10.3*
K	Mixing rate constant; see equations (10.6–10.8)
k	Rate constant of reaction of second order
L	Size of reaction zone
N	Impeller rotational speed
P	Power drawn by impeller
R	Radius of reaction zone
T	Tank diameter, absolute temperature
t	Time
t_c	Mean circulation time
t_M	Mean mixing time
t_R	Mean residence time in reactor
V	Liquid volume
x	Length coordinate
X	Degree of unmixedness; see equation (10.2)
X_S	Measure of product distribution; see equation (10.24)

Greek symbols

β	Proportionality constant; see equation (10.10)
ε	Power input per unit mass of fluid
λ_K	Kolmogoroff velocity microscale
μ	Dynamic viscosity
ν	Kinematic viscosity
ρ	Fluid density

References

1 BEEK, J. and MILLER, R. S. (1959) *Chem. Engng. Progr. Symp. Ser. No. 25*, **55**, 23.
2 HILL, J. C. (1976) *Ann. Rev. Fluid Mech.*, **8**, 135.
3 BRODKEY, R. S. (ed.) (1975) *Turbulence in Mixing Operations*, Academic Press, New York.
4 KRAMERS, H. BAARS, G. M. and KNOLL, W. H. (1953) *Chem. Engng. Sci.*, **2**, 35.
5 HOLMES, D. B., VONCKEN, R. M. and DEKKER, J. A. (1964) *Chem. Engng. Sci.*, **19**, 201.
6 VONCKEN, R. M., HOLMES, D. B. and DEN HARTOG, H. W. (1964) *Chem. Engng. Sci.*, **19**, 209.
7 VONCKEN, R. M., ROTTE, J. W. and TEN HOUTEN, A. T. (1965) *A.I.Ch.E.–I.Chem.E. Symp. Ser. No. 10*, 24, I. Chem. E., London.

8   PROCHAZKA, J. and LANDAU, J. (1961) *Coll. Czech. Chem. Comm.* **26**, 2964.
9   LANDAU, J., PROCHAZKA, J. and POLASEK, F. (1963) *Coll. Czech. Chem. Comm.*, **28**, 1093.
10 KHANG, S. J. and LEVENSPIEL, O. (1976) *Chem. Engng. Sci.*, **31**, 569; (1976) *Chem. Engng.*, 11th October, p. 141.
11 GRAMLICH, S. and LAMADE, S. (1973) *Chem. Ing. Tech.*, **45**, 116.
12 ZLOKARNIK, M. (1967) *Chem. Ing. Tech.*, **39**, 539.
13 THYN, J., NOVAK, V. and POCK, P. (1976) *Chem. Engng. J.*, **12**, 211.
14 BIRD, R. B., STEWART, W. E. and LIGHTFOOT, E. N. (1960) *Transport Phenomena*, Wiley, New York, Chapter 18.6.
15 SUTER, K. (1974) *Diss. No. 5356*, ETH Zurich.
16 HUGHES, R. R. (1957) *Ind. Eng. Chem.*, **49**, 947.
17 BRENNAN, D. J. and LEHRER, I. H. (1976) *Trans. Inst. Chem. Engrs.*, **54**, 139.
18 BIRD, R. B., LIGHTFOOT, E. N. and STEWART, W. E. (1960) *Transport Phenomena*, Wiley, New York, Chapter 11.1.
19 ATKINS, P. W. (1978) *Physical Chemistry*, Oxford University Press, Chapter 27.
20 TOOR, H. L. (1962) *A.I.Ch.E.J.*, **8**, 70.
21 SCHWARZ, G. (1981) *Diss. No. 6804*, ETH Zurich.
22 RICE, A. W., TOOR, H. L. and MANNING, F. S. (1964) *A.I.Ch.E.J.*, **10**, 125.
23 MAO, K. W. and TOOR, H. L. (1970) *A.I.Ch.E.J.*, **16**, 49.
24 NAUMAN, E. B. (1975) *Chem. Engng. Sci.*, **30**, 1135.
25 GUERDEN, J. M. G. and THOENES, D. (1972) *Proc. Second Int. Symp. Chem. React. Engng.*, Paper B1–35, Elsevier, Amsterdam.
26 BRADLEY, J. N. (1975) *Fast Reactions*, Oxford University Press, Chapter 7.
27 CHANCE, B. (1974) in *Techniques of Chemistry, Vol. 6* (Hammes, G. G., ed.), Wiley, New York, Chapter 2.
28 HANNA, S. B., *et al.* (1969) *Helv. Chim. Acta*, **52**, 1537.
29 PAUL, E. L. and TREYBAL, R. E. (1971) *A.I.Ch.E.J.*, **17**, 718.
30 ZOULALIAN, A. and VILLERMAUX, J. (1974) *Adv. in Chem. Ser.*, **133**, 348, Am. Chem. Soc., New York.
31 NABHOLZ, F. and RYS, P. (1977) *Helv. Chim. Acta*, **60**, 2937.
32 BOURNE, J. R., CRIVELLI, E. and RYS, P. (1977) *Helv. Chim. Acta*, **60**, 2944.
33 BOURNE, J. R., RYS, P. and SUTER, K. (1977) *Chem. Engng. Sci.*, **32**, 711.
34 BOURNE, J. R. and KOZICKI, F. (1977) *Chem. Engng. Sci.*, **32**, 1538.
35 BOURNE, J. R., MOERGELI, U. and RYS, P. (1978) *Proc. Sec. Europ. Conf. on Mixing*, Paper B-3, BHRA Fluid Engng., Cranfield, UK.
36 TRUONG, K. T. and METHOT, J. C. (1976) *Can. J. Chem. Engng.*, **54**, 572.
37 ANGST, W., BOURNE, J. R. and KOZICKI, F. (1979) *Proc. Third Europ. Conf. on Mixing*, Paper A-4, BHRA Fluid Engng., Cranfield, UK.
38 NABHOLZ, F., OTT, R. J. and RYS, P. (1977) *Helv. Chim. Acta*, **60**, 2926.
39 BOURNE, J. R., *et al.* (1981) *Chem. Engng. Sci.*, **36**, 1643

Chapter 11

Mixing of high-viscosity fluids

J C Godfrey

Postgraduate School of Chemical Engineering, University of Bradford

11.1   Introduction

There are three main points to be considered in the design of mixing equipment for viscous blending duties:

1. mixer application;
2. mixing rate;
3. power consumption.

As performance is sensitive to fluid properties this has necessitated the development of many types of mixer. The development has been of an empirical nature and the amount of engineering knowledge of high viscosity mixing processes is quite limited, especially for non-Newtonian fluids. The topic of power consumption is well understood but the amount of information at present available on mixing rate and mixer application is small. In some instances it may be necessary to estimate the nature of non-Newtonian mixing phenomena on the basis of Newtonian observations. Only a limited number of mixer designs have been examined in detail.

The simplest mixer characteristic is that of power consumption, with the successful correlations for Newtonian fluids having been extended to some non-Newtonian fluids. Consequently the prediction of power consumption is well documented for a number of mixer types and even for those designs not yet investigated the principles are established. The situation is far less satisfactory in the case of mixing rate; little is known about the nature of the mixing process or the effect of non-Newtonian properties on performance. Because of the lack of mixing rate data, the selection of a suitable mixer is difficult to make on a scientific basis. Fortunately there are sufficient experimental comparisons available to allow some generalizations to be made about the relative performance of a number of mixer types.

There are very many designs of mixer available, a wide range of geometries being necessary for the extremes of viscosity which may be encountered. The impellers used for low viscosity materials are usually small, operating at high speed and turbulent flow conditions. As viscosity rises the machinery required is more complicated and massive, with the flow regime generally laminar.

11.1.1   Small impellers

The most common of the small impellers are the turbine and propeller as used for the turbulent mixing of low viscosity Newtonian fluids, see Chapter 8. Both are regarded as versatile and can be used for mixing fluids of moderate viscosity or non-Newtonian properties. The propeller and turbine operate best under turbulent flow conditions but they are used at times for high viscosity materials where the flow is not fully turbulent—viscosities up to 20 poise being quoted for propellers and up to 600 poise for turbines.

Research work has shown that the mixing of high viscosity[1] and non-Newtonian[2] materials with these small impellers is wasteful in terms of power consumed. Part of the reason for this is the change in flow pattern that occurs when these materials are mixed, flow being confined to regions near the impeller. However, power costs can be a secondary consideration when compared with the capital costs of the more efficient large impeller systems.

The flexibility of small impellers is also attractive in that the same equipment can be used for both high viscosity and low viscosity fluids. Because of these advantages small impellers are extensively used in high viscosity applications and specialist types have been developed, e.g. the dispersing impeller shown in *Figure 11.1* and the high shear rotor-stator type shown in *Figure 11.2*. The flexibility of the small impeller designs is particularly advantageous in those batch processes where mixing is accompanied by a substantial increase in viscosity from a low viscosity starting point.

11.1.2   Large impellers

For pseudoplastic fluids of quite low apparent viscosity the use of small impellers can be quite inefficient in terms of energy consumption. This is because the low shear regions of the mixer, distant from the impeller, have apparent viscosities very much higher than those in the impeller region. To minimize such stagnant regions larger impellers are used; in many cases with only a small clearance at the vessel wall. The anchor agitator, *Figure 11.3*, is a widely used simple, close-clearance, design for which a considerable amount of data is available.

There are many other large impeller designs (paddle, gate, leaf) and these all provide more extensive flow than a small impeller. The close-clearance designs give a degree of wall scraping which has several advantages, i.e.

1. Eliminates the build-up of unmixed materials at wall.
2. Provides a region of high shear for dispersing aggregates and lumps.
3. Improves the wall heat transfer coefficient.

The operating speeds of these impellers are quite low but the power requirements are high. Although these large impellers provide extensive flow it is largely rotational and the axial flow characteristic—as produced by propellers in low viscosity fluids—is absent. This means top-to-bottom mixing is poor and, for this reason, more complex designs such as the helical screw

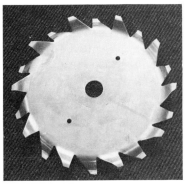

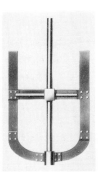

Figure 11.1   *Figure 11.2*   *Figure 11.3*

Figure 11.1   Mastermix high-speed serrated disc impeller for dispersion preparation
(T ≈ 3D), see also Figure 11.6 (Courtesy, Mastermix)

Figure 11.2   Silverson rotor-stator assembly (Courtesy, Silverson)

Figure 11.3   Lightnin anchor impeller (Courtesy, Lightnin Mixers)

(large clearance) and helical ribbon (small clearance, *Figure 11.4*) were
introduced.

Experimental work has shown that these axial mixing impellers are more
efficient than the simpler large impeller designs. A considerable amount of
data is available for the helical ribbon[3,4,5] and also for the combined ribbon
and screw[6,7], *Figure 11.5*.

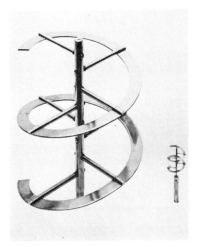

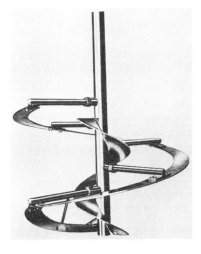

Figure 11.4   Large and small helical
ribbon impellers (D = 250 mm; D =
38 mm) used for experimental
studies[4]

Figure 11.5   Lightnin combined
helical ribbon and screw impeller
(Courtesy, Lightnin Mixers)

11.1.3   Baffles and draught tubes

Various stationary elements have been used in the development of mixer design in an endeavour to improve performance. The conventional vertical baffles used in conjunction with small impellers for the turbulent mixing of low viscosity fluids are detrimental to the mixing of high viscosity Newtonian fluids[1] and even low apparent viscosity pseudoplastic fluids[2].

Alternative arrangements have been used in conjunction with large impellers to increase shear, e.g. vertical baffles located to give close clearance with respect to the inner edges of an anchor impeller.

Axial flow can be produced in viscous or non-Newtonian fluids by the use of impeller and draught tube combinations. Small impellers and helical screws have been used in this way[1] to increase axial flow and to channel the mixer contents through a high shear zone in the impeller region.

11.1.4   Complex motion impellers

For high viscosity fluids, a variety of impeller designs have been developed to reduce the possibility of stagnant regions. An extension of the impeller-baffle combination discussed above is to use moving baffles—effectively a second, contra-rotating, impeller. Combinations of large and small impellers are also used (see *Figure 11.6*).

Planetary arrangements with single, e.g. Nauta (*Figure 11.7*), or multiple impellers have been developed to ensure that all of the vessel contents are subject to shear at the same time. Travelling impellers as well as moving vessels have been used, including combinations of moving vessel with non-rotating impellers.

11.1.5   Intensive mixers

These are used for processing materials of very high apparent viscosities, usually in excess of 1000 poise. The main characteristic of these mixers is the large size of the impellers which give almost total sweeping of the mixer bowl. The typical machine has two impellers mounted on parallel horizontal shafts; the impellers contra-rotate in a bowl shaped to keep blade to wall clearances low. In the Banbury mixer the blades of the impellers are completely enclosed to provide maximum shearing. Similar construction is used in the laboratory Brabender Plastograph which is instrumented for torque measurement.

Other machines (e.g. two arm kneader, sigma blade mixer, *Figure 11.8*) are open at the top and are used for less arduous processes than the completely enclosed designs. Examples of materials processed in high shear mixers are pvc, rubber, cereals, clay, brake linings, dog food, polyethylene.

There are continuous mixers which perform similar duties; these may have a single 'impeller' as in the well-known extrusion machines or multiple 'impellers' (e.g. the 'Ko-kneader'). The diameter of the impeller in a continuous machine is usually very much less than axial flow length. There are also

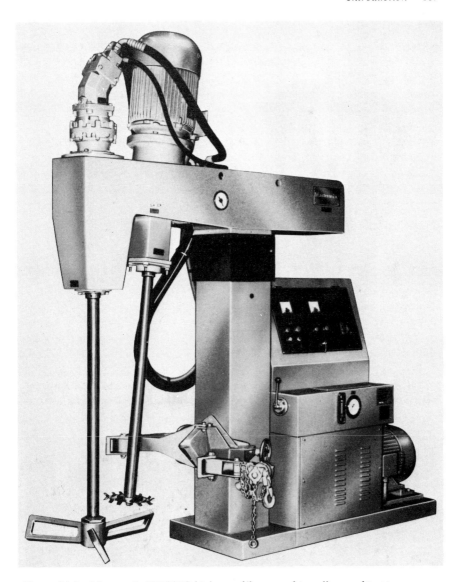

*Figure 11.6   Mastermix HVS/TS high-speed/low-speed impeller combination
(Courtesy, Mastermix)*

continuous mixers—'in-line mixers'—in use which have no moving parts but
rely on mixing elements in a tube for their mixing effect. These mixers are used
for a wide range of mixing processes and the energy is provided by pumps
which force the process fluids through the tube. There is further discussion of
in-line mixers in Chapter 13.

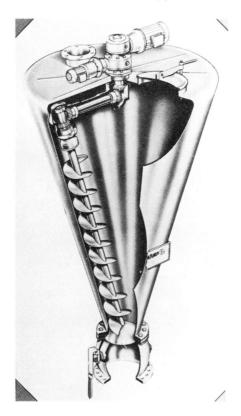

Figure 11.7   Nauta conical mixer with orbiting arm and rotating screw (Courtesy, Hosokawa Nauta, UK)

Figure 11.8   Winkworth Z-blade mixer (Courtesy Winkworth Machinery)

11.2   Power consumption

This is the simplest characteristic of high viscosity mixing equipment and the greater part of work conducted to date has been on this topic. Measurements can be made of shaft torque using conventional transducers although, with small mixing equipment, there may be problems, due to the lack of transducers for very small torques. Very small scale measurements can be made in modified rotational viscometers, as with the small impeller of *Figure 11.4*. In most cases measurements of electrical power consumption are not sufficiently closely related to torque to provide an alternative to the torque transducer.

Although the characteristic of power consumption is simple the information is quite valuable. In addition to the obvious value in motor selection it is perhaps even more useful for gearbox and shaft design for high viscosity mixing equipment where torques are usually much higher than encountered with the higher speed impellers used for low viscosity mixing. Power consumption also provides a useful basis for comparison in the evaluation of different mixer designs for mixing rate and efficiency[3].

11.2.1   Newtonian fluids

Because of the complex motion in any mixing process, dimensional analysis is widely used to establish simple relationships between power consumption and the controlling variables. For a fixed mixer geometry the expression used is (see Chapter 8):

$$Po = f[Re, Fr] \tag{11.1}$$

where Po is the dimensionless power number, analogous to a drag coefficient, and is defined:

$$Po = \frac{P}{\rho N^3 D^5} \tag{11.2}$$

Re is the impeller Reynolds number:

$$Re = \frac{ND^2\rho}{\mu} \tag{11.3}$$

Fr is the Froude number:

$$Fr = \frac{N^2 D}{g} \tag{11.4}$$

Considerable experimental work has been conducted in the evaluation of power numbers for many different mixing conditions. For laminar mixing conditions the relationship is of the same general form for large and small impellers:

$$Po = K_p Re^{-1} \tag{11.5}$$

This is identical in form to the well-known relationship for propellers and turbines with K_p being specific to the impeller geometry tested but independent of scale. For small variations in geometry for a particular impeller design equation (11.5) can be expanded by including various geometric ratios and exponents:

$$Po = K'_p Re^{-1} (G_1)^a (G_2)^b \tag{11.6}$$

This form of relationship has been successfully demonstrated[4] and it appears that the values of K'_p, a, ... are independent of equipment size. It is important to note that:

$$K_p = K'_p (G_1)^a (G_2)^b \tag{11.7}$$

and therefore K_p and K'_p are not equal. The relationship has been evaluated[8] for a wide range of published data:

Helical ribbon

$$Po = 150 Re^{-1} \left(\frac{C}{D}\right)^{-0.28} \left(\frac{p}{D}\right)^{-0.53} \left(\frac{h}{D}\right) \left(\frac{W}{D}\right)^{0.33} N_b^{0.54} \tag{11.8}$$

Anchor

$$Po = 85 Re^{-1} \left(\frac{C}{T}\right)^{-0.31} \left(\frac{h}{D}\right)^{0.48} \tag{11.9}$$

11.2.2 Non-Newtonian fluids

A simple relationship has been shown to exist between Newtonian and much non-Newtonian power consumption data in the laminar regime, using the Metzner and Otto[9] correlation based on calculations of apparent viscosities for mixers. The basis of the method is the assumption that there is an average shear rate for a mixer which describes power consumption and that this shear rate is directly proportional to impeller speed:

$$\bar{\gamma} = k_s N \tag{11.10}$$

where k_s is the mixer shear rate constant. The average shear rate $\bar{\gamma}$ defines an apparent viscosity for power prediction for non-Newtonian fluids. The apparent viscosity is determined from viscometric data at the appropriate shear rate and used directly in the usual Newtonian power number—Reynolds number correlation. To test this hypothesis, measurements are made of the power number—Reynolds number characteristics of Newtonian fluids and power number—impeller speed characteristics for non-Newtonian fluids. For a measured value of power consumption for a non-Newtonian fluid a power number and the corresponding Reynolds number can be calculated. From the Reynolds number:

$$Re = \frac{ND^2 \rho}{\mu_a} \tag{11.11}$$

μ_a can be determined. The shear rate corresponding to this apparent viscosity can be found from apparent viscosity—shear rate data from the non-Newtonian test fluid, see *Figure 11.9*. The mixer shear rate constant, k_s can then be determined from equation (11.10). The values of k_s determined by Metzner and Otto[9] were relatively constant for a range of impeller speeds and fluid properties, and an average value could be used in the prediction of non-Newtonian power consumption data. The use of average shear rate and apparent viscosity also shows potential for the prediction of power consumption for thixotropic fluids[10].

For power law fluids k_s may be incorporated into a power number—Reynolds Number equation. For the power law fluid:

$$\tau = K(\bar{\gamma})^n \tag{11.12}$$

and apparent viscosity is:

$$\mu_a = \frac{\tau}{\bar{\gamma}} = K(\bar{\gamma})^{n-1} \tag{11.13}$$

If $\bar{\gamma} = k_s N$, then:

$$\mu_a = K(k_s N)^{n-1} \tag{11.14}$$

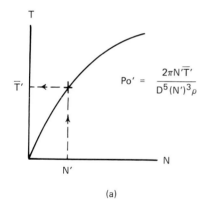

(a)

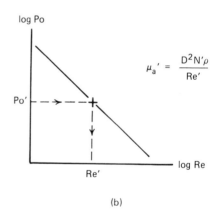

(b)

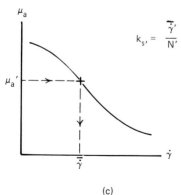

(c)

Figure 11.9 Determination of shear rate constant (k_s) from experimental data. (a) Non-Newtonian power consumption; (b) Newtonian power consumption; (c) non-Newtonian viscometry

This value of μ_a may be incorporated in the Reynolds number:

$$Re = \frac{ND^2\rho}{\mu_a} = \frac{N^{2-n}D^2\rho}{K(k_s)^{n-1}}$$ (11.15)

The power number equation may be rewritten as:

$$Po = K_p \left[\frac{N^{2-n}D^2\rho}{K(k_s)^{n-1}} \right]^{-1}$$ (11.16)

where the value of K_p is valid for Newtonian and non-Newtonian fluids. It is important to note the presence of the mixer shear rate constant in the power number equation and correlations which do not include this constant or some equivalent should be carefully checked before application.

The simple concept of an average mixer shear rate has been widely used in laboratory and industrial work and in most applications it has been assumed that the shear rate constant, k_s, is only a function of impeller type. Research is continuing on the possible influence[11] of flow behaviour index and elastic properties and also on procedures necessary to describe power consumption for dilatant[12] fluids. It should be noted that in all aspects of power prediction and data analysis, power law models (equation 11.12) should only be used with caution. Apparent variability of k_s may be due to inappropriate use of power law equations; when calculations are made it should be ascertained that the average shear rates of interest ($\bar{\gamma} = k_s N$) lie within the range of the power law viscometric data.

11.2.3   Impeller geometry

It is also assumed, and seemingly with success, that the shear rate constant, k_s, is independent of equipment size (no scale-up problems) and small variations in equipment geometry. Some investigation has been made of the influence of wall clearance for helical ribbon and anchor impellers[8].

For the helical ribbon:

$$k_s = 34 - 114\left(\frac{C}{D}\right)$$ (11.17)

for $0.026 < \dfrac{C}{D} < 0.164$

For the anchor:

$$k_s = 33 - 172\left(\frac{C}{T}\right)$$ (11.18)

for $0.02 < \dfrac{C}{T} < 0.13$

In most practical cases the range of values of k_s for a particular agitator design is not large. The major influence of geometry is on the power number—Reynolds number relationship as illustrated by equations (11.8) and (11.9). The influence of geometry described is applicable to both Newtonian and non-Newtonian fluids[4,8].

11.3 Mixing rate

The measurement of mixing rate is a much more complex topic than power consumption and for this reason the amount of data and theory available for rate studies is comparatively small. Practical problems exist with respect to the description of mixture quality and mixing rate and the tedious and time-consuming nature of many of the available measurement techniques.

There is still considerable debate regarding the differences in rate associated with Newtonian[3,5], pseudoplastic[5,11], dilatant and viscoelastic[7,11] properties and also regarding the suitability of the various mixing rate test procedures[3].

11.3.1 Measurement techniques

Many studies of mixing rate have been conducted by simple visual observation of the incorporation of a coloured tracer into a transparent fluid. These experiments give a single value of mixing time which may be studied as a function of mixer design, impeller speed, etc. These measurements are quickly and easily made, the results obtained are useful for initial screening work but are not a basis for design or research because of the subjective nature of the experiment.

Many observations[3,5] suggest that the approach to the fully mixed condition is asymptotic, so that there is no one value of mixing time and that errors in estimation of 'mixing time' are potentially large, see *Figure 11.10*. The situation is improved by using a more quantifiable technique, for example coloured tracer monitored by colourimeter. Further information is gained if some method for following the progress of mixing with time is used.

Some of the combinations of tracer and instrument used to quantify mixedness or mixing are listed below (see also Chapters 8 and 9):

Tracing methods[3]	*Monitoring*
1. Dye[13,14]	Visual/optical
2. Acid-base reaction with indicator[16]	Visual/optical
3. Dye decolourizing[3]	Visual/optical
4. Conducting tracer[5]	Conductivity cell
5. Optically different tracer[15]	Schlieren system
6. Heated tracer[1]	Thermocouple

These methods offer varying degrees of reproducibility but all are affected by a number of variables:

1. Volume of tracer;
2. Location of tracer addition;

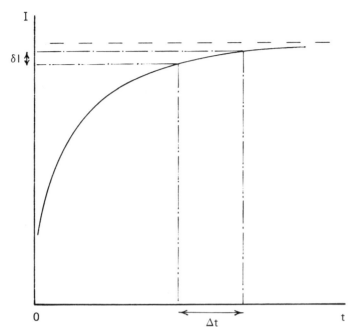

Figure 11.10   Progress of mixing in terms of arbitrary mixing index (I). *As fully mixed condition is approached small errors in mixing index* (δI) *lead to large errors in estimated mixing time* (Δt)

3. Scale of scrutiny (e.g. size of monitoring device; minimum size of non-uniformity visible);
4. Location of monitors;
5. Definition of end-point.

It is rare to find the concepts of scale of scrutiny and scale of segregation used quantitatively in the evaluation of mixing rate data.

11.3.2   Newtonian fluids

For fluids of high viscosity, a turbulent regime is not easily established and the turbine and propeller are inefficient[1,17] at low Reynolds number due to the limited flow profile which exists in that regime. The increased mixing rates at high viscosity achieved with larger impellers are partly due to increased power consumption and partly to improved flow patterns; the 'positive turnover' mixers (impeller and draught tube combinations, off centred screw, helical ribbon) are found to be more efficient. For these impellers Nagata[17], Hoogendoorn and den Hartog[1] and other workers have found that viscosity and impeller speed have very little effect on the number of impeller revolutions, Nt_m, required to achieve mixing. Also it appears that circulation rate is directly proportional to impeller speed and the cube of impeller diameter so that the group Q/ND^3 is constant.

The term $t_m ND^3/V$ is sometimes used as an index of mixer performance: the number of circulations of the mixer contents required for mixing. The number of circulations required by these mixers, approximately 3, is understandably more than is required for a turbine operating in the turbulent regime, about 1.5.

At a more quantitative level experiments which have followed the progress of mixing confirm the observations made above regarding the influence of viscosity and the number of impeller revolutions required for a particular mixture quality.

Measurements of colour change with blending[5] and also of dye decolourizing[3] for the helical ribbon impeller suggest that the same relationship between quality of mixing and number of elapsed impeller revolutions (Nt_m) is always observed regardless of impeller speed. Thus the number of revolutions required to achieve a specified mixture quality appears to be independent of Reynolds number, this observation being supported by the lack of influence of Newtonian viscosity on the progress of mixing[5].

The relationship between variance and impeller revolutions can be approximated by a simple first order relationship, see *Figure 11.11*, providing a useful description of the progress of mixing which is also applicable[18] to the planetary Nauta mixer and the intensive sigma blade mixer. A similar observation was made regarding the relationship between residual dye concentration and impeller revolutions in the dye decolourising experiment except that a finite number of impeller revolutions is required before decolouring becomes apparent[3].

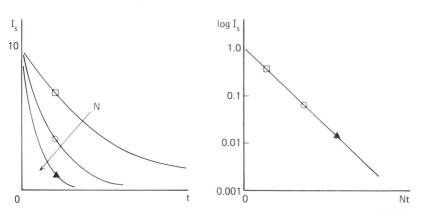

Figure 11.11   Progress of mixing in terms of intensity of segregation $(I_s = \sigma^2/\sigma_0^2)$. *(a) Influence of impeller speed* (N) *on mixing rate; (b) approximate first-order relationship for all speeds in terms of impeller revolutions* (Nt)

11.3.3   Non-Newtonian mixing

The first question that arises is: 'How does the mixing performance of the non-Newtonian liquid differ from that of the Newtonian?' In a number of circumstances Newtonian mixing rate in terms of impeller revolutions has been found independent of Reynolds number, and the circulation rate independent of viscosity. Also Bourne[19] has observed that circulation rate is independent of material properties for both Newtonian and non-Newtonian fluids. It would seem to follow that the mixing rate is independent of material properties. However, there are reasons to doubt that this convenient assumption is justified. Measurements of the mixing rate of pseudoplastic fluids with a turbine impeller[16] gave rates up to 50 times lower than would be expected from similar Newtonian correlations. This is probably due in part to less extensive turbulence in the mixing vessel—as is suggested by examination of non-Newtonian power consumption data—but is unlikely to be relevant to considerations of laminar mixing with large impellers. For laminar mixing the importance of non-Newtonian properties is indicated, as discussed earlier, by the findings of studies concerned with the selection of mixers for various types of fluid[2].

More quantitative measurements of the progress of mixing have shown both similarities and differences in the comparison of Newtonian and non-Newtonian mixing processes. There are similarities in that the progress of mixing can be described in terms of impeller revolutions for a range of impeller speeds and that the relationship between variance and impeller revolutions is approximately first order[5]. However, there is an influence of flow properties on mixing rate which falls as the test fluids become more non-Newtonian. Unfortunately this cannot be taken as the basis of a simple generalization as there is evidence that with other test fluids that the influence of flow properties is less[7]. Other experimental studies give a variety of results for different mixer designs (helical ribbon, helical ribbon and central screw, Nauta mixer, sigma blade mixer) and for different fluid properties.

Although there is not sufficient evidence to establish generalizations it seems important to take into account fluid elasticity. Many of the experimental studies of mixing rate have used test fluids which were polymer solutions and which could be expected to have both elastic and pseudoplastic properties. As different fluids show a different balance of elasticity and pseudoplasticity it is potentially misleading to make judgements on the basis of flow behaviour index only. There is a possibility that the influence of viscoelasticity is different in different mixer types, with reports of mixing rate decreasing with helical ribbons[7] and increasing with sigma blades[14,18] as elasticity is increased.

As there seems to be no single influence that can be associated with non-Newtonian properties it is not realistic to look for explanations which cover all the effects observed. However, it may be useful to consider behaviour that is related to the individual characteristics of pseudoplasticity and viscoelasticity. Pseudoplastic fluids have a lower viscosity in regions of high shear, as near an impeller, and a higher viscosity in regions of low shear, as distant from an impeller. It is possible that this characteristic leads to mixing behaviour different from that observed for Newtonian fluids, in particular for fluid motion to be more than usually confined to the impeller region.

Many phenomena have been associated with the observation of the mixing behaviour of viscoelastic fluids: secondary flow, Weissenberg effect, flow pattern dampening and reports of both decreased and increased mixing rates have been made. It seems therefore that viscoelastic properties affect the performance of different mixer designs in different ways and that there may be some opportunities to exploit viscoelastic properties in mixer design. It also seems that operating conditions, e.g. direction of impeller rotation[20], can also have an important effect when viscoelastic fluids are being mixed.

11.4 Discussion

Much of the preceding description has related to the characteristics of large impellers and reflects the concern of the research literature for these designs. In the present state of knowledge we have simple correlations for both Newtonian and non-Newtonian power consumption characteristics and a number of qualitative observations on the influence of viscosity and non-Newtonian properties on mixing rate. There are few data available for specialized impeller designs, particularly for smaller impellers. There seem to be two major areas where more information would be valuable in developing the present state of our appreciation of mixer performance:

(i) Laboratory trials of the many popular mixers which are at present only sketchily documented. Existing experimental and data processing techniques would provide very useful information. Further consideration of turbulent mixing of less viscous non-Newtonian fluids would be desirable.
(ii) Research studies which improve the understanding of the influence of non-Newtonian properties on both mixing rate and power consumption. For quantitative evaluation our existing knowledge only provides a useful background and is very much lacking in detailed analysis.

The large number of impellers and fluids and the comparatively small amount of quantitative data means that the question of efficiency is difficult to explore. Some attempts have been made to relate power, mixing time and viscosity[1,2,15] and also to use K_p and impeller revolutions[3] for comparisons of efficiency. From the latter work and also from other observations it seems, for the helical ribbon at least, that impellers which use more power tend to be more efficient. Until more data on both mixing rate and power become available it will be very difficult to make efficiency comparisons between different impeller types.

It is clear, from the discussion of power consumption and mixing characteristics above, that there are not enough data available at present to allow mixer selection to be made on a quantitative basis. The shortage of data is particularly evident for non-Newtonian fluids and also for specialized impellers. However, some guidelines are available regarding the selection of mixer type. For optimum performance in terms of mixing time and power consumption, large impellers seem best for high viscosity or non-Newtonian fluids, particularly designs which include both rotational and axial flows.

Small impellers are more attractive when flexibility is required, for example when large changes of viscosity occur during the mixing process. Although small impellers are likely to be inefficient in terms of power the higher rotational speeds used usually mean lower torques and hence less demand on shaft and gearbox construction.

11.5   Conclusions

For high viscosity or non-Newtonian fluids there are a number of simple concepts which are useful for the analysis or design of research problems in laminar mixing. For large impellers and laminar flow these concepts may be summarized as follows:

The relationship $Po = K_p Re^{-1}$ is capable of describing both Newtonian and pseudoplastic power consumption data. The apparent viscosity of a pseudoplastic fluid for purposes of power consumption is defined by an average mixer shear rate, $\bar{\gamma} = k_s N$. The value of the shear rate constant, k_s, is independent of non-Newtonian properties.

The progress of mixing is dependent only upon the number of impeller revolutions, Nt, so that rate is directly proportional to impeller speed. The relationship between quality of mixing and mixing time is approximately first order. Mixing rate is independent of viscosity but may be reduced by either pseudoplastic or viscoelastic properties.

These concepts provide a simple framework for the analysis of data or the planning of experimental or pilot programmes. It is necessary that these assumptions be tested wherever possible and deviations from the simple concepts noted. However, the framework provided allows simple data analysis and economical experimental programmes. To advance the understanding of the mixing of high viscosity fluids it would be advantageous if the simple concepts outlined above were applied to the description of a wider range of mixing equipment and if the concepts themselves were subject to a more thorough analysis.

Some of the fundamental mixing mechanisms of shear, extension and distribution which feature in the mixing of highly viscous materials are treated in detail in Chapter 12.

Notation

C	clearance between impeller and wall
D	impeller diameter
Fr	Froude number, $N^2 D/g$
G_1, G_2	geometrical ratios
g	gravitational acceleration
h	height
I	mixing index
K	consistency index of power law

K_p	constant
k_s	shear rate constant
N	impeller speed
N_b	number of blades
n	flow behaviour index of power law
P	power
Po	power number, $P/\rho N^3 D^5$
p	pitch
Q	circulation rate
Re	Reynolds number, $\rho N D^2/\mu$ or $\rho N D^2/\mu_a$
T	vessel diameter
$\bar{T}$	torque
t	time
t_m	mixing time
V	volume
W	blade width
ρ	density
τ	shear stress
μ	viscosity
μ_a	apparent viscosity
$\dot{\gamma}$	shear rate
$\bar{\dot{\gamma}}$	average shear rate
σ^2	variance
σ_0^2	initial variance

References

1. HOOGENDORN, C. J. and DEN HARTOG, A. P. (1967) *Chem. Eng. Sci,* **22**, 1689.
2. ULLRICH, H. and SCHREIBER, H. (1967) *Chem. Ing. Tech.* **39** (5/6), 218.
3. KAPPEL, M. (1979) *Intl. Chem. Eng.*, **19** (1), 96.
4. HALL, K. R. and GODFREY, J. C. (1970) *Trans. I. Chem. E.*, **43**, T201.
5. HALL, K. R. and GODFREY, J. C. (1971) *Proc. Chemeca (Australia),* **70**, 111.
6. CHOWDHURY, R. and TIWARI, K. T. (1979) *Ind. Eng. Chem. Proc. Des. Dev.*, **18** (2), 227.
7. CHAVAN, V. V., ARUMUGAM, M. and ULBRECHT, J. J. (1975) *A. I. Ch. E. J.*, **21** (3), 613.
8. EDWARDS, M. F. and AYAZI-SHAMLOU, P. (1983) in *Low Reynolds Number Flow Heat Exchangers*, Hemisphere Publishing Corpn.
9. METZNER, A. B. and OTTO, R. E. (1957) *A. I. Ch. E.*, **3** (1), 3.
10. EDWARDS, M. F., GODFREY, J. C. and KASHANI, M. M. (1976) *J. non-Newtonian Fluid Mech.*, **1**, 309.
11. CARREAU, P. J. (1981) *Proc. 2nd World Cong. Chem. Eng. (Canada)*, IV, 379.
12. EDWARDS, M. F. and MacSPORRAN, W. C. (February 1983). Paper presented at 'Emulsion and Colloid Science', I. Chem. E., Chester.
13. JENSEN, W. P. and TALTON, R. T. (1965) *A.I.Ch.E.—I. Chem. E. Symposium Series No. 10*, 82.
14. HALL, K. R. and GODFREY, J. C. (1968) *Trans. I. Chem. E.*, **46**, T205.
15. ZLOKARNIK, M. (1967) *Chem. Eng. Tech.*, **39**, (9/10), 539.
16. GODLESKI, E. S. and SMITH, J. C. (1962) *A. I. Ch. E. J.*, **8**, 617.
17. NAGATA, S., YANAGIMOTO, M. and YOKIYAMA, T. (1956) *Mem. Fac. Eng. Kyoto Univ.*, **18**, 444.
18. HALL, K. R. and GODFREY, J. C. (1979) *Proc. Nat. Conf. Rheol. (Australia),* **77**.
19. BOURNE, J. R. and BUTLER, H. (1969) *Trans. I. Chem. E.*, **47**, T11.
20. CARREAU, P. J., PATTERSON, I. and YAP, C. Y. (1976) *Can. J. Chem. Eng.*, **54**, 135.

Chapter 12

Laminar flow and distributive mixing

M F Edwards

Schools of Chemical Engineering, University of Bradford

12.1 Introduction

Many liquids, pastes, doughs and melts, particularly those found in the
plastics, rubber and food industries, are too highly fiscous to be mixed and
blended in conventional mechanically agitated vessels. Such materials are
processed using Z-blade or Banbury-type mixers, single or twin screw
extruders or roll mills. These units are characterized by the massive machinery
having a high power input and relatively small effective working volumes in
which the fluids are mixed.

Frequently the liquids will be non-Newtonian in rheology and sometimes,
under certain operating conditions, the materials may exhibit significant
elasticity. The high apparent viscosity of such fluids leads to laminar flow
conditions. Although this will result in long processing times it is not feasible
to induce turbulence because the power requirements would be prohibitively
high and the stresses on the machinery would be too great.

The duties to be achieved in processing these highly viscous materials are
blending and/or the incorporation of fine solids. In blending operations two or
more materials, often having widely differing viscosities and volume ratios,
have to be mixed to yield a product of desired uniformity or mixture quality.
For the dispersion of fine solids into viscous liquids the objective is to produce
a final mix with an acceptably low level of agglomerates of the basic individual
particles.

In the absence of rapid mixing which can be achieved in turbulent situations
it is the laminar flow which must now be exploited. The basic mixing
mechanisms are:

(a) 'Laminar shear'—in which relative motion between streamlines across the
    flow brings about deformation of fluid elements which yields increased
    interfacial areas between the liquids to be mixed and reduced striation
    thicknesses.
(b) 'Elongational or extensional flow'—here the stretching effect due to
    changes in flow channel geometry or due to fluid acceleration can produce
    reduced striation thicknesses and hence improved mixture quality.
(c) 'Distributive mixing'—the laminar flow through equipment, e.g. in in-line

static mixers, produces a stream-splitting and recombination effect, again reducing the striation thickness and increasing the interfacial area between components.

(d) 'Molecular diffusion'—in viscous liquids the molecular diffusion coefficients are small. Thus mixing at the molecular level is slow and only becomes significant once the mechanisms of shear, extension and distributive mixing have reduced the striation thickness down to a sufficiently low level.

(e) 'Stresses in laminar flow'—for the dispersion of fine solids into viscous liquids it is the stresses generated within the liquid during laminar flow which are responsible for breaking and rupturing agglomerates of particles to ensure that they are dispersed uniformly as individual particles. High shear stresses for such duties can be achieved in roll mills and the close-clearance regions of Banbury mixers. In order to achieve high shear stress levels it is advantageous to have high shear rates combined with high apparent viscosities. Such conditions inevitably lead to significant viscous heating and it is advantageous to apply cooling in such instances to avoid the temperature rise leading to a reduction in apparent viscosity and a consequent lowering of the stress level.

The assessment of the 'process result' following the mixing of high viscosity fluids in equipment such as roll mills, extruders, kneaders, etc. is an area of considerable difficulty and this will be discussed in a later section. At present it is sufficient to note that, of the mechanisms discussed above, laminar shear, elongational flow and distributive mixing produce a reduction in the 'scale of segregation'. In the absence of molecular diffusion these three mixing processes serve to reduce the size and scale of the unmixed clumps. Once the scale of segregation reaches a sufficiently low level, molecular diffusion can act to reduce the 'intensity of segregation', i.e. the differences in concentration between the components of the mixture are reduced. In the case of solids dispersion, when no mass transfer occurs between the solid and liquid phase, the mixing process can be assessed by the 'scale of segregation'.

Thus the quantities of scale and intensity of segregation are useful concepts in the understanding of high viscosity mixing. However, difficulties are encountered in the measurement of these parameters to give quantitative assessments of mixture quality.

A further important consideration is the 'scale of scrutiny' and this can be illustrated by reference to a familiar duty in the plastics industry involving pigment dispersion. Here it is common practice to add the fine pigment particles in the form of a 'masterbatch' which is a pre-compounded high concentration mixture of the pigment in the base polymer. In extrusion processes therefore, the base polymer granules are mixed and melted with granules of 'masterbatch', the latter containing the necessary quantity of pigment for the final extruded mix. When the mixing of the molten polymer in the extruder is viewed on the scale of the polymer granules it is a laminar blending operation of material of dissimilar colour. However, on the scale of

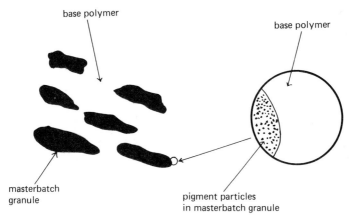

base polymer

base polymer

masterbatch
granule

pigment particles
in masterbatch granule

Figure 12.1   Effect of scale of scrutiny. (a) Polymer and masterbatch granules on scale of granules; (b) on scale of pigment particles

the pigment particles the operation is one of solids dispersion, see *Figure 12.1*.

In the following sections a more detailed analysis is presented of the mixing processes described above. The application of these analyses to mixing machinery is considered and the assessment of resulting mixture quality is discussed further.

12.2   Laminar shear

An idealized laminar flow field, i.e. tangential flow in a coaxial cylinder arrangement, can be used to illustrate some of the principles of laminar shear mixing which are relevant to more complex flows in real mixing machinery. Consider the situation in *Figure 12.2* in which a line of coloured tracer (minor component) is laid down and its motion is observed as the outer cylinder is rotated in an anti-clockwise direction.

If the minor component is aligned, as in *Figure 12.2a*, parallel to the plane of shear no mixing results (if molecular diffusion is neglected). However, when the tracer is placed perpendicular to the plane of shear, see *Figure 12.2b*, mixing occurs due to the velocity field. If h is the radial distance measured between adjacent lines of tracer then

$$h = L/(n+1) \tag{12.1}$$

where L is the gap between cylinders and n is the number of rotations.

After a large number of rotations $(n \gg 1)$ then

$$h \approx L/n \tag{12.1a}$$

and clearly after an infinite number of rotations h approaches zero and the mixing becomes complete.

A more detailed analysis of this flow field is presented by Middleman[1] for

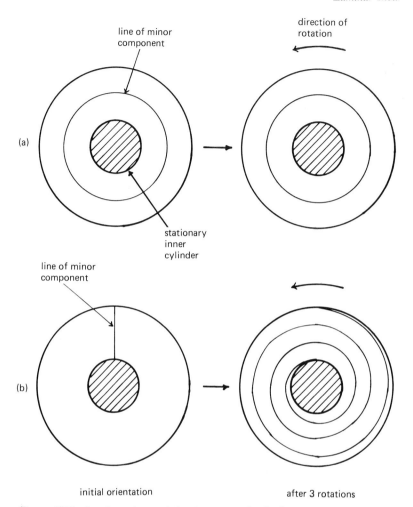

Figure 12.2   Laminar shear mixing in a coaxial cylinder arrangement

the case of the streak minor component having a width *w* and aligned initially
as in *Figure 12.2b*. The result of this treatment is shown in *Figure 12.3* where
the reduced streak width *w′* is plotted against the number of revolutions *n* for
various values of the cylinder diameter ratio, *K* (inner/outer). Such results
indicate the very significant reductions in streak width that can be achieved
after a small number of revolutions in this very idealized mixing situation,
provided the initial orientation is appropriate.

It is also worthy of note that the flow paths in a coaxial cylinder
arrangement can be reversed by changing the direction of rotation of the non-
stationary cylinder. Thus, if in *Figure 12.2b* the direction of rotation were to be
reversed for three rotations after the initial three rotations counterclockwise,
the result would be to re-form the original line. Thus reversal of the flow can
produce an apparent 'unmixing' effect.

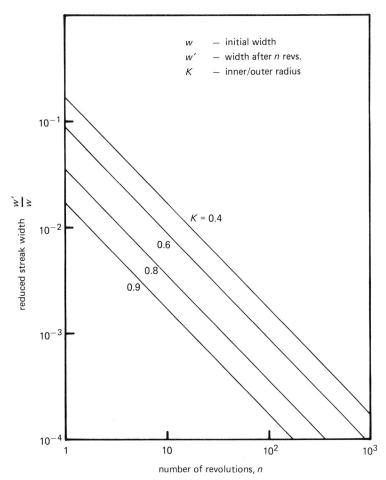

Figure 12.3   Effect of number of revolutions on streak width for a coaxial cylinder arrangement[1]

Consider now the one dimensional laminar flow between parallel plates as illustrated in *Figure 12.4* where the lower plate is stationary and the upper plate moves with velocity V' establishing a linear velocity profile across the gap. The minor component is initially in the form of a rectangle of length, l, and width, w, and after shearing for time t it is deformed in the laminar flow field to yield a streak of length l' and width w', see *Figure 12.4*. It is readily shown that the ratios of final to initial length and final to initial width are given by

$$\frac{l'}{l} = (1 + [\dot{\gamma}_s t]^2)^{1/2} \tag{12.2}$$

and

$$\frac{w'}{w} = (1 + [\dot{\gamma}_s t]^2)^{-1/2} \tag{12.3}$$

where $\dot{\gamma}_s$ is the shear rate in the gap V'/h.

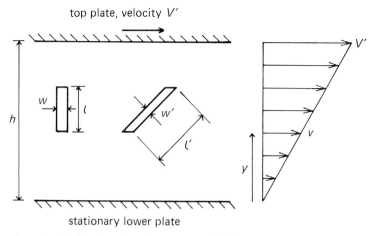

top plate, velocity V'

stationary lower plate

Figure 12.4 Laminar shear between parallel plates

Equations (12.2) and (12.3) show that the streak becomes elongated and reduced in width, i.e. a mixing process has taken place. Furthermore, it is the product of the shear rate, $\dot{\gamma}_s$, and time t, i.e. the total strain which determines the extent of this laminar shear mixing process. In order to enhance mixing, a combination of a high shear rate and a large time of shearing is desirable. This is equivalent to the requirement of a large number of rotations in the coaxial cylinder flow described earlier.

A further observation which applies to the parallel plate situation is that the linear velocity profile applies regardless of the rheological behaviour of the material and thus this idealized mixing behaviour is independent of rheology. Of course, in situations when the velocity field does vary with material rheology the viscous properties of the liquids become relevant. For example in the coaxial cylinder the velocity profile will depend upon the viscous behaviour of the fluid and the results in *Figure 12.3* are only valid for Newtonian liquids.

A more general analysis of mixing in unidirectional laminar shear flow was developed by Spencer and Wiley[2] who evaluated the change in interfacial area between the major and minor components. The process is illustrated in *Figure 12.5* and it can be shown[2] that the ratio of final to initial areas is given by

$$\frac{S}{S_0} = [1 - 2\gamma_s \cos \alpha \cos \beta + \gamma_s^2 \cos \alpha]^{1/2} \qquad (12.4)$$

where γ_s is the strain ($\dot{\gamma}_s t$), α is the angle between the normal n to the original plane and the x axis and β is the angle between the normal and the y axis.

The influence of orientation is brought out by equation (12.4):

(a) for surfaces initially parallel to the x, y plane

$$S/S_0 = 1 \qquad (12.5)$$

that is, no improvement in mixture quality is achieved;

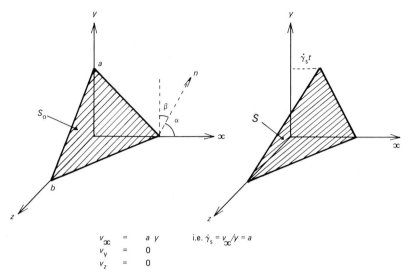

Figure 12.5   Effect of unidirectional laminar shear on interfacial area

(b) for surfaces initially parallel to the x, z plane

$$S/S_0 = 1 \qquad (12.6)$$

and again no mixing occurs;

(c) for surfaces initially parallel to the y, z plane, i.e. perpendicular to the plane of shear

$$S/S_0 = (1 + \gamma_s^2)^{1/2} \qquad (12.7)$$

giving an increase in interfacial area between components and hence an improvement in mixing.

Furthermore, the role of the total strain, γ_s, in laminar shear mixing is again highlighted in equation (12.4).

In a striated system, see *Figure 12.6*, the 'striation thickness' δ is defined as the distance between like interfaces. Thus for long streaks, neglecting the ends, the total system volume V is related to the mean striation thickness δ and the interfacial area S by

$$V = \frac{\delta S}{2} \qquad (12.8)$$

Mohr *et al.*[3] considered the unidirectional, laminar shear of cubes which were randomly distributed, but with faces parallel to the co-ordinate axes, in order to relate the striation thickness, δ, and the total strain γ_s. Basing the analysis on equations (12.4) and (12.8) Mohr *et al.*[3] derived, for large strain ($\gamma_s > 10$), the following expression:

$$\delta = l_2/\gamma_s V_2 \qquad (12.9)$$

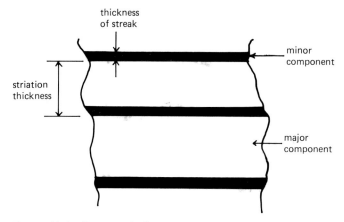

Figure 12.6   Striation thickness as a measure of mixture quality

where l_2 is the original cube length of the minor component and V_2 is the volume fraction of the minor component. Thus these authors were able to relate the total strain $\gamma_s\ (=\dot{\gamma}_s t)$ which is a measure of the processing to which a mixture has been subjected, to the striation thickness which can be used to measure mixture quality. Mohr *et al.*[3] state that in most practical mixing applications the requirement for equation (12.9) that $\gamma_s > 10$ is normally satisfied. It is interesting to note the similarity between equation (12.9) and the prediction of streak width at large strain given by equation (12.3).

All the above analyses have considered the laminar blending of components of identical viscosities. This restriction was removed by Mohr *et al.*[3] who postulated that continuity of stress must be maintained across the interface between components and thus derived a modified version of equation (12.9), i.e.

$$\delta = \tfrac{1}{2}/(\gamma_s)_1 V_2(\mu_1/\mu_2) \tag{12.10}$$

Here the total strain is calculated as the product of shear rate and residence time for the major component and μ_1 and μ_2 are the viscosities of the major and minor components respectively. It can be deduced from equation (12.10) that large striations (i.e. poor mixture quality) are favoured by a small volume fraction of the minor component and a high viscosity of the minor component. In other words, it is more difficult to mix by laminar shearing small volumes of a high viscosity minor component into a major component than it is to mix equal volumes of equal viscosity components.

For the laminar shearing of components of equal viscosity Hold[4] has considered the deformation and mixing which occurs when simple shearing is applied in two consecutive steps, the second being at right angles to the first. This is illustrated in *Figure 12.7* where it is seen that the change in direction of the applied shear is beneficial to the mixing process.

It is therefore advantageous to reorientate the minor component during shear rather than to apply a unidirectional shear field continuously without

Velocity fields

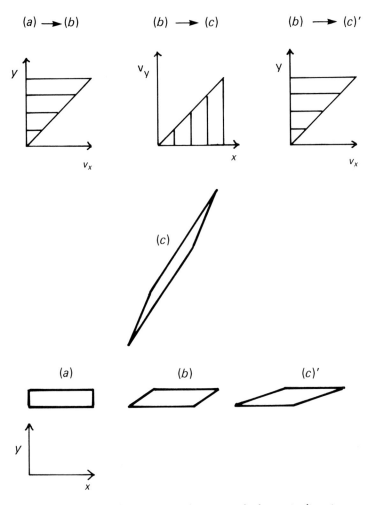

Figure 12.7   *Effect of consecutive shearing with change in direction*

any change in orientation. Clearly, this approach provides evidence of the need for redistribution of the minor phase to accompany the shearing process if mixing is to be achieved efficiently. Thus it may be anticipated that a combination of laminar shear mixing and distributive mixing will provide effective blending.

The concept of redistribution has been developed by Erwin[5] who showed that the generation of surface area can be optimized by changing the direction of the instantaneous shearing action so that it always maintains angles $\alpha = \beta = 45°$ with the normal to the interface, see *Figure 12.5*. Although it is not proposed by Erwin that such a mixer exists with this flow characteristic, the concept of an upper bound on performance is both interesting and useful. This concept will be developed further in the later sections on elongational flows.

The significance of the total strain, γ_s, in the laminar shearing process has been illustrated in several of the developments above and this quantity can clearly be used as a measure of mixing. For any fluid element the total strain is evaluated by summing the product of the shear rate and the residence time as the material moves through the shear field. In the general case fluid elements will be subjected to a variety of shear rates and residence times. For such cases it is useful to calculate the 'weighted average total strain' (WATS) which gives a measure of the mean strain to which the mixture has been subjected.

As a simple example of the calculation of the 'weighted average total strain' (WATS) consider the laminar flow of a Newtonian fluid in a pipe of length L and radius R. The velocity profile gives the local velocity v at radial position r as:

$$v = 2\bar{v}\left\{1 - \left(\frac{r}{R}\right)^2\right\}$$

(12.11)

where $\bar{v}$ is the mean velocity. From equation (12.11) the local shear rate $\dot{\gamma}_s$ at position r is given by:

$$\dot{\gamma}_s = 4\bar{v}r/R^2$$

(12.12)

Further, the residence time t at position r is:

$$t = L/v$$

(12.13)

and the local total strain $\dot{\gamma}_s t$ is given by:

$$\gamma_s = \dot{\gamma}_s t = 4\bar{v}rL/R^2 v$$

(12.14)

Since the mass flow rate of material between radial positions r and $r + dr$ can be written as $2\pi\rho v r dr$ the weighted local strain is $2\pi\rho v r dr (4\bar{v}rL/R^2 v)$. Finally the weighted average total strain follows as:

$$\text{WATS} = \int_0^R (8\pi\rho L\bar{v}/R^2)r^2 dr / \rho\bar{v}\pi R^2$$

(12.15)

i.e.

$$\text{WATS} = 8L/3R$$

(12.16)

Although this is a simple example it serves to illustrate the method of calculating the weighted average strain and this will be utilized in subsequent sections dealing with realistic mixing equipment. It will be clear, however, from this example of pipe flow that it is a straightforward procedure to evaluate the residence times and strains and hence WATS if the velocity profile is known.

12.3   Elongational (or extensional) laminar flow

As was the case for laminar shear flows it is useful initially to consider some idealized flow situations[4] [8]. In 'uniaxial extension' in the x direction, compression occurs in the y and z directions at half the strain rate in the x direction. For an interface arranged as shown in *Figure 12.8* the ratio of final to initial length is given by:

$$\frac{l'}{l} = \exp{(\gamma_E)}$$ (12.17)

the final to initial width:

$$\frac{w'}{w} = \exp{(-\gamma_E/2)}$$ (12.18)

and the final to initial area ratio becomes:

$$\frac{S}{S_0} = \exp{(\gamma_E/2)}$$ (12.19)

where γ_E is the strain in the x direction.

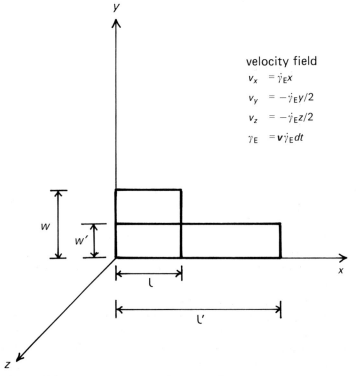

velocity field
$$v_x = \dot{\gamma}_E x$$
$$v_y = -\dot{\gamma}_E y/2$$
$$v_z = -\dot{\gamma}_E z/2$$
$$\gamma_E = v\dot{\gamma}_E dt$$

Figure 12.8   *Increase in area under uniaxial extension*

For 'planar extension', elongation in the x direction is accompanied by compression at the same rate in the y direction with no deformation in the z direction. In this case, see *Figure 12.9*, the ratio of new to initial area for the initial orientation shown becomes:

$$\frac{S}{S_0} = \exp(\gamma_{PE}) \tag{12.20}$$

where γ_{PE} is the strain in the x direction[7,8].

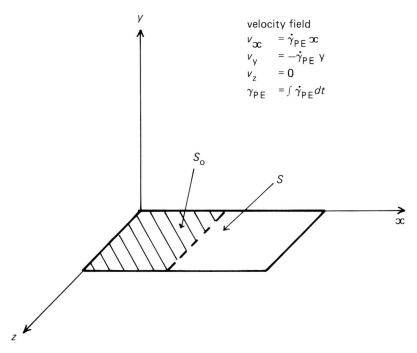

Figure 12.9   Increase in area under planar extension

In *Figures 12.8* and *12.9* the initial surfaces are located in the 'optimum' positions to give the greatest area increases and equations (12.19) and (12.20) apply to these orientations[7].

It is now possible to compare the changes in interfacial area produced in unidirectional shear, equation (12.7), uniaxial elongation, equation (12.19) and planar extension, equation (12.20). However, as pointed out by Cheng[7], this comparison needs to be done on a rational basis. For example, it is possible to examine the area ratios at equal strain defined by $\gamma_s = \gamma_E = \gamma_{PE}$ and it is clear from this viewpoint that effectiveness in generating new surface area increases in the order simple shear, uniaxial extension to planar extension. It is clear that at large strains the advantage is heavily in favour of the extensional flows. If a value of strain equal to 10 is considered (Mohr[3] states that most shear mixers exceed this value) then the area ratios S/S_0 are given in *Table 12.1*. However,

Table 12.1 Comparison of surface area generation[7] in shear and elongational flows ($\gamma_s = \gamma_E = \gamma_{PE} = 10$)

	Simple shear	*Uniaxial extension*	*Planar extension*
Area ratio (S/S_0)	10.05	1.484×10^2	2.202×10^4

this criterion of equal 'strain' is not the only basis for comparison. Cheng[7] gives an extensive survey based on several criteria including equal energy dissipation which measures the power consumption in the mixing process. Typical data on the basis of equal dissipation are presented below for a Newtonian fluid.

Table 12.2 Comparison of surface area generation[7] in shear and elongational flows ($\gamma_s = 10$ and extensional flows give same dissipation)

	Simple shear	*Uniaxial extension*	*Planar extension*
Area ratio (S/S_0)	10.05	17.93	148.4

From his detailed analysis, Cheng[7] concludes that it cannot always be said that extensional flows are more effective for mixing purposes than simple shear flow. Care is needed, particularly for flows involving non-Newtonian liquids. The position is rather complicated and Cheng states that at small strains, extensional flows are on the whole more effective than simple shear, but only marginally so, at sufficiently large strains extensional flows are very much more effective. However, at some intermediate strains simple shear may be more effective particularly when the fluid is non-Newtonian.

The analyses presented so far in this chapter have been concerned with the analysis of blending operations in idealized flows. It is instructive at this stage to consider how such flows may be achieved. For example, flow in a coaxial cylinder arrangement, *Figure 12.2*, is readily achieved and the parallel plate arrangement, *Figure 12.4*, can be approached in a coaxial arrangement with large diameter and small annular gap. Such configurations can operate continuously and therefore generate large strains.

In *Figure 12.10* a simplified version of an internal mixer is shown. The material from the bulk volume of the mixer (a) flows into the high shear region and in this stage elongation will be present. Once in the high shear region (b) the flow approximates to simple shear whilst in region (c) the flow is again elongational. There are clearly regions where both shear and elongation occur simultaneously and similar considerations apply to the flow between rollers in mixing mills. The elongational flows approach the ideal case of planar extension considered earlier.

In order to achieve the idealized elongational flows of uniaxial and planar extension it is necessary to depart from simple flows which can be related to

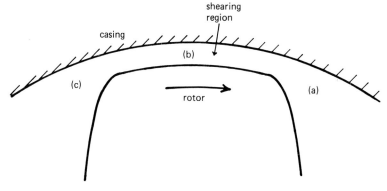

Figure 12.10   Simplified flow of rotor past wall

conventional mixing equipment[4,5,7,8]. For example, planar extension can be achieved in a four roll arrangement[7], see *Figure 12.11*. However, uniaxial extension is more difficult to achieve and recent studies in rheometry using convergent channels with lubricated walls have pursued this problem[7]. It is clearly not possible to continuously elongate material and in any real flow the elongational (accelerating) region must be followed by a compression (decelerating) zone. It is therefore important to ensure that the benefits in extension are not subsequently lost in compression.

Although these comments relating the idealized flows to real blending

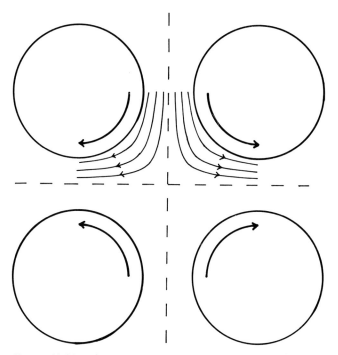

Figure 12.11   Planar extension in four-roll arrangement

equipment appear to be discouraging, it is the concepts which have been laid down through the simple flows which are important if mixer design is to be improved and optimized. For laminar blending the role of shear and the benefits of redistribution are important. The implications of orientation and the potential of elongational flows must be understood by the design engineer.

12.4   Distributive mixing

Blending by spacial redistribution due to 'slicing and replacing' is illustrated by the idealized mechanisms shown in *Figure 12.12*. In the situations illustrated there is an increase in interfacial area (and a corresponding reduction in striation thickness, from equation (12.8)) between the two

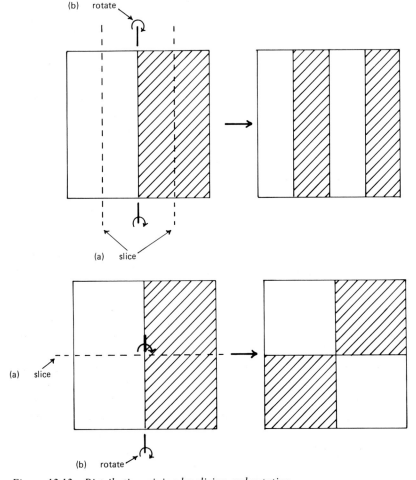

Figure 12.12   Distributive mixing by slicing and rotating

components. If such distributive mixing mechanisms can be applied repeatedly there will be an improvement in mixture quality after each individual action.

It is clear from *Figure 12.8* that the 'slicing and replacing' mechanism occurs in isolation and is not accompanied by either shearing or extension, both of which would deform the fluid elements. However, it is possible to create this distributive mixing action during laminar flow over appropriately designed elements inserted in tubes. For example, the Kenics-Chemineer in-line static mixer achieves laminar blending in this way (see Chapter 7, *Figures 7.7* and *7.8*, and also Chapter 13). For this particular design of in-line static unit the striation thickness is reduced by each element in turn and for the idealized mechanism shown in *Figure 7.8* the following equation relates striation thickness δ to the number of elements n and the tube diameter D:

$$\delta/D = 2^{-n} \tag{12.21}$$

Other in-line mixers generate 'slicing and replacing' actions which are similar in principle but different in detailed geometry and are not therefore described by equation (12.21), see Chapter 13. However, there is usually a power relationship, as in equation (12.21), between the striation thickness and the number of elements[9].

It is interesting to note that equation (12.21) and equivalent expressions for other mixers predict that the laminar mixing action is essentially independent of liquid flow rate and also of rheological properties of the liquid. However, care must be taken in applying these considerations to viscoelastic liquids which tend to resist this type of mixing action[10].

Whilst in-line static mixers act on the distributive mixing principle for their effectiveness in laminar blending, it has been mentioned earlier in this chapter that laminar shear mixing can be enhanced by redistribution, see *Figure 12.7*. Thus redistribution of the two components to be mixed between periods of laminar shear or elongation is very beneficial.

For example, in the mixer shown in *Figure 12.13*, the region between the rotor and the casing will have regions of high extension and shear. It is in this region that the greatest mixing action occurs. However, the relatively slow movements in the remainder of the mixing chamber serve to redistribute fluid elements before they next enter the high shear and elongational zone.

Many mixers handling highly viscous liquids (e.g. kneaders, internal mixers, planetary mixers, helical ribbon units, roll mills) have very complex overall flow patterns. However, they all incorporate regions of high shear (usually with elongational flows) and zones where the fluids are redistributed. It is often not possible to analyse in a quantitative manner these operations using the idealized mechanisms illustrated in this chapter. Nevertheless, an understanding of the roles of shear, extension and redistribution is essential if superior designs are to be evaluated in the future. Some further comments on the possibility of analysing real mixing machinery using fundamental principles are presented in a later section of this chapter.

From the discussion so far it is clear that when a mixer has zones of high

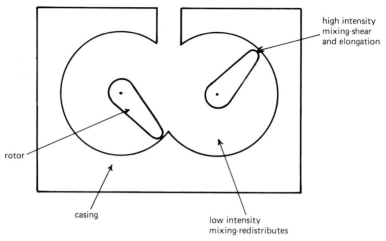

high intensity
mixing-shear
and elongation

rotor

casing

low intensity
mixing-redistributes

Figure 12.13   Redistribution in mixers

shear and/or extension, the components should pass through these zones as often as possible. The role of the redistributive zones is to present the components to the high shear/extensional regions with a fresh orientation for each pass. All circulation paths in the mixer should pass through the high shear/extensional zones. Furthermore, it will be advantageous if the residence time in these intense regions is not too prolonged prior to a redistribution step.

So far the analyses in this chapter have centred upon laminar blending operations. In the following section the stresses generated in laminar flows will be considered and related to the dispersive mixing in solid/liquid and liquid/liquid systems.

12.5   Dispersive mixing in laminar flows

For mixing duties involving the dispersion of fine solids into liquids it is usually necessary to break up clusters of particles which may be held together by either strong or weak inter-particle forces. Often the solid/liquid system will be highly viscous and usually the mixture will exhibit marked non-Newtonian characteristics. In such cases the break-up of the clusters will be brought about by the stresses set up in the laminar flow of the mixture. Rupture will occur when the stresses exceed the bonding forces between the particles[4,7,11,12] and for present purposes it will be assumed that break-up by collision and impact is not significant because of the viscous nature of the mix.

For the creation of fine emulsions similar considerations will also be relevant. Often the emulsion will be viscous and non-Newtonian in rheology and therefore will be processed under laminar flow conditions. Again, neglecting effects such as impact, it will be the stresses set up during this laminar flow which will deform and eventually break up large droplets to create a finer emulsion[7,13,14]. As mixing proceeds a state will be reached when, as the droplet size is reduced, surface tension forces become sufficiently strong to withstand the deforming forces generated in the laminar flow without rupture occurring.

The dispersive mixing processes described above for viscous solid/liquid and liquid/liquid duties will be carried out in a wide range of mills, internal mixers and emulsifiers. A detailed analysis of such equipment to evaluate the stress fields which are responsible for the mixing action is very complex. In general the flow will involve regions of high shear and probably elongation together with zones where redistribution can take place. Furthermore, the rheology of the mixtures will usually exhibit an apparent viscosity which is dependent upon shear rate. It is not uncommon for fine solid/liquid and liquid/liquid systems to exhibit viscoelasticity and thixotropy. Therefore it is not feasible in this section to present more than a very simple assessment of some idealized flow fields, e.g. elementary shear or elongational flows. Further considerations of the dispersive processes are given in Chapters 6 and 15.

It has been stated many times[7,12,13,15] that elongational flows are more effective for dispersion than simple shearing flows. For example, it has been demonstrated[16] that a drop of high viscosity relative to the suspension medium would not rupture in simple shear but could be easily elongated and burst into droplets in planar extensional flow. In addition, for rigid, spherical particles in contact in an idealized dumb-bell arrangement, the maximum separating force in elongational flow is twice that in shear flow at the same rate of deformation in a Newtonian liquid[12]. These simplified concepts point to the effectiveness of elongational zones in mixing equipment[13,15]. However, whilst it is a relatively simple matter to design equipment with zones of extremely high shear rate by the use of high relative speeds and small gaps between moving and stationary parts, it is not easy to generate high rates of extension[12].

From the foregoing discussion it is clear that effective dispersive mixers will exhibit regions of high shear and probably elongation. This will generate high stresses for the dispersive process and will result in correspondingly high power inputs. Thus there is likely to be significant heat generation resulting in a temperature rise of the mixture. Analyses of dispersive mixing processes based on isothermal considerations can give misleading results[11]. Furthermore, in many processes the rheology of the mix will change as the dispersion mechanisms proceed. Detailed studies of dispersive mixing should take account of such factors.

12.6   Applications to blending and dispersing equipment

Single-screw extruders are used extensively in the plastics industry for melting polymers, incorporating additives (for example fillers, colouring pigments, stabilizers) and delivering a homogenized melt at high pressure to a die prior to shaping processes. The granular or powdered feed from a hopper is fed onto the screw and there is an initial region, the 'solids conveying zone', before any melting commences. This is followed by a 'melting zone' where the melting of the polymer occurs gradually. After this melting region the helical channel between the screw and the barrel is filled completely with melt and in the final 'metering zone' the melt is pumped to the die. Here the pressure increases as the die is approached. Thus in the metering section the relative motion between the screw and the barrel creates a 'drag flow' towards the die. This is opposed by the 'pressure flow' from the die towards the hopper[11,12,17,18]. The net output of the extruder is the difference between the drag flow and the pressure flow.

The mixing which occurs in single screw extruders is often very important in determining the product quality of the extrudate and most analyses have focused upon the laminar shear mixing which takes place in the metering zone. In such studies the mixture quality is inferred from the weighted average total strain, WATS[19]. This approach was outlined earlier in section 12.2 for the simple case of laminar flow in a pipe. Following this calculation procedure it is necessary to evaluate the velocity profile in the helical channel and this can then be used to obtain the local shear rates and residence times. The weighted average total strain is then calculated by averaging the product of shear rate and time (i.e. the strain) across the flow.

Analyses based on this approach have been presented for screws of constant channel depth, tapered screws and for Newtonian and non-Newtonian rheological properties[11,12,17-19]. Typical results are sketched in *Figure 12.14* and these indicate that higher average strains and hence mixture qualities are achieved when the pressure flow becomes significant compared with the drag flow. In the limit when these flows are equal and opposite, the situation corresponds to a completely closed die with no output and hence a maximum die pressure. At the other extreme, as the ratio of pressure to drag flow decreases to zero, the die pressure falls, the throughput is increased but the mixture quality becomes worse. Thus it is possible to increase mixture quality in extruders (inferred from a high value of the WATS) by restricting the output whilst maintaining the same screw speed, thus adding more power per unit mass of polymer. This is not an unexpected result!

Such calculations of the weighted average total strain for the metering sections of single screw extruders must be used in conjunction with a knowledge of the mixing which occurs in the solids conveying and melting zones where mixing effects can be important. For example, if the polymer and additive particles in the hopper are dissimilar in size and/or density it is possible that segregation, rather than mixing, may occur in the solids conveying zone. In addition, there is evidence[20] that the mixing which occurs in the melting zone can be more significant than that in the metering zone. At the present time there is no reliable model available for the mixing which accompanies melting.

The studies of WATS are relevant to the laminar blending which occurs in the shearing motion in the helical channel formed between the barrel and the screw. There is no significant extensional flow nor redistribution and as a result the mixture quality achieved in some single screw machines is

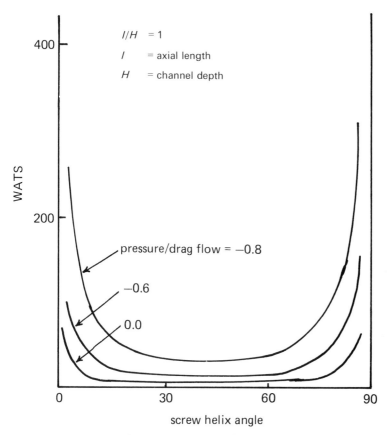

Figure 12.14   Weighted average total strain (WATS) for a single-screw extruder[19]

insufficient to meet the required duties. In such cases an improvement can be achieved by the use of devices which either impart a degree of redistribution and/or force the material through regions of very high shear. Thus various modifications to the simple screw design are possible using mixing pins, interrupted flights[21] etc., see *Figure 12.15*. In addition a variety of additional sections, e.g. the Cavity Transfer Mixer[22], are available, see *Figure 7.17*.

The additional geometrical complexities of these devices for improving mixture quality make any analyses very difficult. However, the principles of redistribution and increased shear rate have been highlighted in the simplified situations presented in previous sections of this chapter. Thus redistribution will enhance the laminar shearing flow which is causing blending[23] and increased shear rates will generate higher stresses which may be necessary to achieve any required dispersive mixing. For a simple single screw machine the only region of extremely high shear is in the small clearance region between the tips of the flights of the screw and the barrel surface.

In the case of twin screw extruders many designs are available including co-rotating, counter-rotating, intermeshing and non-intermeshing

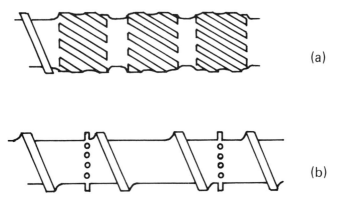

Figure 12.15   Some additional mixing elements for single-screw extruders

units[4,12,21,24,25]. As a result of this wide range of geometries it is not possible to present a unified treatment of mixing in twin screw machines. However, for any given design it is possible to examine the residence time distributions and total strain distributions as the mix moves through regions of high and low shear rate. There is also a redistributive mixing component to take into account in these devices as material moves between low and high shear regions. It is generally considered that twin screw units are capable of providing better mixing and dispersion than single screw extruders because of these high shear regions and the redistributive action.

Many dispersive mixing duties where fine solids are incorporated into viscous liquids are carried out in roll mills. The important region in such machinery is the nip region in the small gap between two rotating rolls, see *Figure 12.16*, and there is a clear resemblance between roll mills, roll coating machinery[26] and calenders[11,12]. Since the gap between the rolls is small compared with the radius of the roll, the analysis of the flow field in the nip region is usually based upon the lubrication approximation. Using this approach it is possible to evaluate the velocity and the stress fields. Unequal roll speeds and purely viscous non-Newtonian behaviour can be taken into account[11,12,26].

In the text of Tadmor and Gogos[12] the dispersive mixing capability of a roll mill is evaluated by comparing the stress distribution in the nip region with the critical stress level which must be exceeded to separate clusters of particles. This is a useful approach but it is based on isothermal conditions with purely viscous flow behaviour. More complex situations involving non-isothermal operation with viscoelastic rheology require sophisticated numerical techniques to solve the governing equations[15]. Considering rubber mixing using roll mills, Funt[15] states that the dispersion of additives in a roll mill is dominated by the shear deformation in the nip region rather than the extensional deformation. However, the stresses due to the stretching flow can exceed the frictional force between the rubber and the metal surface of the roll and in such cases the rubber will not then flow into the nip.

One of the most commonly used mixers in the plastics and rubber industries

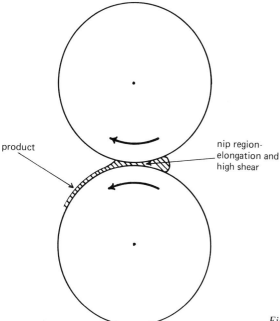

product

nip region-
elongation and
high shear

Figure 12.16   Roll mill

is the internal mixer, see *Figure 12.13*. Here the mixing chamber has regions of high shear with some elongation in the close clearance region between the rotors and the casing and in some designs between the two rotors. The remaining volume in the chamber has a low mixing intensity but, as discussed earlier, this serves to redistribute material prior to a further pass through the intense mixing zones.

Internal mixers have been used for many years and over this time there have been a large number of studies of their mixing characteristics[4,11,12,15]. These range from considerations of simplified geometries[11,12] to more realistic geometries involving non-isothermal operation and viscoelastic effects[15]. A complete and generalized theory for the mixing characteristics of internal mixers is yet to be developed.

In-line static mixing units operate by a redistributive mixing action when used for laminar blending operations. As discussed in section 12.4 this mechanism is essentially independent of flow rate and rheology. However, for dispersion duties it is the stress field which is important and here both flow rate and viscous properties will be critical. Further consideration of these mixing devices is given in Chapter 13.

12.7   Assessment of mixture quality

In order to assess the efficiency of mixing equipment which is used for laminar blending and dispersion duties it is necessary to obtain a measure of the mixture quality of the product. However, such measurements pose severe problems when systems are well mixed. Many mixing indices have been proposed[27], perhaps the most widely quoted being the scale and intensity of segregation, see Chapter 1. The scale of segregation is clearly related to the striation thickness (and hence interfacial area between components) which has been used in this present chapter to assess laminar mixing in shear and extension and also in redistribution processes. However, as mixing approaches completion, striation thicknesses reduce (and surface area increases) to the point where diffusive mixing processes become significant. At this stage the intensity of segregation becomes a relevant parameter.

Practical measurements of the scale and intensity of segregation for well-mixed samples are beyond the scope of the routine analysis which is required by industry. Therefore it is usual to infer the mixture quality of products from laminar mixers by measurements of rheological properties, colour, etc. This difficulty of assessing the process result is one of the major problems which impedes our progress in laminar mixing operations. The advent of sophisticated image analysis equipment capable of scanning rapidly over many samples with a fine resolution should be helpful in future studies of mixture quality for certain blending or dispersing duties with very viscous materials.

Notation

l length
l' length after deformation
L length
n number of rotations, number of elements
r radial position
R tube radius
S area
S_0 initial area
t time
v velocity
V volume
V' plate velocity
w width
w' width after deformation
δ striation thickness
γ strain
$\dot{\gamma}$ shear rate

Subscripts

E   uniaxial extension
PE   planar extension
S   shear
1   major component
2   minor component

References

1   MIDDLEMAN, S. (1977) *Fundamentals of Polymer Processing*, McGraw-Hill Inc.
2   SPENCER, R. S. and WILEY, R. M. (1951) *J. Coll. Sci.*, **6**, 133.
3   MOHR, W.D., SAXTON, R. L. and JEPSON, C. H. (1957) *Ind. Eng. Chem.*, **49**, 1855.
4   HOLD, P. (1982) *Adv. Polymer Tech.*, **2**, 141 and 197.
5   ERWIN, L. (1978) *Polymer Eng. Sci.*, **18**, 738.
6   MOHR, W. D. (1974) in *Processing of Thermoplastic Materials* (Bernhardt, E. C., ed.), Robert E. Kriger Publ. Co., New York.
7   CHENG, D. C-H. (April 1979) *Proc. 3rd European Conference on Mixing*, BHRA Fluid Engineering, Cranfield, Bedford, England, volume 1, p. 73.
8   ERWIN, L. (1978) *Polymer Eng. Sci.*, **18**, 1044.
9   STREIFF, F. A. (April 1979) *Proc. 3rd European Conference on Mixing*, BHRA Fluid Engineering, Cranfield, Bedford, England, volume 1, p. 171.
10 WILKINSON, W. L. and CLIFF, M. J. (1977) *Proc. 3rd European Conference on Mixing*, BHRA Fluid Engineering, Cranfield, Bedford, England, Paper A2.
11 McKELVEY, J. M. (1962) *Polymer Processing*, John Wiley and Sons Inc.
12 TADMOR, Z. and GOGOS, C. G. (1979) *Principles of Polymer Processing*, John Wiley and Sons Inc.
13 VAN DEN TEMPLE, M. (February 1977) *The Chemical Engineer*, 95.
14 MASON, G. (1974) *Rheol. Acta*, **13**, 648.
15 FUNT, J. M. (1980) *Rubber Chem. Technol.*, **53**, 772.
16 RUMSCHEIDT, F. D. and MASON, S. G. (1961) *J. Coll. Sci.*, **16**, 238
17 SCHOTT, N. R. (1979) in *Science and Technology of Polymer Processing* (Suh, N. P. and Sung, N-H., eds.), MIT Press.
18 BIGG, D. M. (1979) in *Science and Technology of Polymer Processing* (Suh, N. P. and Sung, N-H., eds.), MIT Press.
19 PINTO, G. and TADMOR, Z. (1970) *Polymer Eng. Sci.*, **10**, 279.
20 EDWARDS, M. F., GOKBORA, M. N. and ZAYADINE, K. Y. (May 1982) *Polymer Extrusion II*, P.R.I., London, Paper 17.
21 SWANBOROUGH, A. (July 1980) *The Chemical Engineer*, 482.
22 GALE, M. (May 1982) *Polymer Extrusion II*, P.R.I., London, Paper 18.
23 ERWIN, L. (1978) *Polymer Eng. Sci.*, **18**, 572.
24 JANSEN, L. P. B. M. (1978) *Twin Screw Extrusion*, Elsevier, Amsterdam.
25 SCHENKEL, G. (1966). *Plastics Extrusion Technology*, Iliffe Books Ltd., London.
26 BENKREIRA, H., EDWARDS, M. F. and WILKINSON, W. L. (1984) *J. Non-Newtonian Fluid Mech.*, **14**, 377.
27 SCHOFIELD, C. (1974) *First European Conference on Mixing and Centrifugal Separation*, BHRA Fluid Engineering, Cranfield, Bedford, England, Paper C1.

Chapter 13

Static mixers

J C Godfrey

Postgraduate School of Chemical Engineering, University of Bradford

13.1   Introduction

Static mixers, often referred to as motionless mixers, are in-line mixing devices which consist of mixing elements inserted in a length of pipe. There are a variety of element designs available from the various manufacturers but all are stationary in use. The energy for mixing is derived from the pressure loss incurred as the process fluids flow through the mixing elements and additional pumping energy is necessary over and above that normally needed for pumping requirements. The number of elements required in any application is dependent on the difficulty of the mixing duty, more elements being necessary for difficult tasks.

The static mixer presents an alternative to the more traditional agitated vessel. It differs from the mixing vessel in that it is used in continuous processes although it is possible that a static mixer could be incorporated in a batch mixing operation. The static mixer has only been available commercially for a relatively short time but has already been used in a large number of mixing applications.

The range of applications quoted by static mixer manufacturers is very wide and includes both laminar and turbulent processes. Laminar flow mixing proceeds by a combination of flow division and flow re-orientation while, in turbulent flow, the elements cause a higher level of turbulence than in the corresponding empty pipe. The mixers have been used in processes requiring blending, reaction, dispersion, heat transfer and mass transfer.

In addition to the wide range of applications there is also a wide range of designs and suppliers, just as there is for agitated tank mixing equipment, and the same difficulties of comparison and selection of equipment for a particular task arise. As with agitated vessels, a satisfactory process result can often be obtained even if the best design for the job has not been selected. Manufacturers and their agents are often able to supply adequate equipment for a particular task from their records of experience in similar projects. Predictive methods for mixture quality are often claimed but the understanding of the basics which govern static mixer performance is still quite limited.

The flow of fluids in all types of mixer is complex and difficult to describe quantitatively. Consequently, fundamental methods of predicting mixing rate or power consumption are both complex and inaccurate at the present time. However, in this sense, the understanding of the science of mixing is poor for all types of equipment, both static mixers and agitated vessels; there is a great need for design equations and techniques for comparison of performance. The main considerations from the scientific point of view are rate of mixing, power requirements and efficiency, while from the point of view of application, design methods and scale-up criteria are also necessary.

In most mixing investigations it has been found that the prediction of power requirement can usually be dealt with satisfactorily by empirical relationships, usually based on dimensionless groups, derived from experimental results. The description of mixing rate is usually much more difficult, as are the experiments, while scale-up has presented problems in most areas of mixing. For the static mixer the major variables requiring attention in any study are flow rate and viscosity, bearing in mind that in many processes there will be two or more fluids involved and that each inlet stream may have a different flow rate and viscosity.

At the present time static mixer technology is broadly similar to agitated vessel technology with the two areas having many problems in common. Many installations are optimized by trial and error with much depending on previous experience and wide safety margins. Manufacturers of static mixers have attempted to counter the long experience of use of agitated vessels by a considerable involvement in the collecting and publishing of performance data but much still remains to be done.

13.1.1   Mixer types

Static mixers have been in use in one form or another for a considerable time. Perhaps the most common form is the packed column as used in distillation, gas/liquid contacting and liquid/liquid contacting. These traditional chemical engineering operations are, in nearly all cases, countercurrent in nature with the light phase flowing up the column and the heavy phase down. However, there are co-current applications[1] and it is usually co-current mixing that static mixers are concerned with. An early in-line mixer was described by Taber and Hawkins[2] and used a length of pipe packed with metal turnings to mix viscous resins. The use of random packing material was thought unlikely to lead to predictable mixing performance and a number of more ordered packings were developed to improve this aspect of the in-line mixer characteristic.

Nobel[3] described a mixing element which divided a tube flow into two annuli. Each annulus is divided into a number of radial subsections and the mixing action is achieved by splitting and relative rotation between two annuli. The process of division and rotation is also incorporated in the design of Schippers[4] which uses short mixing elements with rectangular ducts. A more sophisticated use of the same principle is described by Ingles[5] where the mixing element is composed of a number of groups of four circular ducts. Each of the four ducts rotates the liquid stream by 90° with the outlet corresponding in position with the inlet of the adjacent duct. Plates were designed with various duct diameters, the larger diameter being used in the early stages of mixing with smaller diameters following.

One of the first studies of mixing quality in static mixing processes was conducted with this design by Hall and Godfrey[6]. In this work mixture patterns were obtained from various combinations of mixing elements using coloured clay pastes as the test fluids. Various mixing indices were reviewed and the data presented in the form of an autocorrelation function as a function of distance. These data have subsequently been further analysed by Cooke and Bridgwater[7] and this analysis is of particular interest with regard to the importance of sample size. A simple form of the duct type static mixer was produced by Dow Badische. The design by Harder[8], known as the Interfacial Surface Generator (ISG), has only four ducts per element but uses these to rotate and translate fluid elements with respect to one another. The inlets to the ducts lie on a diameter of the mixer element and ducts pass through the element appearing on a diameter at right angles to the first with the two outer ducts emerging as the inner pair and vice versa.

A multichannel, low pressure drop mixing element was developed by Sulzer Bros.[9] using their techniques developed for producing multiple element packing for gas/liquid contacting. The intention of the design is to split the fluid into individual streams which meet other streams as they flow transversely through the element. Each element mixes principally in two dimensions and elements are aligned at 90° to their neighbours to give three-dimensional mixing. There are several versions of the Sulzer mixing element available to cover the range from laminar to turbulent mixing, *Figure 13.1*.

Another low pressure drop design is the Kenics mixer, first described by Armeniades[10]. This design also consists of a series of mixing elements aligned at 90°, each element being a short helix of one and a half tube diameters in length. Each element has a twist of 180° and right-hand and left-hand elements are arranged alternately in a tube, *Figure 13.2*. It has been proposed[11] that flow division, flow reversal and radial mixing all contribute to the performance of the Kenics mixer.

In addition to the Sulzer and Kenics designs, there are a number of other low pressure drop designs. Wymbs Engineering offer laminar flow (HV) and turbulent flow (LV) designs, *Figure 13.3*. Charles Ross & Son have two turbulent flow designs (LPD and LLPD, *Figure 13.4*) in addition to the ISG. Lightnin Mixers have an element design which allows a variety of assemblies for different applications, *Figure 13.5*. A large number of static mixers and aspects of their performance have been described by Pahl and Muschelknautz[12].

13.2   Laminar mixing

The laminar distribution process usually produces a mixing effect by random rearrangement of fluid regions in an uncontrolled flow process. This process

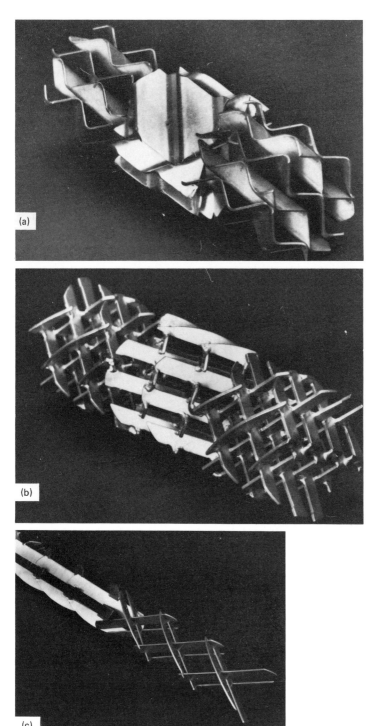

Figure 13.1   Sulzer mixing elements for various applications. (a) SMV for turbulent flow; (b) SMX for laminar flow; (c) SMXL for heat transfer for viscous fluids; for maximum pressure drop (Courtesy, Sulzer UK)

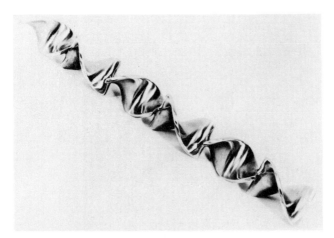

Figure 13.2   Assembly of Kenics states mixer elements (Courtesy, Chemineer-Kenics, UK)

Figure 13.3   Assemblies of Etoflo mixing elements: (left) HV-type for high viscosity and laminar flow; (right) LV-type for gases or low-viscosity fluids in turbulent flow (Courtesy, Wymbs Engineering Ltd)

alone will reduce the volume of individual regions of non-uniformity by continual rearrangement of the mixer contents. The rearrangement process is a major aspect of the performance of static mixers.

Laminar shear dispersion is required in particular mixing applications where the ultimate size of particles or drops of a dispersed phase is to be reduced by the mixing process. In laminar mixing the shear stresses available for dispersion are usually greater when the viscosity of the mix is high. In mixing processes either or both distribution or dispersion may be required and

Figure 13.4 Ross mixing element-assemblies. (a) LPD assembly with 45° baffles for turbulent flow applications; (b) LLPD assembly (in Teflon) with 60° baffles for low-pressure drop turbulent applications (Courtesy, Transkem Plant)

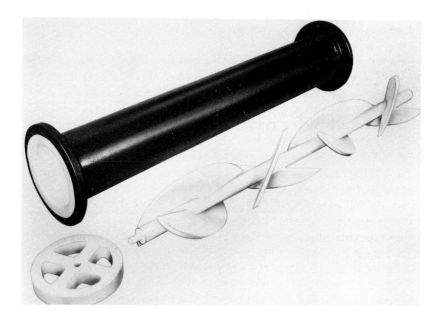

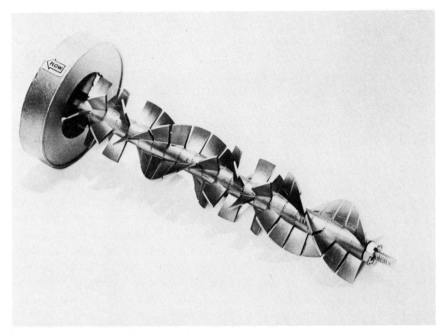

Figure 13.5   Assembly of Lightnin mixer elements—Series 50 Inliner for turbulent flow (Courtesy, Lightnin Mixers and Aerators)

the relative demands of these two aspects need to be considered in mixer selection and design. It needs to be borne in mind that at present a great diversity of agitated tank designs are in use for laminar mixing processes. While there is probably no need for the very large numbers that are available, there are certainly mixers which are very good at specific tasks without being universally applicable.

There seems no reason why a particular static mixer should be well suited to all laminar mixing tasks so there will definitely be some difficulties in choosing the best mixer for a particular task. Smith[13] discusses a modification of the Kenics design for improved contacting of gases with viscous liquids. Streiff[14] presents performance information for three types of Sulzer mixing element and also discusses the relative contributions of 'macro-mixing' and 'micro-mixing' which are relevant to the concepts of distributive and dispersive mixing. In the design of the Lightnin Inliner mixer, flexibility can be achieved by various arrangements of the individual flighted elements with respect to one another. Flexibility in the Sulzer mixing element can be achieved by changing the relative dimensions of channel element and pipe diameter.

In addition to the questions of distributive and dispersive mixing, there are other important variables to be considered in the selection of a suitable mixer—in particular, viscosity ratio and flow ratio. Mixing applications which involve large differences in viscosity or flow rate in the streams are usually significantly more difficult than average. Measurements showing the effect of viscosity ratio and flow ratio are discussed by Wilkinson and Cliff[15]; the effect of viscosity ratio is discussed by Streiff[14].

However, as laminar mixing is usually associated with fluids of high viscosity, it must also be expected that non-Newtonian fluid properties will be encountered in a significant number of cases. As yet, there has been little consideration of the effect of non-Newtonian characteristics on the performance of static mixers, although Wilkinson and Cliff[15] report difficulties with the mixing of viscoelastic polyacrylamide in water solutions and Ottino[31] has attempted to calculate the effect of non-Newtonian properties on static mixer performance.

When the results of any mixing research are examined, it is important to give careful consideration to the techniques used for the measurement of mixing rate. In some instances mixing lengths (or mixing times for agitated vessels) are given and these can be very subjective. A mixing length only has meaning if the quality of mixing achieved at the end of the mixing process is reasonably well defined. Visual methods for deciding when a mixing end point has been achieved are always difficult; the disappearance of the last trace of non-uniformity is not easily followed by eye. Some form of sampling is usually a better approach but the results obtained are very much dependent on the sample size[6,7].

Various techniques have been used in the examination of static mixer performance. Cross-sections of mixtures of materials of two colours have been examined[6,16,18], temperature profiles have been measured for the mixing of a fluid at two temperatures[15] and conductivity profiles[17] for the mixing of two fluid streams identical except for conductivity. The use of the Redox reaction to following mixing has been described[14], photoelectric cells being used to establish the change in colour during a reaction between two streams.

13.2.1   Mixing indices

If a detailed picture of the mixture cross-section can be obtained there are a number of indices which can be used to describe mixedness[6,18]: variance, variance based mixing indices, length of interface, auto-correlation. Variance (σ^2) is extensively used in the definition of mixing indices:

$$\sigma^2 = \frac{\sum (C - \bar{C})^2}{n-1} \tag{13.1}$$

where C is the sample concentration, $\bar{C}$ the mean value of concentration and n the number of samples. The variance based indices of Lacey[19,20] and Danckwerts[21] have been applied to the progress of mixing element by element[6]:

$$I_1 = \sqrt{\left/ \left(\frac{\bar{C}(1 - \bar{C})}{n\sigma^2} \right) \right.} \tag{13.2}$$

$$1 - I_2 = \frac{\sigma^2 - \sigma_R^2}{\sigma_0^2 - \sigma_R^2} \tag{13.3}$$

$$I_s = \frac{\sigma^2}{\sigma_0^2} \tag{13.4}$$

where σ_0^2 is the initial variance before commencement of mixing and σ_R^2 is the variance of a random mixture, all for the same sample size.

Variance is a function of the size of individual samples and sample size is therefore of considerable importance in the analysis of mixture quality. All of the above indices are therefore dependent on sample size and this dependence has been investigated[6,7]. Similarly, the influence of sample size on the estimation of interface length can be made from mixture patterns.

For a complete cross-section of a mixing pattern the auto-correlation coefficient and its relation to sample separation distance provides the most informative basis for mixture analysis. The correlation coefficient for all points separated by distance r is defined:

$$R(r) = \frac{\overline{(C_1 - \bar{C})(C_2 - \bar{C})}}{\sigma^2} \tag{13.5}$$

The plot of coefficient against distance gives a 'correlogram' and examples of this plot are available in *Figure 13.6*[6], and also for commercial designs of static mixer[18]. Further processing of the correlogram data is possible to give the scale of segregation:

$$L_s = \int_0^l R(r)\, dr \tag{13.6}$$

and the similarity coefficient of two correlograms to be compared[6,18] (e.g. mixed and partially mixed conditions):

$$I_{sc} = \frac{\sum_r R(r)_1 R(r)_2}{[\sum_r R(r)_1^2]^{1/2}[\sum_r R(r)_2^2]^{1/2}} \tag{13.7}$$

mixing indices for these calculated values have been proposed[18]:

$$I_3 = \log\left[\frac{L_{so}}{L_{se}}\right] \tag{13.8}$$

$$I_4 = \log\left[\frac{1 - I_{sco}}{1 - I_{sce}}\right] \tag{13.9}$$

Some initial analyses of auto-correlation and related assessments of mixing have been conducted[6,18] and given the capability and low cost of modern image analysis equipment these methods offer a basis for detailed, fine scale analysis of mixture patterns. It is important to re-emphasize the influence of sample size or scanning spot size on such analyses. Ideally, this size should be as small as the smallest sub-division of the best mixed cross-section to be analysed but, failing that, size needs to be standardized for any comparative work.

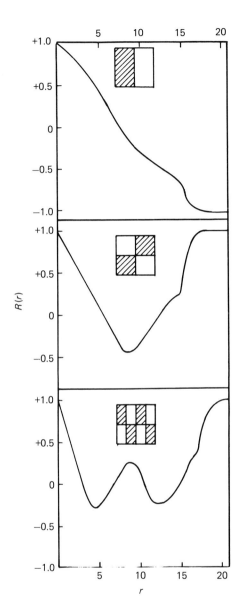

Figure 13.6 Illustration of the relationship between the correlation coefficient (R(r)) and separation distance (r) for non-random mixtures[6]

13.2.2 Mixing rate

The static mixer was almost certainly developed in the first instance with laminar flow blending operations in mind but other characteristics of practical interest have emerged, in particular heat transfer and residence time distribution. Heat transfer coefficients are considerably increased by the use of static mixers and, although the cost in terms of power consumption can be high, this is a considerable advantage in reducing temperature gradients and contact times when processing heat sensitive materials.

The enhanced heat transfer coefficient is due to the increase in radial mixing compared with the empty tube and possibly a contribution due to the 'fin effect' associated with the presence of the metal element—particularly if there is good contact between element and tube wall. It is difficult to relate the heat transfer characteristics of static mixers to empty pipes but low pressure drop mixers (e.g. Kenics, SMX) give increases in heat transfer coefficient of the order of 300% for an increase of pressure drop of the order of 700%. Some simple correlations have been presented for the Kenics mixer[22]

$$(L_t/D_t) < 30: \quad Nu = 4.65 \left[\frac{Re\ Pr\ D_t}{L_t} \right]^{0.33} \tag{13.10}$$

$$(L_t/D_t) > 30: \quad Nu = 1.44 (Re\ Pr)^{0.33} \tag{13.11}$$

and for the Sulzer[23] mixers $(L_t/D_t = 100)$

$$SMX: \quad Nu = 2.6 (Re\ Pr)^{0.35} \tag{13.12}$$

$$SMXL: \quad Nu = 0.98 (Re\ Pr)^{0.38} \tag{13.13}$$

For heat transfer applications the static mixer has an advantage over the empty tube in that it has a much more uniform residence time distribution characteristic, further reducing the possibility of damage to heat sensitive materials. The more uniform residence time characteristic has other advantages, particularly for the design of continuous flow tubular reactors.

Not only is this characteristic a design simplification but it also provides a more controlled environment for reaction with respect to completeness of reaction, minimizing by-products and general uniformity of product. For non-isothermal reactions there is also the opportunity to take advantage of the enhanced heat transfer characteristics and to optimize the reactor design by suitable choice of mixer element type and length.

In all of the work published to date the major concern has been with the description and evaluation of the blending characteristics of static mixers. In early discussions of mixing rate the concept of the number of sub-divisions of flow produced per static mixer element was used as an illustration of mixing rate. This was frequently interpreted as the hypothetical number of 'striations' (n_δ) in a mixture pattern produced from equal volumes of segregated black and white materials. A relationship was proposed[25] for the Kenics mixer.

$$n_\delta = 2^e \tag{13.14}$$

where e is the number of mixer elements.

Wilkinson and Cliff[5] made an experimental study of the production of 'striations' by measuring temperature profiles for the mixing of two fluids of different temperature but used a mixing index incorporating both variance and striation numbers to describe their data:

$$I_S = \left(1 - \frac{1}{n_\delta} \right) (1 - \sigma) \tag{13.15}$$

Charts were produced[25] for the Kenics mixer which related striation thickness and number of mixer elements and some other manufacturers have followed suit. Several relationships have been proposed to describe the number of striations produced by other mixer types:

ISG: $n_\delta = 4^e$ (13.16)

SMV: $n_\delta = m(2m)^{e-1}$ (13.17)

Lightnin: $n_\delta = 3(2)^{e-1}$ (13.18)

Proposals have also been made for scale-up procedures based on the striation concept[26].

Despite wide use in commercial literature, there has been little application of the striation concept to experimental studies. Recent descriptions of mixing rate, both commercial and research, have been presented in terms of the reduction of variation coefficient $(\sigma/\bar{C})$ with number of mixer elements or mixer length. It has been suggested[14] that the variation coefficient is more relevant to the description of commercial mixing processes than the relative standard deviation (σ) or intensity of segregation (σ^2/σ_0^2).

The relationship between the variation coefficient and mixer length is flow ratio dependent, as illustrated by data published for Sulzer mixers[27], this apparently not being the case for the other two indices[14]. There is also the disadvantage, shared by most other variance based indices, of a dependence on sample size which can lead to difficulty in comparing data from different sources.

As a rule of thumb a value of $\sigma/\bar{C} = 0.05$ is taken to estimate the fully mixed conditions although a range of values is being used to specify the requirement of different mixing processes.

In addition to the simple idea of a value of variation coefficient which corresponds to the fully mixed condition, the coefficient has also been used to describe research work on the progress of mixing. In an experimental study conducted at constant flow rate, the sample variance decreases as the fluids travel through the static mixer elements.

At each point along the tube length there is a steady state value of variance which can be determined by sampling and the progress of mixing can be monitored with respect to length. The data obtained are usually presented as $\log(\sigma/\bar{C})$ against 'normalized' length, L_t/D_t. This plot illustrates the exponential nature of the data and the relationship:

$$\sigma/\bar{C} = a \exp(-bL_t/D_t)$$ (13.19)

is used in some cases to describe the data. Plots of data of this sort have been presented by Pahl and Muschelknautz[28] and Alloca[29] for research data and by Sulzer[27] as a performance specification. The Sulzer data are shown in *Figure 13.7* and illustrate the significant effect of mixer design (SMX and SMXL) and the large effect of flow ratio (in terms of $\bar{C}$). The influence of mixer design is also illustrated in the other available data, *Figures 13.8* and *13.9*, there being differences in length up to ten fold to produce the same variation coefficient.

The range of equipment studied in these two works was not the same but

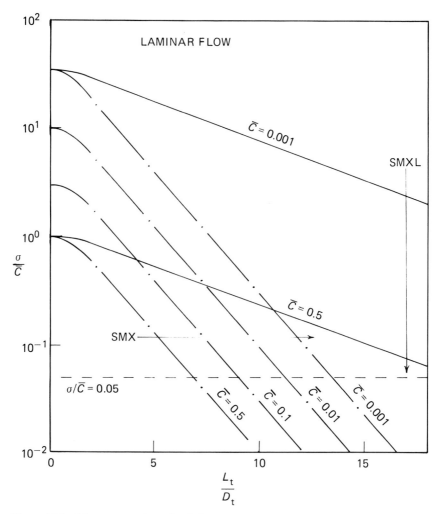

Figure 13.7   Mixing rate chart for Sulzer mixers in terms of variation coefficient
$(\sigma/\bar{C})$ *and dimensionless mixer length* (L_t/D_t)[27]

some designs were included in both studies so some comparison is possible. In both studies the rates of mixing in terms of the variation coefficient and normalized length are in the same order: SMX, Hi-mixer, Kenics, Komax, Lightnin with SMX giving the most rapid rate. This order of rates of mixing does not include any consideration of energy requirements and this point will be discussed in the next section.

In an alternative approach to the analysis of performance of commercial static mixers Heywood, Viney and Stewart[18] have used a range of procedures based on the correlation coefficient to describe mixture cross-sections and hence mixing rate. Once again it is easy to differentiate between the mixing rate of the various designs tested, in this case in terms of mixing per element of static mixer. There are not many mixer types in common with previous work but the

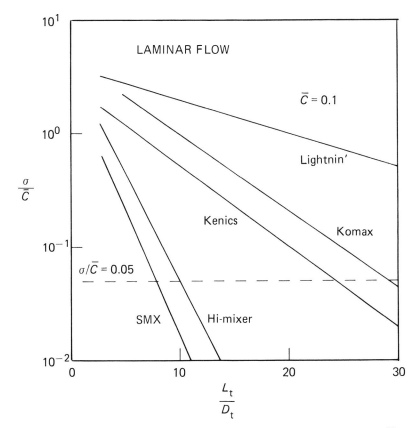

Figure 13.8   Comparison of mixing rates; data of Pahl and Muschalknautz[28]

relationship between the Ross-ISG, SMX and Kenics mixers is as reported elsewhere. Thus, although there are several differences in procedure between the various experimental studies available, there is also a degree of consistency in the results obtained which provides a basis for rating commercial mixers in terms of comparative mixing rate.

13.2.3   Energy and efficiency

An understanding of the energy requirements of static mixers is necessary both with respect to the establishment of installed pressure drop and flow rate requirements and to objective performance comparisons of different mixer types. In laminar flow, the pressure drop–flow rate characteristics of static mixers are simple and analogous to pipe flow.

There is a possible source of confusion in the definitions of friction factor used in different works but in all cases the product of friction factor (ϕ) and Reynolds number (Re) is approximately constant. For the definition of friction factor used by Wilkinson and Cliff[15] and more recently by Pahl and Maschelknautz[28] the empty pipe value of the product $\phi\, Re$ is 64.

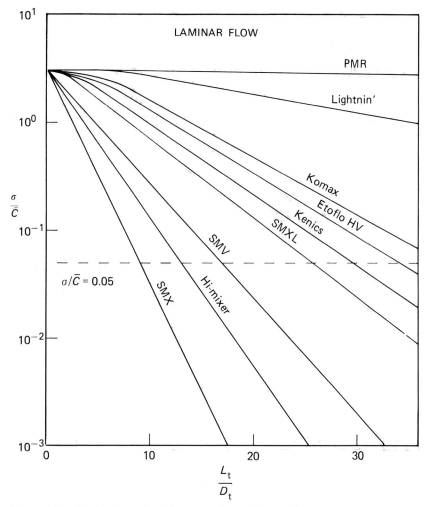

Figure 13.9   Comparison of mixing rates; data of Alloca[29]

$$\phi = \frac{2D_t \Delta P}{L_t \rho v_s^2} \tag{13.20}$$

$$Re = \frac{v_s D_t \rho}{\mu} \tag{13.21}$$

The two dimensionless groups are based on the empty pipe diameter (D_t), including the value of v_s which is a superficial velocity. In principle it only requires one value of ΔP to be measured at one value of v_s to define the product $\phi\, Re$ for any static mixer but a range of measurements can give useful information regarding the slight variability of $\phi\, Re$ and also the upper limit of the laminar flow regime in terms of Re^{28}.

A similar approach is used to describe the pressure drop characteristics of

Sulzer mixers in terms of the Newton number and other dimensionless groups defined in terms of hydraulic diameter (for the SMV element $D_t/D_h \approx 4.2$)

$$Ne_h = \frac{\phi_h D_t}{2\varepsilon_F^2 D_h} \tag{13.22}$$

$$\phi_h = \frac{2D_h \Delta P}{L_t \rho v_s^2} \tag{13.23}$$

Values of $Ne\,Re$ are given for the SMX (≈ 1200) and SMXL (≈ 250) elements, i.e. in terms of an empty pipe Reynolds number. As with the product $\phi\,Re$ the value of $Ne\,Re$ is approximately constant, the empty pipe value being 32.

In more recent work[18,28] a simpler method has been used where the pressure drop characteristics are described as a ratio of mixer pressure drop: empty pipe pressure drop for the same flow rate and diameter:

$$z = \frac{\Delta P_m}{\Delta P_0} \tag{13.24}$$

As both $\phi\,Re$ and $Ne\,Re$ are constant for any mixer or pipe in the laminar regime it follows that z is also constant. Also the relationship between the various parameters is very simple:

$$z = \frac{\Delta P_m}{\Delta P_0} = \frac{(Ne_h\,Re)_m}{(Ne_h\,Re)_0} = \frac{(\phi\,Re)_m}{(\phi\,Re)_0} \tag{13.25}$$

i.e.

$$z = \frac{Ne_h\,Re}{32} = \frac{\phi\,Re}{64} \tag{13.26}$$

Much of the recent data reviewed by Heywood *et al.*[18] are in agreement in these terms. However, as discussed above, all of the parameters used (i.e. z, $Ne\,Re$, $\phi\,Re$) have some small dependence on Reynolds number. This can be accommodated by a simple power law expression[18]:

$$z = A\,Re^B \tag{13.27}$$

where $B \ll 1$, or an Ergun type equation[15]:

$$\phi = \frac{115}{Re} + 0.5 \tag{13.28}$$

or a modified Darcy equation[30]:

$$z = 3.24(1.5 + 0.21\,Re^{1/2}) \tag{13.29}$$

As pressure drop characteristics have been established with a reasonable degree of precision it is possible to re-examine mixing rate data in terms of efficiency. This seems a much more relevant basis for comparison of performance of different mixer types than simple rate data as it has been shown that high mixing rates can be accomplished in some design by very high energy requirements[28].

In most comparisons to date[18,28,29] mixers have been ranked in order of performance according to the criterion chosen. While the subject remains in its semi-quantitative state with respect to experimental data this simple ranking procedure is useful for the reader but a more quantitative approach is worth attempting. To this end it is suggested that the product of the pressure drop ratio (z) and the value of L_t/D_t for $\sigma/\bar{C}=0.05$ would provide a useful quantitative estimate of mixer efficiency. This is less ambiguous than the ratio values used in previous comparative tests and provides a basis for the comparison of data from various sources. The data of Alloca[29] and Pahl and Maschelknautz[28] can be compared in this way as in *Table 13.1*.

The values of L_t/D_t for $\sigma/\bar{C}=0.05$ are consistently larger in the data of Alloca and this suggests that the effective sample size was less than that used by Pahl and Maschelknautz. However, the trends observed are very similar for the seven mixer designs which are covered in both works, especially when it is considered that some of the data are reported[28] to only one significant figure. The same hierarchy of performance in terms of efficiency emerges from both groups.

In this comparison it is convenient that the two sets of data use the same

Table 13.1 Estimates of efficiency in terms of $z\ (L_t/D_t)$ for published laminar flow data

Mixer type	Data of Alloca[29]			Data of Pahl and Muschelknautz[28]		
	z $\left[=\dfrac{Ne_h\,Re}{32}\right]$	L_t/D_t for $\left[\dfrac{\sigma}{\bar{C}}=0.005\right]$	$z\,(L_t/D_t)$	z $\left[=\dfrac{\Delta P_m}{\Delta P_o}\right]$	L_t/D_t for $\left[\dfrac{\sigma}{\bar{C}}=0.05\right]$	$z\,(L_t/D_t)$
Kenics	6.9	29	200	7	25	180
Etoflo HV	5.9	32	190			
Sulzer SMXL	7.7	26	200	7.8*	17	130
Sulzer SMX	39	9	350	38*	8.5	320
Toray Hi-mixer	36	13	470	38	10	380
Bran and Lubbe N-form	17	29	490	17	—	—
Komax	19	38	740	25	29	730
Sulzer SMV	45	18	800			
Lightnin In-liner	9.1	100	910	9	67	600
Ross ISG	300	10	3000	300†	5.5	1700
Prema-technik PMR	16	320	5000			

* Sulzer data: $z=Ne_h\,Re/32$
† Range quoted: 250–300

flow ratio (and therefore $\bar{C}$) and the same value to describe the 'mixed' condition ($\sigma/\bar{C}=0.05$). However, to explore the possibilities of further use of z (L_t/D_t) to describe efficiency, the exponential nature of equation (13.19) allows estimates to be made of L_t/D_t for any value of $\sigma/\bar{C}$. For this estimation it is worth noting the relationship between variation coefficient and relative standard deviation:

$$\frac{\sigma}{\bar{C}} = \frac{\sigma}{\sigma_0} \bigg/ \sqrt{\left(\frac{1}{\bar{C}} - 1\right)} \qquad (13.30)$$

where $\dfrac{\sigma}{\sigma_0} = 1$ when $L_t/D_t = 0$, thus fixing a point on the $\sigma/\bar{C}$ axis and allowing the calculation of $\bar{C}$ from any set of data where this is not specified:

$$\bar{C} = \frac{1}{1 + \left(\dfrac{\sigma}{\bar{C}}\right)_0^2} \qquad (13.31)$$

If it is also assumed that plots of $\sigma/\bar{C}$ are parallel, as suggested by the Sulzer data of *Figure 13.7*, it should also be possible to make estimates for one value of $\bar{C}$ from data obtained at another using equation (13.30). It must be emphasized that all aspects of the comparison in terms of z (L_t/D_t) described above are estimates, particularly as the values of σ obtained experimentally will depend on the equipment used.

Thus a small advance in simple comparisons of data can be made but many other points remain to be investigated. Experimentally some standardization in techniques would be useful to give values of variance which are directly comparable and further study of important variables (flow ratio and flow rates, viscosity and viscosity ratio, scale).

Present use of the plots of $\sigma/\bar{C}$ and L_t/D_t assume that there is no effect of viscosity or flow rate on mixing rate, only on pressure drop. Similarly it is often assumed that the use of L_t/D_t takes account of scale. It also seems worthwhile to look at the more detailed analysis of mixture cross-section by auto-correlation[6] and related indices[18] as this provides a more precise description of the state of mixedness than the determination of variance and also a possibility of exploring the influence of sample size[7]. Other topics needing further investigation are the influence of scale[26] and non-Newtonian fluid properties[15,31].

13.3 Turbulent mixing

For low viscosity process fluids, turbulent pipe flow can usually be established and mixing is then much more quickly and easily achieved than for the laminar mixing of more viscous fluids. In turbulent flow, radial mixing is much stronger and turbulent flow characteristics lead to a rapid reduction in scale of any inhomogeneities present. The characteristics of turbulent flow can also be effectively applied to multiphase processes, such as gas/liquid and liquid/liquid contacting, and it is well known that coefficients of heat or mass transfer are much higher in turbulent flow.

Although the concept of the static mixer was developed with laminar blending in mind there are probably now more applications in turbulent flow. It could be argued that an empty tube could provide a very suitable environment for the turbulent mixing process but a variety of applications have been found for the static mixer. It is often claimed that the static mixer offers a better solution than the empty pipe but the relationship between the two alternatives has not been much explored. For the present it would seem sensible to regard the static mixer as a device for raising the level of turbulence without changing pipe diameter or flow rate.

The static mixer can reduce the contact time required for any particular process at the cost of increased energy consumption. In some cases, e.g. heat transfer, the increase in energy requirements is very large but, equally, mixing energy is not a large cost for many of these processes in the context of overall plant energy requirements.

13.3.1   Mixing rate

The turbulent flow characteristics of the static mixer are even less well documented than those for laminar flow, probably because of the fairly recent emphasis on turbulent applications and the background of development for laminar flow applications. Measurements of mass transfer have been made[32] and data are available for drop size in liquid–liquid dispersions for Kenics, Sulzer and Lightnin mixers.

It has always been difficult to have confidence in the prediction of drop size from correlations and, even though the situation is less complicated than the agitated tank[33], this is also likely to be the case for static mixers. For the Kenics mixer a correlation has been proposed[34] relating drop size to that produced in an empty tube.

$$\frac{(d_{vs})_m}{(d_{vs})_o} = \left[\frac{\phi_o}{\phi_m}\right]^{0.4} \tag{13.32}$$

For the Sulzer mixer a correlation in terms of the Weber number has been developed[35]:

$$\frac{(d_{vs})_m}{D_h} = 0.21\, We^{-0.5}\, Re^{0.15} \tag{13.33}$$

Like all correlations these will only apply for the conditions for which they were developed. For drop dispersions there will be a minimum energy requirement below which a uniform dispersion will not be produced. It is also necessary to consider which of the two phases in a liquid–liquid dispersion is desired to be the drop phase. In very small diameter mixers the wetting relationship between the mixer elements and process liquids may dominate and effectively fix which phase (the non-wetting) will be dispersed[36].

However, since liquid–liquid processes can be very complicated, it is as well to consider the advantages of the static mixer in this application. The combination of uniform turbulence level and uniform residence time characteristics presents a rather simpler operating characteristic than the highly non-uniform conditions existing in agitated tanks. The same would seem to apply to gas–liquid contacting[37] where the agitated tank has a further disadvantage in its variable gas hold-up performance. However, for both gas–liquid and liquid–liquid processes long contact times, i.e. long mixers, limit the range of applicability.

As in the case of laminar flow the static mixer is also used in turbulent flow to enhance blending and heat transfer. A correlation for the Kenics mixer[32]:

$$Nu = 0.078 \, Re^{0.8} \, Pr^{0.33} \tag{13.34}$$

suggests an increase in heat transfer coefficient of a factor of three but the corresponding pressure loss is likely to be one hundred fold or more, so the scope for application is limited.

Data are available for the blending characteristics of the Sulzer SMV mixer[27], see *Figure 13.10*. As with the laminar data for the SMX and SMXL types the data are presented as the relationship between variation coefficient $(\sigma/\bar{C})$ and length (L_t/D_t). The same criterion of mixedness is used, $\sigma/\bar{C} = 0.05$, as

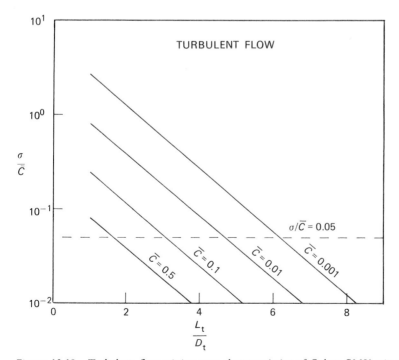

Figure 13.10   Turbulent flow mixing rate characteristics of Sulzer SMV mixer[27]

in laminar flow and a similar dependence on flow ratio ($\bar{C}$) is shown. The data of *Figure 13.10* are for $D_t < 100$ mm where the length of a mixer element used is equivalent to its diameter.

There are other data available, e.g. Kenics[38], but not sufficient to enable the comparison of mixer types that is possible for laminar mixing. The thermal homogenizing of gases is another process where static mixers are used for blending in the turbulent flow regime[39].

13.3.2   Energy requirements

The main value of data describing turbulent energy requirements is in the computation of pressure drop–flow rate characteristics for installed plant but there are also examples of performance evaluation using energy data[39]. As with laminar flow characteristics, although different, those for turbulent flow are relatively simple and easily described in terms of the friction factor–Reynolds number relationship used to describe empty tube.

At high Reynolds number, $Re > 10^3$, the friction factor approaches a constant value. A limited number of comparative[28] values are summarized in *Table 13.2*.

Table 13.2   Turbulent flow friction factors[28]

	Friction factor
Empty tube	0.02
Kenics	3
Sulzer SMV	6–12
Sulzer SMX	12
Toray Hi-mixer	11

Data are also described in terms of the relationship between pressure drop in the static mixer and the equivalent pressure loss in an empty tube. For the Kenics mixer[30] the relationship:

$$\Delta P_m = B K_{OT} \Delta P_o \tag{13.35}$$

is used with B a function of Reynolds number and K_{OT} a function of tube diameter. For the Lightnin mixer the corresponding relationship is:

$$\Delta P_m = 66.5\, Re^{0.086}\, \mu^{0.064}\, \Delta P_o \tag{13.36}$$

13.3.3   Applications

There is a wide range of applications of static mixing equipment. Most of the areas traditionally served by agitated vessels seem to have been considered, including various combinations of gases, liquids and solids. The range of applications and industries is now so extensive that there seems no point in attempting to summarize. There are many useful suggestions for applications in the advertizing material available from manufacturers, much more than is the case for conventional agitated tanks. Of particular interest is the potential for improving existing mixing processes by the addition of static mixers for pre-mixing or post-mixing[24].

There are also many mixing processes where the inherent characteristics of the static mixer are well suited. The static mixer differs from most conventional agitated tank mixer designs in being a continuous rather than a batch mixer and this generally means that the size of a static mixer and the space it requires is less than the corresponding agitated tank. Other advantages are also claimed for the static mixer—flexibility in that one installation could be designed to process a range of different fluids; predictable and repeatable mixing performance; low power requirements; a wide choice of materials of construction. Obviously a number of these points are not yet supported by independent investigation and in some cases the apparent superiority of static mixers is more associated with a fresh approach to mixing problems rather than the inherent characteristics of the mixer itself.

There are a number of potential problems associated with the use of static mixers[40]. An important limitation, which is not very obvious, is that the virtual absence of any axial mixing requires that the feed streams are held at a constant flow ratio—not always easy to achieve in practice. For some applications, e.g. liquid–liquid and gas–liquid dispersions, mixer performance will depend on flow rate and there will be a minimum value below which a dispersion will not be created.

Because of the relatively large amount of wetted area per unit volume in some static mixers, wetting characteristics need to be considered in some processes; particularly liquid–liquid dispersions. There are many small points which need to be considered: possibility of blockage, need for visual observation, the storage and flexibility provided by agitated tanks. Thus, although the static mixer has now been very widely used, it should be considered alongside of its agitated tank counterparts with all of the relevant characteristics in mind.

13.4 Conclusions

There are many designs of static mixer available for use in both laminar and turbulent mixing applications. A very wide range of applicability is claimed by the manufacturers and this seems to be confirmed by the large number and wide range of installations already made. However, from the chemical engineer's point of view the documentation of the static mixer is very poor. The only quantitative calculations that can be performed relate to pressure drop and energy requirements. There seems to be no satisfactory way of estimating mixture quality and it follows that quantitative comparisons of the various designs available is very difficult. Nor is it possible to say at the present time to which applications the static mixer is genuinely suited. This said, it must be admitted that very similar problems exist in design or comparison of agitated tanks.

The main requirement of any item of process equipment is the ability to fulfil the required specification. In this, the technical competence of the manufacturing or installing company is at least as important as inherent characteristics of the equipment being used. A good engineering team can often get better results from a mediocre piece of equipment than *vice versa*; equipment companies are selling technology as much as equipment. It may be that this is the reason that the flow of information on static mixers from the major companies has not increased significantly in recent years.

The static mixer has particular advantages: it is particularly suited to continuous operation; it is potentially of small size; it has no moving parts; there is considerable freedom in the choice of materials of construction. For these reasons alone there must be many operations where a static mixer is the best choice. However, at the moment, the only way a buyer can judge what is the best choice for his process is to balance the cost of the quotation and the reputation of the manufacturer.

Notation

a	constant in equation (13.19)
b	constant in equation (13.19)
C	concentration
$\bar{C}$	mean value of concentration
D_h	hydraulic diameter
D_t	tube diameter
d_{vs}	volume/surface mean diameter
e	number of elements, equations (13.14) to (13.18)
I_1, I_2, I_3 etc.	mixing indices
L_s	scale of segregation
L_{se}	scale of segregation at entrance
L_{so}	scale of segregation at exit
L_t	length of mixer
m	number of channels, equation (13.17)
Ne	Newton number, equation (13.22)
Nu	Nusselt number
n	number of samples
n_δ	number of striations
Pr	Prandtl number
ΔP	pressure drop
Re	Reynolds number
$R(r)$	correlation coefficient
r	distance
v_s	superficial velocity
We	Weber number
z	pressure drop in mixer/pressure drop in open tube

Greek letters

ρ	density
ε_F	voidage, free volume fraction
σ	standard deviation
σ^2	variance
σ_0^2	initial variance
σ_R^2	variance of random mixture
μ	viscosity
ϕ	friction factor, equation (13.20)

Subscripts

h	based on hydraulic diameter
m	mixer
o	open pipe, or initial value

References

1 GODFREY, J. C. and SLATER, M. J. (1978) *Chem. & Ind.* (October 7), 745.
2 TABER, R. E. *et al.* (1959) *US Patent 2,894,732.*
3 NOBEL, L. (1962) *US Patent 3,051,452.*
4 SCHIPPERS, K. H. (1965) *US Patent 3,206,170.*
5 INGLES, O. G. (1963) *International Plastics Engineering*, **3**, 133.
6 HALL, K. R. and GODFREY, J. C. (1965) *A. I. Ch. E.-I. Chem. E. Symposium Series No. 10*, 71.
7 COOKE, M. H. and BRIDGWATER, J. (1977) *Chem. Eng. Sci.*, **32**, 1353
8 HARDER, R. E. *US Patent 3,583,678.*
9 TAUSCHER, W. and SCHUTZ, G. (1973) *Sulzer Technical Review* (2), 1.
10 ARMENIADES, C. D., JOHNSON, W. C. and RAPHAEL, T. (1966) *US Patent 3,286,992.*
11 CHEN, S. J. and MacDONALD, A. R. (1973) *Chem. Eng.*, **80** (March 19), 105.
12 PAHL, M. H. and MUSCHELKNAUTZ, E. (1979) *Chem. Ing. Tech.*, **51**, 347.
13 SMITH, J. M. (1978) *The Chemical Engineer* (388), 827.
14 STREIFF, F. A. (1979) *Third European Conference on Mixing (BHRA)*, 171.
15 WILKINSON, W. L. and CLIFF, M. J. (1977) *Second European Conference on Mixing (BHRA)*, A2: 15.
16 TAUSCHER, W. and MATHYS, P. (1974) *First European Conference on Mixing and Centrifugal Separation (BHRA)*, D3: 25.
17 CHAKRABARTI, A. (1979) PhD thesis, Postgraduate School of Chemical Engineering, University of Bradford.
18 HEYWOOD, N. I., VINEY, L. J. and STEWART, I. W. (1984) *Institution of Chemical Engineers Symposium Series No. 89.*
19 LACEY, P. M. C. (1943) *Trans. I. Chem. E.*, **21**, 53.
20 LACEY, P. M. C. (1954) *J. Appl. Chem.*, **4**, 257.
21 DANCKWERTS, P. V. (1953) *Appl. Sci. Res.*, **A3**, 279.
22 SUNUNU, J. H. *Kenics Technical Report 1002.*
23 Sulzer 'Static Mixing' e/22.07.06—Cgi 50.
24 OLDSHUE, J. Y. (1983) *Fluid Mixing Technology*, McGraw-Hill.
25 CHEN, S. J. (1973) *J. Soc. Cosmet. Chem.*, **24** (September 16), 639.
26 BOSS, J. and CZASTKIEWICZ, W. (1982) *Int. Chem. Eng.*, **22** (2), 362.
27 WILLIAMS, G. D. (1980) *Proc. Eng.* (June), 85.
28 PAHL, M. H. and MUSCHELKNAUTZ, E. (1982) *Int. Chem. Eng.*, **22** (2), 197.
29 ALLOCA, P. T. (1982) *Fiber Producer*, **12** (April).
30 DE VOS, P. (1972) *Inform. Chim.*, **109**, 109.
31 OTTINO, J. M. (1983) *A. I. Ch. E. J.*, **29** (1).
32 MORRIS, W. D. and BENYON, J. (1976) *Ind. Eng. Chem. Proc. Des. Dev.*, **15**, 338.

33 BAIRD, M. H. I. (1977) *Chem. Eng. Sci.*, **32**, 981
34 MIDDLEMAN, S. (1974) *Ind. Eng. Chem. Proc. Des. Dev.*, **13** (1), 78.
35 STREIFF, F. (1977) *Sulzer Technical Review* (3), 108.
36 GODFREY, J. C., SLATER, M. J., WYNN, N. and ZABELKA, M. (1980) *Proc. Int. Solv. Ext. Conf. (ISEC 80)*, Paper 31.
37 GROSZ-ROLL, F., BATTIG, J. and MOSER, F. (1982) *Fourth European Conference on Mixing*, 225.
38 HARTUNG, K. H. and HIBY, J. W. (1972) *Chem. Ing. Tech.*, **44** (18), 1051.
39 TAUSCHER, W. A. and STREIFF, F. A. (1979) *Chem. Eng. Prog.*, **75** (April), 61.
40 PENNY, W. R. (1971) *Chem. Engng.*, **78** (7), 86.

Chapter 14

Mechanical aspects of mixing

R. King

BHRA, Cranfield, Bedford

14.1   Introduction

The process industries have demonstrated quite remarkable growth over the past 30 to 40 years, induced by growing markets, larger and more efficient production units and advances in chemical engineering and allied technologies. Equally remarkable advances were made in product ranges, process technology and environmental impact, largely through a combination of economic and technical factors and new social pressures. During the same period, these process advances were not matched by similar progress in hardware. With a few notable exceptions, there has been little change in the designs of plant and equipment used for carrying out the processes.

Process improvement and plant design evolution should advance together. However, one of the greatest problems facing plant designers and operators is that of scale-up from repeatable and proven, but often very small scale, experiments. In many cases, true scale-up is not achieved and processes frequently run at 10% to 20% below their predicted performance yield. To build a succession of pilot units of gradually increasing size is plainly not cost effective and to be able to scale-up in one step from moderate size to industrial size is considered a matter of urgency. Small scale work can often be extremely instructive, but investigations at near industrial size remove many of the uncertainties. Hall, for example, inventor of the electrolytic process for smelting aluminium, never succeeded in making his original small cell function properly. On larger scale, the problems that had dogged his earlier attempts disappeared 'as if by magic'[1].

In the field of mixing, particularly in agitated vessels, the equations for determining the power consumption of agitators of varying geometry in fluids of widely differing properties have been adapted from work undertaken by Thompson in 1853, who experimented with rotating discs in water[2]. In other attempts to take account of the many different geometrical, chemical and physical properties of components and constituents, complex equations have been generated, often with little theoretical basis.

This chapter discusses the mechanical considerations of mixing processes undertaken in circular cylindrical vessels typically one diameter deep, having an agitator mounted centrally and driven from above through a motor-gearbox arrangement. This is a conventional configuration and there are literally thousands of such vessels in use throughout the process industries, ranging from less than one metre to over four metres in diameter.

The various sections of this chapter are predominantly adaptations of research work undertaken at BHRA and address the impeller, gearbox, seal and shaft dynamics in sequence. To this extent, this chapter represents the philosophy of BHRA and the author, on fluid loading. A design exercise is given, in which two different design criteria are used to determine the shaft size. Advice is given on interpreting results of such exercises. Finally an economic survey is described and examples given of the ramifications of component failure.

14.2   Production of forces

An agitator is used to promote dispersion, heat transfer or mass transfer. It does this by continually renewing the interfaces between the constituents, by the production of turbulence or by the creation of mass circulation. Assume that the matching of agitator to purpose has already been undertaken and that the next steps are to consider the various mechanical aspects associated with supporting and driving the agitator.

It is perhaps appropriate at this stage to review the types of agitator available and used throughout industry. The agitators are shown diagrammatically in *Figure 14.1* and *Table 14.1* describes the various agitators and their functions. However, in the following example, a flat blade paddle (*Figure 14.1*) will be taken as the agitator because, with the exception of the anchor, the comments are sufficiently general to cover the majority of other types of agitator.

Consider a flat plate such as a single blade on an agitator, with a fluid passing about it at a relative velocity v. The dynamic pressure exerted on the plate is theoretically $\frac{1}{2}\rho v^2$ and the theoretical force F_t is this pressure multiplied by the plate area, i.e. $F_t = \frac{1}{2}\rho v^2 A$. The actual force on the plate F_a, is related to the theoretical force by a 'drag coefficient' C_D where $C_D = F_a/F_t$, thus:

$$F_a = C_D \tfrac{1}{2}\rho v^2 A \tag{14.1}$$

Power (P) is force × velocity (force × distance/time), so:

$$P = F_a . v \tag{14.1a}$$

$$P = C_D \tfrac{1}{2}\rho v^3 A \tag{14.2}$$

For a flat paddle agitator, the frontal area A is a combination of the length of the blade and its width, both of which are functions of D; thus, $A \propto D^2$. Velocity is a function of impeller diameter and speed of rotation, thus $v \propto ND$. Combining these gives:

$$P \propto N^3 D^5$$

The 'propeller law', an experimentally proven law derived from dimensional

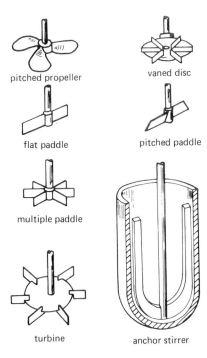

pitched propeller

vaned disc

flat paddle

pitched paddle

multiple paddle

turbine

anchor stirrer

Figure 14.1 Some common types of agitator[4]

analysis, states that the theoretical power requirement P_t is given by

$$P_t = \rho N^3 D^5 \qquad (14.3)$$

and the actual power P_a is related to this through a coefficient, *Po*, i.e.

$$P_a = PoP_t$$

$$P_a = Po\rho N^3 D^5 \qquad (14.4)$$

The power coefficient, *Po*, is also termed the Power Number and, like the drag coefficient, varies with Reynolds number and Froude number. A typical curve of *Po* against Reynolds number for turbine impellers is given in *Figure 14.2*. Similar curves would be obtained from flat blade impellers.

This description of the production of forces and of the subsequent power requirements centres on a single flat blade impeller. If more blades were added, it may at first sight seem that the power would increase in proportion; but this is not so, because the blades are subjected to a diminishing local normal velocity. The wake influence of 'upstream' blades is similar to a sheltering effect and the power increases in proportion to the numbers of blades raised to a factor in the range 0.5–0.8.

Having determined the power requirement of the agitator the next problem is that of designing a suitable method of transmitting this power from the electric or hydraulic motor.

Table 14.1   Impellers in common use[3]

Impeller	Description	Characteristics
Flat paddle	Single flat blade (two arms) usually about 2/3 of vessel diameter long	Usually large low-speed agitators, but capable of producing high intensities of agitation, especially when baffled. Cheap to make, and suitable for construction in timber, plastics, etc.
Turbine	Flat disc with blades attached to periphery. Similar effects are produced with the same number of blades attached directly to a boss (often called multi-blade paddle). Usually 1/3 diameter of vessel (overall)	Generally moderately fast agitators. Particularly suitable for high intensities of agitation and high power inputs. Somewhat more expensive to fabricate than a simple paddle. Very versatile. Recommended for applications where gas dispersion combined with intense agitation is required
Propeller	Marine type propellers, usually less than 1/4 of vessel diameter overall. Wide variety in form possible, from paddles with twisted arms to properly formed marine propellers. No standardization of pitch or number of blades between manufacturers	Usually small high-speed agitators. Cheap to make, but limited to duties where agitation is not very intense, and unsuitable for high viscosities. Much used for relatively small-scale blending operations
Anchor	Agitator following closely the contour of the vessel	A large low-speed agitator, useful where the wall film must be disturbed (e.g. heat transfer to viscous liquids from a jacket), or where build-up of solids on the walls is likely (as in crystallizing). At low speeds has a very gentle action and will prevent caking in the bottom of vessels without vigorous agitation elsewhere. Widely used in enamelled equipment
Vaned disc	A circular disc, usually 1/4–1/2 of vessel diameter with radial vanes 1/6–1/24 of disc diameter deep on its underside	A small or moderately sized high-speed agitator, limited usually to gas dispersion work. The gas is fed under the centre of the disc. It is an effective agitator with or without gas flow, but can be 'flooded' if excessive gas flow is applied, when its agitating effect falls abruptly. The

Table 14.1 (*continued*)

Impeller	Description	Characteristics
		power consumption without gas flow will be much higher than when gas is on, and drives should be adequate to cover the gasless condition
Pitched paddles and turbines	Paddles or turbines with blades twisted to deflect flow upwards or downwards	Similar applications to those of plain paddles and turbines, but larger vertical flow is produced. Power consumption drops with increasing twist, but relatively slightly unless the angle exceeds 45° from the vertical

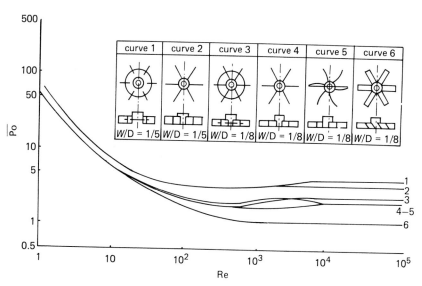

Figure 14.2   Variation of power coefficient P_0 *with Reynolds Number* Re *for six different types of turbine agitators[5]*

14.3   Transmission of power

Before going on to consider details of power transmission it is necessary to make a brief detour into the subject of stress analysis.

Consider a solid shaft of diameter D_{sh} transmitting power P at (variable) speed N rev/s and torque M. The power transmitted is (from equation 14.1a)

$$P = 2\pi N M$$

(14.5)

The torque is constant along the shaft, although the angular displacement of one end relative to the other naturally alters with distance from the driving end. In general, if speed can be varied, low speed is equivalent to high torque and *vice versa* for a given power transmission duty. High torque transmissions almost invariably necessitate large diameter shafts.

The power transmitted depends upon the mixing load at the agitator. As this load is increased so the power supplied by the motor increases. The rated power of the motor should not be exceeded because problems can be encountered in extended running in this condition. Exceeding the rated power will cause the motor to stall or 'pull-out'. Although continual operation in this condition would never be considered, the shaft is designed to accommodate this larger torque if the EEUA[3] approach is followed. In general, it can be taken that the pull-out torque for design purposes is between 1.5 and 2.5 times the rated motor torque depending upon the nature of the mixing duty.

In the absence of any other imposed conditions, the shaft diameter can be sized according to the EEUA method[3]:

$$M_r = P/2\pi N \qquad \qquad \text{(from 14.5)}$$

Let M_r be the rated motor torque and M' be the pull-out torque ($=1.5$ to $2.5 \times M_r$). Then, from standard strength of materials theory: for torsion alone:

$$\frac{M}{J} = \frac{\tau}{r_{sh}} \qquad \qquad (14.6)$$

Substituting

$$J = \frac{\pi}{32} D_{sh}^4 \qquad \qquad (14.7)$$

gives:

$$D_{sh}^3 = \frac{16M'}{\pi\tau} \qquad \qquad (14.8)$$

To determine the working shaft size, τ is replaced by τ_s where τ_s is the safe working shear stress of the shaft material. Typical values of τ_s are given in *Table 14.2*.

Table 14.2 Typical mechanical properties[6]

Material	Ultimate tensile stress ($N\,m^{-2} \times 10^{-8}$)	Yield stress ($N\,m^{-2} \times 10^{-8}$)	0.2% Proof stress ($N\,m^{-2} \times 10^{-8}$)	Allowable working stresses ($N\,m^{-2} \times 10^{-8}$)		
				Bending	Torsion	Combined
0.2% carbon steel	4.0	2.0		1.1	0.5	0.5
Stainless steel	6.5		3.1	2.3	1.1	1.1

The majority of applications have factors of safety of 1.5–3.0; the lower value represents conditions where loads and environments are not over-demanding and can be predicted accurately

In reality, the situation is more complex and a side load could be applied causing pull-out of the motor and bending of the shaft. Reference 3 gives a guideline to shaft design based on a stalling load three-quarters of the way along the paddle blade. Such loading could be caused by the paddle blade running into a solid obstruction (sometimes jokingly referred to as a 'dead cat') in the mixing vessel.

If a side load F_s is given to the shaft at the agitator, the bending moment M_b at the driving end will be the product of force and length of shaft, i.e. $M_b = F_s l_{sh}$. Thus, at the driving end of the shaft, the forces, moments and torques are as shown in *Figure 14.3*.

From standard strength of materials theory:

for bending:

$$\frac{M_b}{I} = \frac{\tau_b}{y} \tag{14.9}$$

for torsion:

$$\frac{M}{J} = \frac{\tau}{r_{sh}} \tag{from 14.6}$$

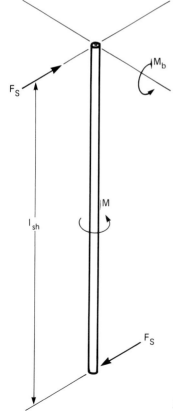

Figure 14.3   Forces on an agitator shaft

The maximum bending stress is at the outer fibres of the shaft ($y = D_{sh}/2$) thus

$$\tau_b = \frac{M_b D_{sh}}{2I} \tag{14.10}$$

For torsion alone, the maximum shear stress also occurs at the outer fibres where $r_{sh} = D_{sh}/2$, thus,

$$\tau = \frac{M D_{sh}}{2J} \tag{14.11}$$

substituting

$$M_b = F_s l_{sh} \qquad I = \frac{\pi}{64} D_{sh}^4 \qquad J = \frac{\pi}{32} D_{sh}^4$$

gives:

$$\tau_b = \frac{32 . M_b}{\pi D_{sh}^3} \tag{14.12}$$

$$\tau = \frac{16M}{\pi D_{sh}^3} \tag{14.13}$$

These stresses can be combined in a variety of ways depending on the criterion of operation. Wide variations exist between answers derived from the different formulae. The maximum strain energy criterion is the most realistic according to at least one author[4]. This results in an equivalent bending moment M_{be} which would produce the maximum strain energy:

$$2M_{be} = M_b + \sqrt{(M_b^2 + M^2)} \tag{14.14}$$

From this, the equivalent axial stress in the outer fibres can be calculated and the direct stress due to the agitator added to it. For the shaft to operate successfully the resulting stress must be less than the yield stress of the shaft material.

The stalling force F_{max} at three-quarters of the distance along a blade is given by:

$$F_{max} = \frac{8M'}{3D} \tag{14.15}$$

This in turn creates a bending moment M_b about the driving end of the shaft given by $M_b = F_{max} l_{sh}$. Recalling equation (14.14) we have the equivalent bending moment M_{be} given by the combination of M_b and M', from which the shaft size can be calculated:

$$D_{sh}^3 = \frac{32M_{be}}{\pi \tau_y} \tag{14.16}$$

where τ_y is the yield stress of the shaft material.

These methods of calculation will be used in the step-by-step worked example in Appendix 1 at the end of this chapter. In the meantime it is necessary to move from the ideal concept of 'steady' forces as depicted by this treatment and introduce fluctuating or time-dependent forces.

14.4   Fluctuating forces, vibrations and fatigue

14.4.1   Fluctuating forces

The previous sections have outlined one design philosophy dealing with the generation of steady forces and the contribution to power and side loading on the agitator shaft. In the majority of practical situations the shaft experiences fluctuating forces of both mechanical and fluid dynamic origin. Some of the fluctuating forces have fairly clearly defined frequencies in a very narrow range of frequencies, whilst others are very broadband.

Forces with clearly defined frequencies originate in out-of-balance mechanical rotating equipment, where the magnitude and frequency of the forces depends on the speed of rotation. Broad-band frequencies can arise from the fluid dynamics of the liquid being stirred, although in certain cases the turbulence contains more power at one frequency than others. Interactions between the mechanical equipment and forces at the dominant frequencies can result in effects similar to those caused by rotating out-of-balance forces. Anyone driving a car with an unbalanced road wheel must have experienced the phenomenon of resonance; as the speed is increased, so the response of the car to the out-of-balance forces becomes more violent. In most cases the violence subsides above a certain speed in much the same way that below a much lower speed the out-of-balance forces are hardly noticeable. The most violent response occurs when one of the frequencies of the out-of-balance forces coincides with a natural frequency of one of the mechanical components or linkages of mechanical components. Where this coincidence of frequencies occurs is defined as resonance.

Sources of time-dependent mechanical forces in mixing vessels are gearboxes (the meshing of irregular teeth) and rumbling of bearings. Time dependent fluid dynamic forces are created by bulk flows passing about solid objects; wakes and vortex shedding lead to a mix of broad-band and discrete frequency forces.

14.4.2   Vibrations

It is not intended to give an intense course on vibrations, but it is essential to fill in a little of the background in order that the remainder of this section can be readily understood.

14.4.2.1   Sway or bending vibrations

Consider an agitator shaft held rigidly at the top, and having an agitator mass m_p at the free end. If the shaft free end were deflected and released it would vibrate at a fixed frequency known as its natural frequency. In general, the shorter the length the higher the natural frequency.

The natural frequency (ω_n) of sway or bending vibrations of the agitator and shaft is given by:

$$\omega_n^2 = \frac{3EI}{m_p l_{sh}^3 + 0.24 m_{sh} l_{sh}^4} \tag{14.17}$$

If $3EI/l_{sh}^3$ is defined as the shaft stiffness K_{sh} and if $m_p + 0.24 m_{sh} l_{sh}$ is defined as the effective mass m_e, then the natural frequency equation reduces to the form characteristic of all vibrational components:

$$\omega_n^2 = \text{stiffness/mass} \tag{14.18}$$

thus

$$\omega_n^2 = K_{sh}/m_e \tag{14.19}$$

If the support is non-rigid, then its flexibility must be incorporated in the overall stiffness K_{sh}. In general, if the support is not infinitely stiff compared with the shaft, the resulting natural frequency will be lowered relative to the ideal rigid support case.

For multiple impellers, intermediate and footstool bearings, the calculation methods would be more complex.

14.4.3   Response to forcing

Resonance was touched on earlier and the purpose of this sub-section is to explain very briefly the selective response mechanism of the agitator-shaft system.

In common with all vibration systems the shaft and agitator will show greater or lesser response (i.e. vibratory motion) to some frequencies than to others depending upon its characteristics and how close the forcing frequency is to the system natural frequency. If the forcing frequency is near zero, then the shaft would deflect as an elastic beam and the deflection would be directly proportional to the forcing load—the static deflection. As the frequency is increased, the deflection increases to some value higher than the static deflection—the dynamic deflection.

Figure 14.4, from ref. 8, shows this effect and the subsequent behaviour very clearly in tests on a laboratory agitator shaft. The dynamic deflection is plotted vertically and frequency is plotted horizontally. At low frequencies the dynamic amplitude is approximately equal to the static amplitude and at very high frequencies the dynamic amplitude approaches zero as all the input energy is converted to kinetic energy at the mass. Between these two frequencies the dynamic response reaches a peak where the forcing frequency is coincident with the natural frequency.

The equivalent response to rotating out-of-balance is shown in *Figure 14.5*.

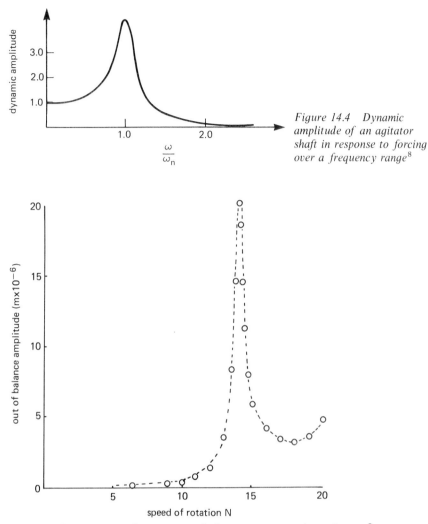

Figure 14.4 Dynamic amplitude of an agitator shaft in response to forcing over a frequency range[8]

Figure 14.5 Response of an agitator shaft over a rotational speed range[8]

Note that static deflection has no meaning since at zero frequency there is no out-of-balance loading.

14.4.4 Whirling of the agitator shaft

The rotation of the agitator shaft is analogous to the rotation of the out-of-balance car wheel referred to previously. For all manufactured components, some degree of imperfection in mass distribution is inevitable and in most cases the effective centre of mass rarely coincides with the geometric centre of the component. Consider the set-up of a slightly bent or eccentric agitator shaft, as shown in *Figure 14.6* adapted from ref. 9.

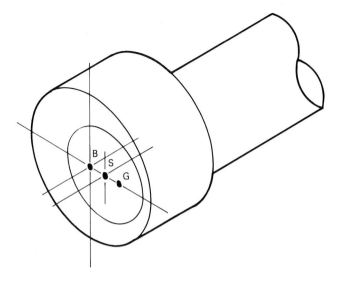

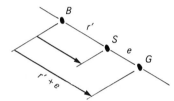

Figure 14.6   A 'bent' shaft

If the eccentricity of the centre of mass is e and the deflection of the shaft is r', then for a vertical shaft we can consider the forces acting on the agitator in terms of these variables. There are two forces: the out-of-balance centrifugal force acting on the centre of mass G, which point is rotating in a circle of radius $(r' + e)$, and the elastic restoring force of the shaft, trying to return S to B. The centrifugal force is defined as the product of effective mass, effective radius and square of rotational speed, i.e. $m_e\omega^2(r' + e)$ and is directed outwards from the centre of rotation. The elastic restoring force depends on the bending stiffness of the shaft (K_{sh}) and is proportional to its deflection, thus the restoring force is $K_{sh}r'$. At a steady whirling speed the two forces must be in equilibrium, there being no others present, i.e.

$$K_{sh}r' = m_e\omega^2 r' + m_e\omega^2 e \tag{14.20}$$

where $\omega = 2\pi N$.

Solving for the shaft deflection r' gives

$$r' = e \cdot \frac{\omega^2}{K_{sh}/m_e - \omega^2} \tag{14.21}$$

It will be remembered that K_{sh}/m_e is the square of the natural frequency of the sway vibrations of the shaft and agitator thus

$$r' = e \frac{\omega^2}{\omega_n^2 - \omega^2} \qquad (14.22)$$

This equation shows how the shaft deflection is related to speed of rotation. For low speeds ($\omega \sim 0$) the radius of whirl ($r' = BS$) is practically zero; at the critical speed, when $\omega = \omega_n$, $r' = BS$ becomes infinite whilst for very much higher speeds B coincides with G. Thus, at very high speeds, the centre of mass remains at rest and the shaft rotates about this point, i.e. the shaft is self-centering. This explains why high speed shafts are usually designed for operation at speeds above the critical whirling speed in order to achieve stability.

However, there are other rather more esoteric unbalancing mechanisms and higher critical speeds above the first or fundamental one discussed here. The reader is referred to Den Hartog[9] for a deeper interpretation of the phenomena.

14.4.5 Fatigue

In many practical cases, stresses are not constant either in magnitude or direction and in such situations the designer must include this departure from direct steady stresses.

The classification of variable stresses is dependent upon the absolute magnitude of the mean stress, which is the average of the maximum and minimum stresses. The variable stress is the increase or decrease in stress above or below this mean stress. If the mean stress is zero the variable stress is classed as a reversed stress; if mean stress is greater than zero and the minimum stress is zero then this is classed as repeated stress; when the mean, maximum and minimum stresses are all positive the variable stress state is classed as fluctuating; lastly, when the mean stress is less than zero and the maximum stress is positive (the minimum stress obviously must be negative) then this is an alternating variable stress.

The endurance limit of a material is determined experimentally by rotating a test specimen whilst it is loaded in bending (i.e. reversed variable stress). The magnitude of the stress is adjusted by altering the loading, and it is found that the number of cycles of reversed stressing that can be sustained by the material is a function of the magnitude of the stressing.

A typical graph is presented in *Figure 14.7* which shows the maximum stress (S') plotted against number of stress reversals (N') necessary to produce fracture. For a typical steel (0.37% annealed carbon steel), the stress decreases rapidly as N' increases but after a million or so reversals there is no longer any appreciable change in maximum stress and the curve levels out at about 2.3×10^8 N m^{-2}. The stress corresponding to such a levelling out is called the *endurance limit* of the material under completely reversed stressing. For the aluminium alloy specimen also shown in *Figure 14.7*, no such well-defined endurance limit is found.

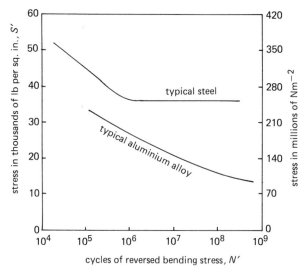

Figure 14.7   Typical fatigue graph (S'–N' curve)

For ferrous materials there is a definite relationship between the endurance limit and the static tensile strength. Let q represent this ratio. For steels, the values of q lie between 0.35 and 0.60 with the majority falling between 0.4 and 0.5. For non-ferrous materials the effective q is lower, being between 0.21 and 0.49 with the majority having a value of 0.37.

Some caution is necessary in applying these criteria because the values obtained from such tests are highly dependent upon surface conditions, the form of the test bar, the atmosphere surrounding the specimen, the temperature, length of test and previous processing of the material. The reader is referred to specialist literature on this topic when in doubt.

The value of the endurance limit obtained in reversed loading will differ from that appropriate to other types of loading. If no test results are available it is possible to construct an approximate design curve for ductile materials by assuming that the endurance limit under reversed loading is $q \times$ the ultimate tensile strength of the material. The vertical axis of *Figure 14.8* shows the variable stress, and the horizontal axis depicts the static or mean stress. The line joining the two end points together (the so-called Goodman or Soderberg line) is the failure stress line.

The diagram shows that the material will fail in direct static loading (horizontal axis crossing point) or under variable stressing (vertical axis crossing point) and that for combination loading, failure occurs at a combination of static stress (horizontal axis) and variable stress (vertical axis). The line parallel with this failure line (the design curve) is the safe stress line, obtained by applying a factor of safety to both the endurance limit and the yield stress. Typically, factors of safety will lie in the range 1.25–3.00 for ordinary design depending upon how well the operating conditions are specified. The selection of factors of safety is a matter for caution and judgement.

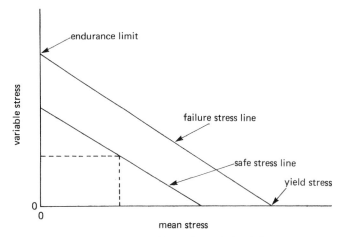

Figure 14.8   Combined variable and steady load design curve

The next stage is the detail design of the blades, bearings and gearbox. It is beyond the scope of this section to cover these points and the reader is referred to Appendix D of ref. 3 for these details.

14.5   Designing to accommodate fluctuating loads

14.5.1   Shafts

In section 14.3 calculations were given to enable agitator shafts to be designed to withstand 'steady' forces using the motor torque characteristic as the item governing agitator shaft sizing. In this section, consideration will be given to accommodating fluctuating loads of fluid-dynamic origin.

Experimental work at BHRA has demonstrated that significant bending loads can be generated during mixing operations. Torque is typically fairly constant. The work has also revealed the dependence of shaft bending loads on the quantity mN/f_n which is a ratio of blade-passing frequency (mN) to shaft whirling frequency (f_n). This has resulted in a fundamentally different approach to the design of shafts, based on a combination of fluid-dynamic loads and torque transmitted[11].

A typical plot of fluctuating loading (expressed in non-dimensional terms as λ) and mN/f_n is given in *Figure 14.9*. This is for a pitched blade impeller. Each impeller type has a characteristic 'loading curve' and tests are under way to determine the curves appropriate to the most commonly used impellers. At present there is no general method of predicting these loadings, and calculation procedures are heavily dependent upon experimentally derived data.

The vertical asix (λ) of *Figure 14.9* is the ratio of $\bar{\mu}/l_{sh}$ to M/D. The quantity $\bar{\mu}$ is the root-mean-square (r.m.s.) bending moment at the gearbox lower bearing. Hence $\bar{\mu}/l_{sh}$ is the apparent r.m.s. bending force applied at the impeller by

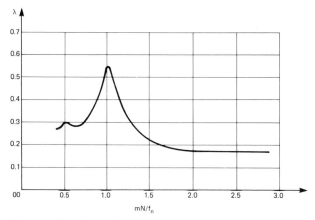

Figure 14.9   Typical design curve for a pitched blade impeller

phenomena of fluid dynamic origin. Similarly M/D is a measure of fluid drag forces producing torsion. Thus λ is the ratio of bending forces to torsion forces, both of which have fluid dynamic origins.

Figure 14.9 shows that the unsteady loading is extremely sensitive to variations in mN/f_n in the region of $mN/f_n = 0.5$ and 1.0, but is a relatively independent function for higher values of mN/f_n. It must be emphasized that these are not true resonance phenomena; at low mN/f_n the shaft rotates at a speed that is far removed from its critical whirling speed f_n. At $mN/f_n = 0.5$ the blade passing frequency is $0.5 \times f_n$; for a four-bladed impeller this would equal 12.5% of the critical speed. Similarly at $mN/f_n = 1.0$, for a four-bladed impeller, the rotational speed is actually only 25% of the critical. Thus, out of balance forces caused by shaft non-uniformity do not contribute to the fluctuating loading, which in this case is therefore fluid dynamic in nature. At high values of mN/f_n the fluid dynamic loading levels off to a constant and for very high values the dynamics of shaft imbalance would need to be considered (see section 14.4.4).

The BHRA calculation procedure[11] generally results in smaller shaft sizes. This has important economic implications as revealed in section 14.6. To determine the influence of the fluctuating fluid dynamic loading the following step-by-step approach is adopted.

Decide on agitator type and make a preliminary guess at rotational speed N and natural frequency f_n. This determines mN/f_n.

From the appropriate design curve read off λ for the assumed value of mN/f_n. Using known values of D, l_{sh} and M (found from equation 14.5) deduce $\bar{\mu}$ from the relationship

$$\bar{\mu} = \lambda . \frac{l_{sh}}{D} . M \tag{14.23}$$

The maximum $\bar{\mu}$ (i.e. $\bar{\mu}_{max}$) likely to be developed is found statistically from

$$\bar{\mu}_{max} = 2.63\bar{\mu} \tag{14.23a}$$

Choose an appropriate working shear stress and determine the minimum shaft diameter D'_{sh} from the relationship:

$$(D'_{sh})^3 = \frac{16}{\pi} \cdot \frac{(\bar{\mu}_{max}^2 + M^2)^{1/2}}{\tau_s} \tag{14.24}$$

Choose the next size up of stocked bar to give the actual shaft diameter, D_{sh}. A check should be made at this stage to ensure that the resonant frequency f_n is close to the guessed value. If there are significant variations, repeat the calculation procedure until close agreement is achieved.

14.5.2   Seals

In those vessels requiring sealing, two common types of seal are employed— packed-gland units and mechanical seals. Selection is usually dictated by application. The cheaper, packed-gland type is often used for low-pressure vessels, whilst at high pressures mechanical seals are preferred. However, when vessel contents are the main consideration, mechanical seals are employed when hazardous materials are under mix, whereas for less hazardous liquors a packed gland is normal.

Reliable sealing is essential when hazardous materials are involved because the escape of toxic process fumes past the seal may necessitate the installation of costly ventilation systems to remove them. In the pharmaceutical industry, for example, the converse problem of maintaining sterility of the process liquor is encountered.

The full details of seal design are beyond the scope of this discussion and are not covered here. However, there are two ways in which the choice of seal type might affect the design of other elements in the drive train and these are now considered. For this, it will be necessary to distinguish between two types of mechanical seal—coaxial and concentric. *Figures 14.10* and *14.11* schematically illustrate these two types; all parts left unshaded rotate with the shaft, although the clamping mechanism is not shown. The shaded elements are attached to the seal housing, which is in turn attached to the vessel.

Both types of seal have a pressurized sealant, usually at 1 to 2 bar above vessel pressure which, along with relatively weak springs, forces a pair of sealing elements against other elements. The particular type of coaxial seal illustrated is a double 'back-to-back' unit, so termed because it consists of two sub-units, whose sealant pressures act away from each other. This means that there is no nett force acting on either the shaft or the seal housing due to the sealant pressure. In the concentric seal, by contrast, the sealant pressure acts in the same direction for both sub-units, with the result that there is a nett axial thrust imparted to the shaft (and an equal and opposite thrust on the housing).

With vessel pressures, and hence sealant pressures, commonly over 10 bar, the thrust developed over the area concerned may become significantly large. Axial thrust forces generated in this way, and from operational features may, under certain circumstances, affect the size of gearbox selected.

The second point to consider regarding seals is the amount of shaft

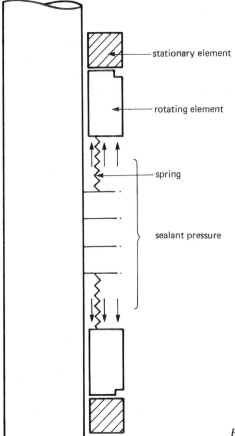

Figure 14.10   A coaxial back-to-back double seal

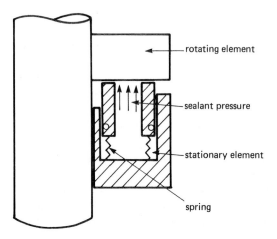

Figure 14.11   A concentric double seal

deflection that may be tolerated while still providing effective sealing. It has been reported that concentric seals can tolerate deflections of about 0.76 mm, whilst coaxial seals and packed-gland units can withstand only 0.076 mm. Thus, in sizing the shaft for a given application, it is essential to consider not only the power transmission capacity, but also the maximum deflection that the seal can tolerate. This latter limitation may well be the deciding factor.

14.5.3   Gearboxes

The components of the gearbox have to withstand the hydraulic moments and forces transmitted from the mixer shaft. The most important of these are:

(i) bending moment;
(ii) torque;
(iii) axial thrust.

The first two of these have been discussed previously. The major components of axial thrust are:

(a) the weight of the impeller and shaft;
(b) upthrust or downthrust if the impeller generates axial flow;
(c) upthrust if concentric mechanical seals are used;
(d) static pressure forces, arising from the difference in vessel and atmospheric pressures, acting on the cross-sectional area of the shaft.

The gearbox will, in general, be sized on the most important combination of the above three moments and forces, and this combination has a strong influence on cost. General-purpose gearboxes, for use in any application requiring speed reduction, are usually sized on the basis of torque alone, without consideration being given to high bending moments that are found when long, overhung mixer shafts are used. For this reason, a number of bearing failures have been recorded associated with the installation of such gearboxes.

Three designs of gearbox have been developed to date, for use when bending moment is the controlling factor for gearbox sizing, and these are shown diagrammatically in *Figures 14.12* to *14.14*.

The first solution (*Figure 14.12a*) involved simply increasing the bearing spacing, and so reducing the bearing loads. The drawback with this design, however, is that a high bearing spacing results in a large deflection of both the final drive shaft of the gearbox, and the mixer shaft itself. Large deflections of the gearbox shaft may result in poor gear meshing, with consequent wear, whilst deflections of the mixer shaft may cause premature wear of the vessel seal.

The solution depicted in *Figure 14.12b* seeks to isolate the main gearbox components from mixer shaft flexure. A hollow drive shaft is mounted on short-span bearings to minimize deflection, whilst the mixer shaft which passes through this is mounted on large-span bearings. The two shafts are coupled by a unit that is torsionally stiff, but flexible in bending. By this means, the high

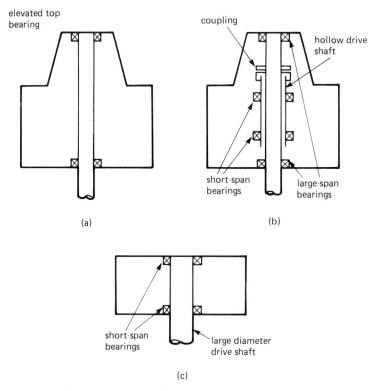

Figure 14.12   *Three designs of gearbox*

bending loads produced on the overhung mixer shaft produce tolerably low reaction forces on the large-span bearings, whilst the final drive shaft undergoes little flexing, leading to improved gear life. Naturally, this arrangement is more expensive than the first, and the problem of high shaft deflection at the vessel seal is still present.

The third design (*Figure 14.12c*) includes a large diameter final drive shaft with short-span bearings. Here, shaft deflection is reduced to a minimum, whereas bearing loads are high. The assumption is that the heavy-duty bearings used are capable of withstanding these loads.

It is sometimes possible, however, to arrange the design so that the large bending moments present on the mixer shaft are not wholly transmitted through to the gearbox. Mounting the gearbox on a stool and installing a third bearing (*Figure 14.13*) greatly reduces the reaction forces on the gearbox bearings and the bending moment on the gearbox final drive shaft, and so may allow gearboxes to be sized on torque and thrust (when important).

Further isolation of the gearbox is achieved in the design shown in *Figure 14.14*. Here, a thrust bearing is added to the design of *Figure 14.13*, so leaving shaft torque as the dominant moment on which the gearbox is sized. A flexible coupling is also shown to cater for any shaft misalignment. Such an arrangement might well prove more economic than any of the other designs

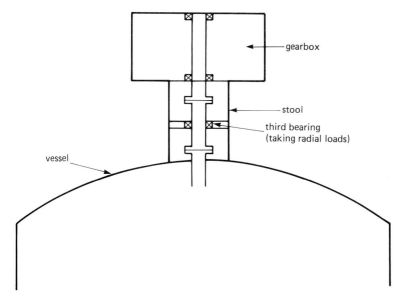

Figure 14.13 *Reducing bending loads on a gearbox by installing a third, radial-load bearing in a stool housing*

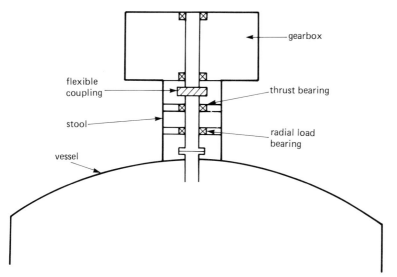

Figure 14.14 *Reducing bending and axial loads on a gearbox by installing a radial-load bearing and a thrust bearing in a stool housing*

considered, in those cases where the vessel is under high pressure, and a concentric mechanical seal is employed, particularly if the impeller develops downward axial flow. In such instances components (b), (c) and (d) of the axial force may become so great that a larger size gearbox is needed. By catering for axial forces with an external thrust bearing, a smaller unit may be used.

14.5.4 Impellers

No other item in the drive train has attracted such a wide diversity of designs as the impeller itself. The main impeller types, paddles, disc turbines, propellers, anchors and gates are described in *Table 14.1* and shown in *Figure 14.1*. For paddle and turbine types in particular, a large number of variations on the basic theme have been introduced by mixer manufacturers and by process designers in industry and in the academic world. Unfortunately, reliable figures are not available for their relative popularity or efficiency in industry. The Fluid Mixing Processes (FMP) project set up at BHRA has, as one of its objectives a programme of quantitative assessment of the performance of the various impeller types available.

The impeller type has an influence on bending loads on two counts. Firstly, its size and shape make it susceptible to turbulent buffeting by the process liquor. Secondly, its mass, and associated 'added mass' of liquor, determines, along with the stiffness of the shaft, the critical speed of the agitator. The critical speed is of prime importance in determining out-of-balance mechanical loads. It is now apparent, from the work carried out by BHRA that the critical speed is also of importance in determining hydraulic bending loads which may be the dominant design consideration.

Although the effect of the mass of the impeller is easy to take account of, little notice has been paid to the associated 'added mass' of liquor, which is the amount of liquid that effectively vibrates with the impeller, so increasing its inertia and reducing its natural frequency (and critical speed). The added mass of an impeller has been determined as a coefficient C_m multiplied by the volume of liquid contained within a cylinder of diameter equal to the impeller diameter and of length equal to the blade width. C_m is typically within the range 0.2–0.5.

Some manufacturers include on the underside of their impellers a short ring known as a stabilizer; two distinct uses for this device have been claimed. It is said that when running completely submerged, the large surface area of the stabilizer provides damping forces against any lateral vibration of the shaft. The increase in damping then allows the agitator to run closer to its critical speed for the same value of shaft lateral displacement. The benefits claimed for the device appear dubious in this context because the addition of stabilizers must increase hydraulic forces associated with turbulent buffeting by the liquor.

The other benefit claimed for this device involves those cases where the agitator must be kept running whilst the vessel is drained. It is suggested that hydraulic forces could become particularly high when the impeller just skims the liquor surface, and that large shaft deflections are possible. It is supposed that the stabilizer, being still submerged at this point, suppresses these deflections by providing resistance to sideways movement. This argument seems rather more plausible than the former claim.

14.6 Economic considerations

For users in general, whether they supply their own equipment or not, the dilemma of high capital outlay versus risk of mechanical problems is present, but is less critical to their operation than for suppliers. Given the choice, most users would tend to invest more in capital outlay, as an insurance against possibly ruinous mechanical problems. The difficulty for most users, however, who do not become involved technically in the problem is that they have no reliable rational basis on which to decide what constitutes an over-design nor what the safety factor is. They therefore rely on the expertise of the equipment vendor.

Losses incurred due to mechanical problems can be put into two categories—*remedial losses* and *marketing losses*[10]. The latter loss is potentially much greater than the former, and will in any case figure more highly in the thinking of user companies. Each of these categories is now considered in some detail and examples given of the costs associated with the various types of loss typical in the majority of industrial installations.

14.6.1 Remedial losses

Remedial losses can be divided into:

1. Loss of process liquor in the plant at the time of failure.
2. Replacement of failed parts.
3. Man-time involved.
4. Additional remedial measures.

Item 1 refers to costs of liquor in the vessel or at other stages of the process that are irretrievably lost because of mechanical problems. In some processes, if agitation is stopped, the contents of the vessel degenerate and cannot be re-used when agitation re-starts. It is also possible that stoppages may have similar repercussions at earlier stages of the process. In other processes, by contrast, the liquor may suffer no ill effects due to lack of agitation, and may be re-used. If a product is lost in this way, the costs incurred may be negligibly small for some materials (bulk low-cost chemicals, for example) or extremely high in other cases (perfumes or antibiotics).

The second and third items naturally depend on the scale of the plant, the type of failure experienced, the materials in use and the effort required to remedy the problem.

The fourth item covers various other costs resulting from a lack of mechanical reliability. An example of costly remedial measures being employed as a result of minor mechanical difficulties has already been quoted in the section on seals, in which extensive ventilation systems were installed as a result of leakage of toxic fumes. As another example in this category, the user may be forced to consider installing additional spare equipment, or additional facilities for stockpiling product, if particularly severe and prolonged problems are experienced.

14.6.2 Marketing losses

Marketing losses can be divided into:

1. Penalties for missed delivery dates.
2. Lost production.
3. Loss of future orders.

The first item might refer either to cash losses arising from penalty clauses in the contract, or to special costs involved in avoiding these losses. Such special costs might cover, for example, chartering an express transportation service to make up lost time or supplying the product in question from an alternative source.

The critical factors in determining marketing losses are:

(i) downtime resulting from failure.
(ii) the level of demand for the product.
(iii) the levels of stockpiled product.
(iv) whether all production passes through a small number of large vessels, so that mechanical problems result in a major loss of capacity, or whether a large number of small vessels are used.

At one extreme, if the level of demand is high and no stockpiles exist and the plant is running at full capacity, then even minor stoppages will result in lost production. If the problems are major or prolonged ones, then delivery dates may not be met, with consequent penalties. If the market is a competitive one, future contracts may be lost to other producers. On the other hand, if demand is low or stockpiles high, and if production can be made good in other vessels, even major problems will result in little or no marketing losses.

The following examples illustrate various features referred to above.

14.6.3 Example 1: Seal problems

A fermentation vessel holds $40 \, \text{m}^3$ of liquor that must be kept sterile. Wholesale price of the product is 200 pounds sterling/m^3. Because of the consequences of contamination, an expensive mechanical seal, costing 3000 pounds sterling, is installed. Dynamic run-out at the seal is greater than expected, however, and occasionally the product becomes contaminated. This happens on average once a year and a fresh seal cartridge costing 2100 pounds sterling, is installed. Lost production is quickly made up.

Remedial losses: *Pounds sterling*

1. Loss of product: one batch of $40 \, \text{m}^3$ is lost per annum. Assuming raw material costs are 10 pounds sterling/m^3, annual losses are   400
2. Parts replacement: one seal cartridge per annum   2100
3. Man-time: negligible
4. Additional remedial measures: nil

Marketing losses: *Pounds sterling*

1. Lost production: one batch is lost each year, with a loss of 8000 pounds sterling in terms of wholesale prices. Assuming a 10% profit margin, annual production losses are: 800
2. Penalties for missed delivery: nil
3. Loss of future orders: nil

Total losses: 3300 pounds sterling per annum

14.6.4   Example 2: Gear and bearing problems

In order to break quickly into a rapidly developing market, a 10-year-old plant with three vessels each of $20\,m^3$ capacity, is modified to handle the new process. Both fluid type and shaft speed have been significantly altered. High demand has stretched capacity to the limit, so that any downtime means lost revenue, and has kept prices high at 400 pounds sterling/m^3, of which 150 pounds sterling/m^3 is profit. Four batch cycles are achieved in each vessel per day.

The plant experiences mechanical problems, however, with gear and bearing failures. On average, each gearbox requires a major overhaul every 18 months, each overhaul taking 2 days to complete. Because the gearbox is of an old design, it is not possible to buy a standby unit.

Remedial losses: *Pounds sterling*

1. Loss of product: nil
2. Parts replacement: two gearboxes require overhauling each year. Assuming each costs 4200 pounds sterling, annual costs are   8 400
3. Man-time: cost of fitters and administrative staff   4 000
4. Additional remedial measures: nil

Marketing losses:

1. Lost production: with two overhauls each year, each lasting 2 days, a total of 16 batches are lost per annum, with a profit loss of   48 000
2. Penalties for missed delivery: nil
3. Loss of future orders: nil

Total losses: 60 400 pounds sterling per annum

14.6.5   Example 3: Shaft failure

A single large vessel, 7 m diameter and 12 m overall depth, contains $350\,m^3$ of a highly corrosive liquor. The double bank of impellers and the overhung shaft are of hastelloy and cost in total 35 000 pounds sterling. The vessel, too, is lined with hastelloy. Three batch cycles are completed each day, and the final product price is 100 pounds sterling/m^3, 10 pounds sterling/m^3 of which is profit.

After 6 months of operation, the shaft fatigues and fails, resulting in a buckled impeller blade and severe damage to the vessel lining. The batch concerned is ruined, with a raw materials loss of 15 000 pounds sterling. Production is halted for 3 weeks whilst repairs are carried out on the shaft and vessel, and a delivery date is missed by 15 days, with a 1000 pounds sterling per day penalty. As a precaution against further losses, a spare shaft is purchased. Because of lack of confidence, a customer places a major contract, for 200 000 m^3 of product, with a competitor.

	Pounds sterling
Remedial losses:	
1. Loss of product: one batch lost, with raw material losses	15 000
2. Parts replacement: assuming materials cost of shaft repair and vessel repair are 8400 pounds sterling and 11 200 pounds sterling respectively	19 600
3. Man-time involved in repair, liaison with equipment vendors and with customers	10 000
4. Additional remedial measures: one spare shaft	35 000
Marketing losses:	
1. Lost production: 21 days production equivalent to 63 batches are lost, with a profit loss of	220 000
2. Penalties for missed delivery	15 000
3. Loss of future orders: profit on 200 000 m^3 is	2 000 000

Total losses: 2 314 600 pounds sterling

14.6.6 Summary

From the above examples it may seem that one answer to many of the typical industrial problems is simply to install larger equipment at the outset. However, the alarming increase of capital costs with size (*Table 14.3*) is often seen by organizations as a reason for opting for the smallest equipment acceptable.

Frequently decisions of purchase are made on the basis of lowest competitive tender. It is only by appreciating fully the force and power transmission capabilities of equipment that objective decisions on size and cost be made. Minimum installed cost can often turn out to be far from cheapest in the longer term.

Table 14.3 1983–84 Prices of the main items in the drive train[10]

Shaft diameter (in)	Gearbox cost (pounds sterling)	Most used motor cost (pounds sterling)	Seal cost (pounds sterling)	Shaft cost per unit length (pounds sterling/ft)
2	1 631	74	543	42
$2\frac{1}{2}$	2 471	98	823	74
3	2 793	155	931	105
$3\frac{1}{2}$	3 381	167	1127	133
4	4 046	526	1348	168
$4\frac{1}{2}$	4 725	711	1575	200
5	8 491	711	2831	231
$5\frac{1}{2}$	10 710	1117	3570	255
6	12 936	1117	4312	294

14.7   Overall conclusions

This chapter has shown how forces are created by, and on, agitator blades and describes the various criteria for ensuring that these forces can be accommodated in the mechanical design of equipment. Unfortunately not all the necessary data are to hand and where gaps exist these have been indicated and methods suggested for at least allowing for the lack of factual design codes.

The conventional designs of agitator shafts based on motor torque characteristics are extremely conservative and result in large heavy installations. On the other hand, the BHRA approach relates shaft loads to the detailed characteristics of the flow forces and generally results in smaller shaft sizes. There is a tendency to assume that massive equipment is absolutely rigid and thus unlikely to be excited into vibrational motion during service. It is hoped that this notion has now been dispelled and that the design guidelines given will at least give a feel for how far each piece of equipment is from being unstable, or failing.

The financial considerations illustrated in the examples showed the cost penalties associated with mechanical failure and the importance of reliable data at the design stages.

In considering objectively the design dilemma experienced by users and suppliers, questions arise as to whether the possibility of suppliers becoming more competitive by reducing equipment size is a beneficial market force.

Clearly, the ideal situation is for the correct shaft size for a given application to be specifiable, and for this design information to be available to all parties concerned, rather than to a few users and a few suppliers. In such a situation, the users would be assured of a smooth-running process, whilst relying on competition amongst suppliers in terms of efficiency of operation, to keep prices low. For the suppliers, the fear of mechanical problems would be reduced.

It might be argued that such widespread information would seriously undermine the market position of those high-technology suppliers who have invested heavily in research in this area, and who generally command higher prices for their equipment, on the basis of improved reliability. Whilst this is certainly true in the context of mechanical aspects of mixing, the whole area of process performance of equipment would remain unaffected. Thus the question of impeller design, selection and operation would still be open to competition among suppliers, and would leave a sound basis for the high-technology companies.

A step towards this ideal situation would be a higher level of technical involvement on the part of the users. It is clear that if technical expertise lies solely with equipment suppliers, this knowledge would remain partial, because it would refer only to their equipment types, and for commercial reasons would not be widespread. Hence, there would still be a fear of mechanical problems, with penalties to all concerned. The active involvement of large numbers of users, by contrast, would ensure that the maximum industrial applications would be covered, since most users see little harm in other users sharing knowledge that does not directly involve competition for their own primary products (chemicals and process products).

This is the philosophy of Fluid Mixing Processes (FMP) referred to previously. Industrial users and equipment suppliers have collaborated to fund generic research work into fluid-mechanical aspects of mixers and mixing processes in a long-term programme of work to be undertaken largely at BHRA. The results of this research are currently confidential but will be released for general publication in the future.

Acknowledgements

The author would like to thank the Director and Council of BHRA for permission to publish this work, which is based on industrially relevant mixing research at BHRA. Particular thanks are extended to Dr G. J. Pollard for allowing much of his Doctoral thesis work to be adapted for this chapter.

Notation

A	area of paddle blade $(= W \times h)$ (m^2)
C_D	drag coefficient
C_m	coefficient of added mass
D_{sh}	diameter of agitator shaft (m)
D'_{sh}	minimum shaft diameter (m)
D	diameter of agitator (m)
e	eccentricity of shaft (m)
E	Young's modulus of elasticity (N m^{-2})
f_n	natural frequency of system (Hz)
F_a	actual force (N)
F_{max}	stalling force on impeller (N)
F_s	side force on impeller (N)
F_t	theoretical force (N)

g	gravitational constant (m s^{-2})
G	modulus of rigidity (N m^{-2})
h	length of agitator blade (m)
I	second moment of area of shaft about a diameter (m^4)
J	polar second moment of area of shaft (m^4)
K_{sh}	shaft stiffness (N m^{-1})
l_{sh}	length of impeller shaft (m)
m_e	effective mass (kg)
m_{am}	added mass (kg)
m_p	mass of agitator (kg)
m_{sh}	mass of shaft/unit length (kg m^{-1})
M	torque (N m)
M'	pull-out torque of motor (N m)
M_b	bending moment about driving end of shaft (N m)
M_{be}	equivalent bending moment (see equation 14.14) (N m)
M_r	rated torque of motor (N m)
N	rotational speed of agitator (s^{-1})
N'	number of stress reversals
P	power (W)
P_a	actual power (W)
Po	power number
P_t	theoretical power (W)
q	endurance limit/static tensile strength
r'	shaft deflection (m)
r_{sh}	shaft radius (m)
Re	Reynolds number
S'	maximum stress (N m^{-2})
t_m	mixing time (s)
T	vessel diameter (m)
v	velocity of flow (m s^{-1})
W	width of paddle blade (m)
y	distance from shaft centreline (m)

Greek symbols

ρ	density of liquid (kg m^{-3})
ρ_s	density of steel (kg m^{-3})
ν	kinematic viscosity (m^2 s^{-1})
ω	generalized frequency of rotation ($= 2\pi N$) (s^{-1})
ω_n	natural frequency of sway vibrations ($= 2\pi f_n$) (s^{-1})
τ	shear stress (N m^{-2})
τ_b	bending stress (N m^{-2})
τ_d	direct stress due to agitator weight (N m^{-2})
τ_s	working stress (N m^{-2})
τ_y	yield stress of shaft material (N m^{-2})
$\bar{\mu}$	r.m.s. shaft bending moment (N m)
$\lambda = \dfrac{\bar{\mu} D}{l_{sh} M}$	ratio of r.m.s. force producing bending to the steady force producing torsion

References

1   EDGEWORTH JOHNSTONE, R. and THRING, M. W. (May 1957) *The Scaling-Up of Chemical Plant and Processes, Introductory Paper, Joint Symposium on Scaling Up*, Institution of Chemical Engineers, London.
2   NIENOW, A. W. (March 1982) *Stirred Not Shaken: Mixing Studies*, Inaugural Lecture delivered at University of Birmingham.
3   EEUA (1962) *Agitator Selection and Design*, Handbook 9, Constable and Co. Ltd, London.
4   SACHS, G. (September 1974) *A Survey of Agitation*, Paper A1, *First European Conference on Mixing and Centrifugal Separation*, BHRA, Cranfield, Bedford.
5   UHL, V. W. and GRAY, J. B. (1966) *Mixing Theory and Practice*, published in two volumes, Academic Press, Vol. 1.
6   BRITISH STEEL CORPORATION (1974) *Iron and Steel Specifications*, SSD 823 30.6.74, Market Promotions Division.
7   BEVAN, T. (1965) *The Theory of Machines*, Longman.
8   POLLARD, G. J. (September 1979) *Fluid Loadings on Mixing Equipment. Introduction, Part 1. The Design of the Test Facility*, BHRA Report RR 1567.
9   DEN HARTOG, J. P. (1956) *Mechanical Vibrations*, McGraw-Hill, 4th ed.
10   POLLARD, G. J. (October 1981) *Fluid Loadings on Mixing Equipment, An Industrial and Literature Survey*, BHRA Report RR 1751.
11   POLLARD, G. J. (February 1984) *A Guide to the Design and Selection of Agitator Drives*, FMP Report 002 (FMP is a service offered by BHRA).
12   POLLARD, G. J. (April 1982) *Hydraulic Bending Loads on Mixer Shafts*, Paper K3, pp. 383–398, *Fourth European Conference on Mixing, Leuwenhurst, Netherlands*, BHRA, Cranfield, Bedford.
13   MIDDLETON, J. C. (April 1979) *Measurements of Circulation within Large Mixing Vessels*, *Proc. 3rd European Conference on Mixing*, BHRA, Cranfield, Bedford, Vol. 1, Paper A2, pp. 15–36.

Appendix 1: Numerical example in sizing an agitator shaft

In this example, the problem to be solved is that of transmitting the shaft power necessary to achieve complete mixing of the tank contents in a specified time, using the two methods of calculation described in the text.

Tank details

Diameter $(T) = 3$ m
Water depth $= 3$ m   (fully baffled)

Impeller/shaft details

Pitched blade impeller diameter $(D) = 1$ m
Projected blade width (W) $= 0.2$ m
Power number (Po) $= 2.0$
Length of shaft (l_{sh}) $= 3$ m
Mass of agitator (m_p) $= 15$ kg
Young's modulus (E) $= 2 \times 10^{11}$ N m^{-2}
Density of steel (ρ_s) $= 8 \times 10^3$ kg m^{-3}
Added mass coefficient (C_m) $= 0.3$

Process details

Liquid assumed to be water; mixing time required $= 2$ minutes.
Middleton[13] correlates mixing time t_M and agitator speed N through the correlation below:

$$t_m = 2.5(\text{Volume})^{0.3}(T/D)^3 N^{-1} \text{ s}$$

Thus, for a mixing time of 2 minutes, we have:

$$120 = 2.5 \times \left(\frac{\pi}{4} \times 3^2 \times 3\right)^{0.3} \times 3^3 \times N^{-1}$$

from which,

$$N = 1.41 \text{ rev/s}$$

A1.1   Conservative approach

$$P_a = Po\rho N^3 D^5 \qquad \text{(equation 14.4)}$$

and

$$Po = 2.0.$$

Thus,

$$P_a = 2.0 \times 10^3 \times 1.41^3 \times 1^5 \qquad \text{(W)}$$

$$\underline{P_a = 5.6 \text{ kW}} \qquad (\sim 7\tfrac{1}{4} \text{ horsepower})$$

$$M_r = P_a / 2\pi N$$

$$\therefore M_r = 633 \text{ N m}$$

and assuming a light duty

$$M' = 1.5 M_r$$

$$M' = 949 \text{ N m}$$

From equation (14.15) the stalling force F_{max} is:

$$F_{max} = \frac{8M'}{3D}$$

i.e.

$$F_{max} = \frac{8 \times 949}{3 \times 1} \text{ N}$$

$$\underline{F_{max} = 2531 \text{ N}}$$

$$M_b = F_{max} l_{sh}$$

$$M_b = 2531 \times 3$$

$$M_b = 7593 \text{ N m}$$

From equation (14.14), inserting M' for M we have:

$$2M_{be} = M_b + \sqrt{M_b^2 + (M')^2}$$

$$2M_{be} = 7593 + \sqrt{7593^2 + 949^2}$$

$$M_{be} = 7623 \text{ N m}$$

The stress corresponding to this is (equation 14.16):

$$\tau_b = \frac{32M_{be}}{\pi D_{sh}^3}$$

The direct stress due to agitator weight is

$$\tau_d = 4\frac{m_p g}{\pi D_{sh}^2}$$

When added together, these two must be less than the yield stress τ_y.

$$\tau_b + \tau_d < \tau_y$$

i.e.

$$\frac{32M_{be}}{\pi D_{sh}^3} + \frac{4m_p g}{\pi D_{sh}^2} < \tau_y \qquad (\text{where } \tau_y \sim 2.0 \times 10^8 \text{ N m}^{-2})$$

$$\frac{32 \times 7623}{\pi D_{sh}^3} + \frac{4 \times 15 \times 9.81}{\pi D_{sh}^2} < 2.0 \times 10^8$$

$$\frac{7.765 \times 10^4}{D_{sh}^3} + \frac{1.874 \times 10^2}{D_{sh}^2} < 2.0 \times 10^8$$

Solving for D_{sh} by trial and error gives:

$$\underline{D_{sh} = 73 \text{ mm}}$$

(note that the direct loading contribution is very small).

It is now necessary to check that the shaft will withstand the torque developed as the driving motor accelerates through its peak torque during start-up, i.e. ensure that the shear stress is less than the safe working stress.

$$D_{sh}^3 = \frac{16M'}{\pi \tau_s} \qquad (\text{equation 14.8})$$

Substituting the appropriate values for D_{sh} and M' gives $\tau_s = 1.24 \times 10^7$ N m^{-2} which is well within the working stress quoted in *Table 14.2* (5×10^7 N m^{-2}); thus the shaft is strong enough to transmit the start-up torque.

Next, the whirling speed of the agitator/shaft assembly must be checked.

Equation (14.17) gives a method of assessing the natural frequency of the shaft and agitator *in air*.

In water, the 'added mass' must be included.

$$\text{Added mass } (m_{am}) = C_m \cdot \rho \frac{\pi}{4} D^2 . W$$

$m_{am} = 47$ kg

$m_p = 15$ kg

$m_{sh} = \rho_s \dfrac{\pi}{4} D_{sh}^2$ kg m^{-1}

$m_{sh} = 33.46$ kg m^{-1}

$m_e = m_p + m_{am} + 0.24 m_{sh} l_{sh}$

$\underline{m_e = 86.09}$ kg

$K_{sh} = 3EI/l_{sh}^3$

$E = 2 \times 10^{11}$ N m^{-2}

$I = \dfrac{\pi}{64} D_{sh}^4$

$\underline{K_{sh} = 30\,929}$ N m^{-1}

From equation (14.19)

$\omega_n^2 = 30\,929/86.09$

$\omega_n = 18.95$ rad s^{-1}

$\underline{f_n = 3.02}$ Hz

Thus, the natural frequency of the agitator/shaft system is 3.02 Hz; this is numerically equal to its critical whirling speed. Since the specified, or derived running speed is $N = 1.41$ Hz (or rev/s, effectively the same) then there is no danger of exciting resonance due to shaft imbalance.

In general, shafts should not be run within $\pm 30\%$ of the critical whirling speed (see section 14.4.3 and *Figure 14.7*). If the calculated critical speed range includes the specified operating range then remedial measures are necessary; these include making the shaft hollow, reducing the mass of the agitator blades, fitting a steady bearing or, if all else fails, varying the process requirements of mixing time. The following example illustrates one remedial option available to the designer.

A1.2 Using the BHRA approach

The step-by-step method is given in section 4.5.1 and this example illustrates its use relative to that of the conservative approach on page 281.

(i) From before (see page 281) $M_r = 633 N_m$
(ii) Guess a critical speed of $f_n = 3$ H

$$\dfrac{mN}{f_n} = \dfrac{4 \times 1.41}{3} = 1.88$$

(iii) From *Figure 14.9*, $\lambda = 0.17$.

(iv) $\bar{\mu} = \dfrac{\lambda l_{sh} \cdot M_r}{D}$

$\bar{\mu} = \dfrac{0.17 \times 3 \times 633}{1}$

$\bar{\mu} = 322.83$ N m

$\bar{\mu}_{max} = 2.63 \bar{\mu}$

$\bar{\mu}_{max} = 849$ N m

(v) $(D'_{sh})^3 = \dfrac{16}{\pi} \left(\dfrac{\bar{\mu}_{max}^2 + M_r^2}{\tau_s} \right)^{1/2}$

$D'_{sh} = 48$ mm

(vi) Check now for natural frequency:
From equations (14.17)–(14.19):

$\omega_n^2 = K_{sh} / m_e$

$K_{sh} = \dfrac{3EI}{l_{sh}^3}$

$m_e = m_p + m_{am} + 0.24 l_{sh} \cdot m_{sh}$

by substitution:

$f_n = 1.40$ Hz

It will be noted that the natural frequency is almost exactly equal to the shaft rotational speed: this is an operational condition that must be avoided from resonance considerations (see section 14.4.3 and *Figure 14.5*).

To overcome this, a larger diameter shaft is selected such that

$N < 0.7 f_n$; $f_n \sim 2.0$ Hz

thus $mN/f_n = 2.8$, λ remains unaltered at $\lambda = 0.17$ thus

$D_{sh} = 60$ mm

By comparing this with the size of shaft resulting from the conservative approach (73 mm diameter) it will be seen that a saving of over 20% on diameter has been made. This has important advantages on the installed costs of seals and shaft as demonstrated in section 14.6.1 and *Table 14.3*.

Chapter 15

Dynamics of emulsification

R Donaldson

Unilever Research Laboratory, Port Sunlight, Merseyside

15.1 Introduction

In preparing emulsions, many factors concerning their end-use have to be taken into account. Often, the preparation of emulsions is merely an enabling step towards, for example, formation of disperse polymer systems, or enhancing a liquid/liquid extraction process. However, in many situations the emulsion is itself the end product. Abundant examples of where this is the case are to be found in the pharmaceutical, food, paint, agrichemical, cosmetic and detergents industries. In these situations, as opposed to those where emulsification is an intermediate step, we are invariably more closely concerned with two key properties—rheology and stability.

To the consumer, consistency of product properties represents quality, together with an assurance that the product is functioning in the regularly expected way. Quite apart from this marketing appeal, there can be a very real danger from incorrect function in an unstable product. This is recognized in law in many countries. For example cosmetics and pharmaceutics may be legally required to have a specified shelf-life during which they show no physical changes.

Accepting that rheology and stability are critical properties, how can large-scale processes be designed and operated to achieve a given product specification? The approaches open are either (a) to adopt an empirical approach using experimental designs for formulation and processing which allow the 'best' combination to be selected or (b) to use an approach based on understanding the basic physical and chemical phenomena operating within the systems, and hope to build these into a formulation/process. There are advantages and disadvantages to both. The former will generally yield a compromise solution but the gaining of that solution offers little insight into how new products and processes might be developed. On the other hand, the latter should yield a solution, plus an expertise for future developments, but only with a considerable investment in skill and equipment.

Of course, neither approach is exclusive and many 'empiricists' make excellent use of basic technology; similarly, no process has yet been designed purely from a set of equations. The purpose of this chapter is to try and bridge the gap by presenting, in a semiquantitative way, some of the concepts which are essential to a fuller understanding of emulsification. The approach taken does not purport to provide answers, but instead to draw attention to some fundamental issues worth considering when designing processes for making emulsions. The emphasis is placed on hydrodynamic and rheological effects; for discussion of the detailed role of emulsifiers the reader is referred to the wide literature on this subject.

Firstly, the aspects of an emulsion which are important to its rheology and stability are briefly considered.

15.2   Rheology and stability

Most emulsions of interest show shear-thinning rheological behaviour, and many show other non-Newtonian features such as elasticity, yield stress or time dependent effects. All of these can be explained qualitatively, and sometimes quantitatively, by a relatively simple model (*Figure 15.1*) which takes into account:

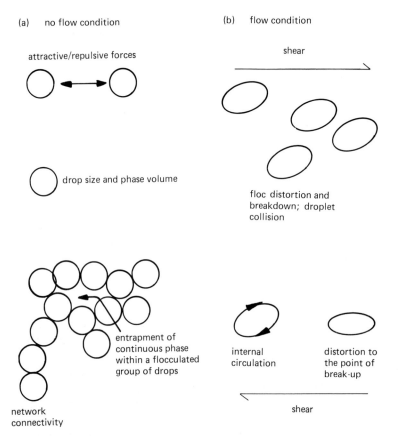

(a)   no flow condition

attractive/repulsive forces

drop size and phase volume

entrapment of continuous phase within a flocculated group of drops

network connectivity

(b)   flow condition

shear

floc distortion and breakdown; droplet collision

internal circulation

distortion to the point of break-up

shear

Figure 15.1   Diagrammatic representation of the parameters which are important to emulsion rheology

Droplet size;
Droplet phase volume;
Conservative (colloid interaction) forces;
Hydrodynamic forces;
Interfacial properties.

Thus, shear-thinning behaviour is the result of either droplet distortion and alignment with the flow, or due to increasing shear stress causing breakdown of weak flocs with a consequent decrease in effective phase volume (*Figure 15.1b*).

If a system is to possess elasticity, it must possess a physical mechanism for storing energy. In flow, droplet distortion causes an increase in surface free energy which is released on cessation of flow, manifesting as e.g., recoverable shear compliance. The magnitude of the (dimensionless) drop distortion is proportional to the Weber number, *We*, which is the ratio of the deforming stress ($\mu_c \dot{\gamma}$) and the restoring Laplace stress (σ/d_d). Thus,

$$We = \frac{\mu_c \dot{\gamma} d_d}{\sigma}$$

where μ_c is the continuous phase viscosity, $\dot{\gamma}$ is the local shear rate, d_d is the droplet diameter and σ is the interfacial tension. Barthes-Biesel and Rallison[1] have shown how these droplet-size dependent elastic stresses contribute to the dilute emulsion rheology.

The property of yield stress arises in the case where the droplets form a continuous network throughout the system, for instance under the influence of van der Waals forces[2]. In such a case a system containing a network of deformable droplets has a yield stress, τ_y, given by

$$\tau_y \propto \frac{A}{H_0^2} \cdot \frac{\phi}{d_d}$$

where A is the composite Hamaker constant, ϕ the droplet phase volume and H_0 the surface–surface distance of separation of adjacent droplets. If a stress is applied to such a system, less than the yield stress, the network deforms elastically with a modulus, G, given by

$$G \propto \frac{\sigma \phi}{d_d}$$

Time and stress dependent flow properties are generally modelled[3] via the making and breaking of interdroplet bonds, such that at any given stress level, there is a distribution of singlets, doublets, triplets etc. and the rate of change of distribution on changing the stress level reflects the thixotropic change noted experimentally. Clearly, both droplet–droplet collision, essential for 'making' bonds, and multiplet breakdown (controlled by attractive forces) are strong functions of droplet size.

The stability of an emulsion depends largely on the balance of attractive and repulsive forces (*Figure 15.2*). If only attractive forces exist then the droplets

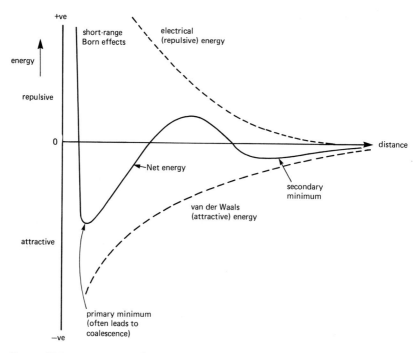

Figure 15.2   Attractive and repulsive energy as a function of surface–surface distance of separation. The net force is zero at distances corresponding to the primary and secondary minima

coalesce, thereby lowering the system free energy (except in the event of extremely small particles where the entropic energy term can dominate). On the other hand, if repulsive forces dominate then even very small density differences are sufficient to give rise to creaming, for example. What is required for stability is a flocculation of droplets which does not give rise to coalescence but forms a network throughout the system.

In *Figure 15.2* we have shown a secondary minimum stabilizing mechanism where stability is achieved simply through an appropriate balance of attractive and electrical repulsive forces. Other stabilizing mechanisms exist such as the use of adsorbed species which present an entropic hindrance to close approach of droplets, or the use of bridging polymers or finely divided solids. Each mechanism leads to a different strength of interaction and, generally, a different response to stress.

It is in this area that the art of the selection of emulsifier and making process reaches its peak of creativity, but it is also an area where fundamental knowledge is beginning to rationalize and order empirical experience. For instance Bancroft's rule[4] can be understood in terms of molecular geometry, and the effects of the materials of construction[5] of the processing vessel on the emulsion type can be understood in terms of critical wetting energy. However, irrespective of the mechanism, such aspects as the number of droplet–droplet

Table 15.1 Some key properties of emulsions and the basic parameters which control them

Property	Basic parameters
Viscosity	d_d, μ_c, ϕ
Yield stress	d_d, ϕ, A, H_0
Stability	d_d, A, H_0
Appearance	d_d, ϕ

junctions, and the surface area available for emulsifier action, are strongly dependent on droplet size.

Through consideration of the factors which can be seen to be important to rheology and stability *and* over which some control is possible during processing, it can be concluded that droplet size is a key parameter (*Table 15.1*). In the following sections, some fundamental aspects of droplet formation are examined, in order to yield some insight into the design and operation of processes for controlled droplet formation.

15.3 Droplet formation

In very general terms, droplets are formed by stress being imparted to a primary drop, causing elongation of all or part of it, followed by development of surface wave growth to the point where the primary drop breaks into droplets and, often, smaller satellite droplets. The factors important to this process are:

(a) the viscous and elastic properties of the disperse and continuous phases;
(b) the interfacial properties;
(c) the flow conditions.

There are a number of difficulties in examining the roles of these factors either experimentally or theoretically. The key difficulty is that, in a practical sense, emulsification does not take place under steady conditions, but under dynamic conditions where time scales may be of the order of seconds down to 10^{-6} s. However, it can be reasonably assumed that the *direction* of effects is independent of time scale. Then, a combination of steady state effects with a knowledge of how time scales might modify their magnitude can be used.

15.3.1 Deformation and break-up in steady flows

This problem was first studied by Taylor[6] and more recently by others[7-9]. Basically, the approaches that were made consider how the hydrodynamic forces give rise to a pressure field around the droplet, distorting it and leading to a counteractive Laplace pressure or, equivalently, a counteractive surface tension. The analysis is complex, even for simple systems. Barthes-Biesel and

Rallison[1] have examined the time-dependent deformation of a capsule with an elastic membrane, in a shear flow, and point out that their analysis must be extended to the region of larger deformation and for membranes which show a viscoelastic response.

In the area of emulsification the latter is critically important because the surface tension of a newly dilated surface relaxes back towards its equilibrium value due to further emulsifier adsorption from the bulk or to spreading of molecules in the interface from emulsifier-rich regions.

In simple terms, however, the extent of deformation of a droplet depends on the Weber number, as described above, and the ratio (R) of the viscosities of the disperse (μ_d) and continuous (μ_c) phases,

$$R = \frac{\mu_d}{\mu_c}$$

Under limiting conditions of $We \ll 1$ (e.g. low flow rate) and with $R \approx 1$, then the deformation, D^1, is given by

$$D^1 = \frac{L-B}{L+B} \propto We$$

where L and B are defined as the major and minor axes of the resulting ellipsoid. Under limiting conditions of $R \gg 1$ and $We \approx 1$, then

$$D^1 \propto \frac{1}{R}$$

Typically, for oil/water mixing in a batch vessel we expect the former condition to prevail (although for turbulent mixing the Weber number has to be reformulated as shown below).

The break-up of droplets in steady flow is generally approached through consideration of the hydrodynamic stability of liquid cylinders, which are assumed to be a precursor of break-up. The critical factor for the *rate* of break-up of a given cylinder is the rate at which instabilities grow, whereas for the ultimate droplet *size*, it is the wavelength of the instability which is determining. Again it is found, just as with deformation, that the controlling parameters are the viscosity ratio and the Weber number.

In principle a continuous spectrum of wavelengths can exist in the earliest stages of wave growth. However, one will become dominant and Chin[9] has shown that, for a given cylinder of Newtonian liquid, both the wavelength of the ultimately dominating instability and its rate of growth are a function of the viscosity ratio. Higher values of R lead to lower growth rate and a longer dominating wavelength (i.e. leading to larger droplets).

The wavenumber, α, may be expressed as:

$$\alpha = \frac{\lambda}{2\pi r_0}$$

where r_0 is the radius of the initial cylinder. If the wavenumber, α_m, is that associated with the dominating wavelength, λ_m, and further it is assumed that a cylinder of length $n\lambda_m$ breaks up into n equal droplets, then it follows that:

$$d_d = (12\pi\alpha_m)^{1/3}r_0$$

Thus d_d is a weak function of α_m but a strong function of the initial cylinder radius. Therefore, although viscosity ratio and Weber number have an influence on break-up of the cylinder once formed (through their effects on α_m) the dominant role of these parameters is in their effect on the initial deformation of a drop, i.e. their effect in determining r_0.

For most real systems, where σ is a function of time and deformation, it would be necessary to couple the surface behaviour with the hydrodynamic equations, but this is a complex procedure[10].

In non-steady flow it becomes necessary to consider the response time of the droplet compared with the time scale of the fluctuating stresses. For viscous drops subject to a high frequency field, but with a time-average stress of zero, or subject to a short residence time in a directional stress field, it can be expected that the drop will not have time to deform. An experimental observation of this phenomenon is discussed later. Clearly there are implications for process design but these have not, to the author's knowledge, been applied in any systematic manner.

15.3.2 Dynamic effects

The most important practical dynamic effects are associated with rapid stretching of a droplet, i.e. surface dilation and bulk elastic effects. The former is important where surface active materials are present. In the equilibrium state, the droplet will have an adsorbed (mono) layer of surfactant and in consequence the interfacial tension will be lowered. On stretching, the interfacial tension rises because the surfactant molecules cannot respond instantaneously, and therefore become depleted in terms of the number/unit area.

In this way, the Weber number falls and deformation is less than might be expected from a knowledge of the equilibrium interfacial tension alone. As time passes, further adsorption from the bulk may occur, or the droplet may contract back to a spherical shape under the influence of its new-found high interfacial tension. The exact circumstances will depend strongly on surfactant properties and concentration. These effects are discussed in detail elsewhere[11].

If the droplet material has viscoelastic properties, characterized by a relaxation time λ_t, then the behaviour of the drop will depend on the timescale of the applied deforming stress. For very short time ($\ll \lambda_t$) the drop will appear as an elastic solid whereas at long times ($\gg \lambda_t$) the viscous nature will dominate. A knowledge of the material rheology across a range of timescales is therefore important for a fuller understanding.

15.3.3   Turbulence

So far steady flow conditions have been considered, but in practice emulsification is often carried out under turbulent conditions. The same physical concepts apply but this time in the context of a rapidly changing flow field. Under such conditions the Weber number cannot be defined using a steady-state shear rate and is reformulated as:

$$We' = \rho_c u^2 (d_d) \frac{d_d}{\sigma}$$

where ρ_c is the continuous phase density and $u^2(d_d)$ is the mean square of the difference of velocities at distance d_d.

A reasonably successful approach to finding the maximum stable drop size, $(d_d)_{max}$, in turbulence has been made by Arai *et al.*[12]. For drop sizes much larger than the Kolmogoroff length scale, i.e.

$$(d_d)_{max} \gg \left[\frac{\mu_c^3}{\varepsilon_T \rho_c^3} \right]^{1/4}$$

(which has a value of order 10^{-4} m for a low viscosity system in a typical 1 ton batch mixer) they showed that theoretically

$$\frac{\rho_c \varepsilon_T^{2/3} (d_d)_{max}^{5/3}}{\sigma [1 + f(N_{vi})]} = \text{const}$$

ε_T is the energy dissipation rate per unit mass ($\alpha N^3 D^2$ for fully developed turbulent flow in baffled vessels) and $N_{vi} = (\mu_d \varepsilon_T^{1/3} (d_d)_{max}^{1/3})/\sigma$ allows for the effect of a finite disperse phase viscosity. $f(N_{vi})$ has to be determined experimentally.

In attempting to confirm their expression experimentally, Arai *et al.* noted a strong interaction between drop viscosity and interfacial tension. Thus, for a low viscosity drop, one might expect a rapid response to the periodic turbulent stress, i.e. the droplet would rapidly adjust its shape. On the other hand, viscous drops having a response time longer than the periodicity will be subject to enhanced deformation by consecutive stress cycles.

For drops smaller than the Kolmogoroff length scale, shear forces rather than periodic inertial forces dominate the break-up mechanism and theoretically[13]

$$\frac{(d_d)_{max} (\mu_e \varepsilon_T \rho_c)^{1/2}}{\sigma f(R)} = \text{const}$$

In order to arrive at a drop-size distribution, it is necessary to take coalescence into account and to ascribe an energy spectrum to the turbulent flow. Some success in this has been achieved and is detailed elsewhere[14].

15.4   Implications for process design

A strong semi-quantitative link has been made between rheology/stability and drop size, and between drop size and process conditions. Control of product properties through control of the process therefore requires a study of both links. Some of the important aspects of the first link have already been outlined and the reader is referred to a review of the subject[15] which, although devoted to dispersion, enlarges on some of the key aspects. The roles of the process variables in controlling drop size are outlined below. For a full description the reader is referred to Nagata[14].

15.4.1   Batch processing

The variables available in a typical baffled batch mixer of standard geometry are impeller geometry and speed, temperature, time and the entry point of the second phase. Each of these is considered below.

Impellers are available which provide for different distribution and stress patterns. The former is quantified by the discharge rate, Q, which is the flow rate perpendicular to the impeller discharge area:

$$Q = N_Q N D^3$$

where N_Q is the discharge coefficient, which is essentially the ratio of discharge rate to swept volume rate and depends on impeller type.

The stress level may be represented by the power number or, useful in some cases, by the power consumption per unit volume. A high stress to discharge rate ratio leads to small drops, and *vice versa*, through the high (turbulent) Weber number. The relationship between power, impeller diameter and rotation rate can be summarized for various types of impellers in power correlation curves (see Chapter 8) from which calculations can be made to support process design work. Use can also be made of the many empirical equations available to correlate equilibrium drop size with impeller geometry and other operational variables. However, in using such equations, care must be taken to ensure that the process under design, and the liquid systems for which it is to be used, have parameters which lie within the range over which the empirical equation was found to hold.

In order to overcome difficulties associated with a broad distribution of stress history within a batch mixer, such as a slow approach to equilibrium, recirculation devices are often used. Generally these are high stress devices and the role of the batch mixer impeller is reduced to ensuring adequate gross mixing. A further alternative is to obtain gross mixing, to some defined scale of scrutiny, and then pump out the mixer through a high stress device. This narrows the distribution of stress history still further and offers a cost-effective way to control drop size.

Temperature may play a strong role in changing the viscosity ratio of the two phases. For two Newtonian fluids obeying the Arrhenius-type activated flow law, with temperature-independent activation energies E_d and E_c, we find:

$$\frac{d \ln R}{dT} = \frac{E_c - E_d}{R_g T^2}$$

where T is absolute temperature and R_g is the gas constant.

For many systems, either or both of E_c and E_d may be strongly temperature dependent. Such would particularly be the case where a phase change occurs. Again, use may be made of such phenomena to control ultimate drop size. Temperature effects will be more complicated where emulsifiers are present because of temperature-dependent changes in adsorption, configuration at the interface, interfacial tension, phase structure etc. and these effects must be studied separately.

Time of mixing is important from the points of view of (a) ensuring adequate gross mixing, (b) ensuring equilibrium drop-size distribution is achieved and (c) avoidance of overprolonged mixing with its penalties of higher energy costs, capacity or throughput problems, and possible damage to the product. For these reasons it is important to have an idea as to the limits to which a product can be 'pushed' in terms of intensity/time of mixing together with a good knowledge of the mixing time for systems of comparable rheology. For the latter, model studies using tracers, dyes, etc. are widely used[16].

The entry point of the second phase can have a pronounced effect. For rate-of-mixing purposes and uniformity of drop-size distribution, it is claimed that entry close to the impeller tip is advantageous. This effect becomes particularly important where the temperatures of the two liquids are not equal and where advantage may be taken of temperature effects on the viscosity ratio.

15.4.2 Continuous processing

It is probably in continuous processing that a deeper knowledge of the fundamental physicochemical molecular effects pays most dividends. Generally, a continuous emulsification process consists of two essential elements: a proportioning pump to ensure correct formulation, and a shearing device to ensure correct dispersion of the phases.

This apparent simplicity belies the true complexity of such a process. For instance, where an emulsifier is present in one or both of the liquid phases, the state of adsorption at the point of shear will play an important role in determining the resulting drop size. Thus a dependence can be clearly seen on the contact time (and contact area) available between the pump and the shearing device. This will depend on pipe dimensions, on the use of in-line devices to produce a coarse initial dispersion, and on the flowrate which, typically for a positive displacement pump, is pulsed!

These are some of the factors upstream of the shearing device; downstream, one has to consider the flow conditions pertaining immediately after the stretching zone, and the relationship between pipe dimensions and flow rate and the coalescence or further breakup of droplets.

Clearly, if the promised advantages for continuous processes, of flexibility of throughput, ease of cleaning, space saving and tightness of dosing and temperature control are to be realized, then their design and operation should be made in the light of such fundamental knowledge.

Notation

A	Hamaker constant (J)
B	breadth of an ellipsoidal drop (m)
D	impeller diameter (m)
D^1	deformation, dimensionless
d_d	droplet size (m)
$(d_d)_{max}$	maximum droplet size (m)
E	activation energy for viscous flow (J mol^{-1})
G	network modulus (Pa)
H_0	surface–surface distance of separation between drops (m)
L	length of an ellipsoidal droplet (m)
N	impeller speed (rev/s^{-1})
N_Q	discharge coefficient, dimensionless
N_{vi}	viscosity number, dimensionless
Q	impeller discharge rate (m^3 s^{-1})
R_g	gas constant (J mol^{-1} K^{-1})
r_0	diameter of a liquid cylinder (m)
R	$\dfrac{\mu_d}{\mu_c}$, dimensionless
T	absolute temperature (K)
$u^2(d_d)$	mean square of the difference in velocity over distance d_d (m^2 s^{-2})
We	laminar shear Weber number, dimensionless
We^1	turbulent Weber number, dimensionless
α	$\lambda/2\pi r_0$, dimensionless wavenumber
α_m	$\lambda_m/2\pi r_0$, dimensionless wavenumber
ε_T	mean turbulent energy dissipation rate (W kg^{-1})
ϕ	phase volume of disperse phase, dimensionless
$\dot{\gamma}$	laminar shear rate (s^{-1})
λ	wavelength of an instability (m)
λ_m	dominating wavelength (m)
λ_t	relaxation time (s)
μ	viscosity (Pa s)
ρ	density (kg m^{-3})
σ	surface tension (N m^{-1})
τ_y	yield stress (Pa)

Subscripts

d	dispersed phase, or droplet
c	continuous phase

References

1 BARTHES-BIESEL, D. and RALLISON, J. M. (1981) *J. Fluid Mech.*, **113**, 251.
2 VAN DEN TEMPEL, M. (1961) *J. Coll. Sci.*, **16**, 284.
   PAPENHUIZEN, J. M. P. (1972) *Rheol. Acta*, **11**, 73.

3   MICHAELS, A. S. and BOLGER, J. C. (1962) *Ind. Eng. Chem. Fund.*, **1**, 153.
    HUNTER, R. J. (1982) *Adv. Coll. Int. Sci.*, **17**, 197.
4   BANCROFT, W. D. (1915) *J. Phys. Chem.*, **17**, 501 (1913); **19**, 275 (1915).
5   BECHER, P. (1965) in *Emulsions: Theory and Practice*, Rheingold Publishing Co., New York, 2nd ed.
6   TAYLOR, G. I. (1934) *Proc. Roy. Soc.*, **A146**, 501.
7   RUMSCHEIDT, F. D. and MASON, S. G. (1961) *J. Coll. Sci.*, **16**, 238.
8   TORZA, S., COX, R. G. and MASON, S. G. (1972) *J. Coll. Int. Sci.*, **38**, 395.
9   CHIN, H. B. (1978) PhD dissertation, Polytechnic Institute of New York.
10 VEDOVE, W. D. and SANFELD, A. (1981) *J. Coll. Int. Sci.*, **84**, 318.
11 LUCASSEN-REYNDERS, E. H. (1981) in *Anionic Surfactants: Physical Chemistry of Surfactant Action* (Lucassen-Reynders, E. H., ed.), Marcel Dekker, Inc., New York.
12 ARAI, K., KONNO, K., MATANUGA, Y. and SAITO, S. (1977) *J. Chem. Eng. Jap.*, **10**, 325.
13 SPROW, F. B. (1967) *Chem. Eng. Sci.*, **22**, 435.
14 NAGATA, S. (1975) in *Mixing—Principles and Applications*, John Wiley and Sons, Kodansha, Tokyo.
15 MEWIS, J. and SPAULL, A. J. B. (1976) *Adv. Coll. Int. Sci.*, **6**, 173.
16 FORD, D. E., MASHELKAR, B. A. and ULBRECHT, J. (1972) *Process Technology Int.*, **17**, 803.

Chapter 16

The suspension of solid particles

A W Nienow

Department of Chemical Engineering, University of Birmingham

16.1 Introduction

Agitators are used in stirred tanks containing particulate solids and liquids in many different applications. They may be used in flocculation where gentle agitation is provided to produce low intensity turbulence leading to frequent contact of colloidal particles with each other and with the large, amorphous flocs. The agitation must not produce regions of high shear sufficient to break down the weak flocs.

There are other applications in which a high concentration ($> \sim 40\%$ by volume) of small size solids produces a solid–liquid mixture with non-Newtonian and possibly time-dependent characteristics.

The widest range of problems is covered by suspensions in which the particles are present in sufficiently low concentration that they can be considered to have a negligible effect on the fluid viscosity. In addition, the stability of the particle is generally unaffected by the agitation. Within this class of problems would fall dissolution, chemical reaction including catalysis, ion-exchange and adsorption, crystallization and precipitation, as well as the mechanical problem of slurry withdrawal from a vessel. In all these cases, the fluid motion is highly turbulent. This chapter is concerned specifically with this latter class of problem, namely, the suspension of solids, present in relatively low concentration, under turbulent flow conditions, i.e. impeller Reynolds numbers $> \sim 10^4$.

16.2 Definitions of states of suspension

16.2.1 Complete suspension

Complete suspension exists when all particles are in motion and no particle remains on the tank base for more than a short period, e.g. 1–2 s. Under this condition, all the surface of the particles is presented to the fluid, thereby ensuring that the maximum surface area is available for chemical reaction, heat or mass transport. Most work in the literature pertaining to particle suspension gives the minimum agitator speed, N_{JS}, or power, $(\bar{\varepsilon}_T)_{JS}$, required to achieve this condition. As indicated in Chapter 8, agitation can be classified as mild, moderate or severe, depending on whether the energy dissipation rate (power input/unit mass) is about 0.05 W/kg, 0.2 W/kg or 1 W/kg respectively. Very often, complete particle suspension is a severe agitation operation though different geometries require significantly different energy dissipation rates. For instance, disc turbines in tanks of Rushton dimensions can only just completely suspend 100 μm particles of sand in water at about 1 W/kg.

16.2.2   Homogeneous suspension

Homogeneous suspension exists when the particle concentration and, for a range of sizes, the size distribution is constant throughout the tank. This condition is particularly desirable when a continuous and representative flow of solids from the system is required.

In general, a considerably higher speed than that giving complete suspension is necessary.

16.2.3   Bottom or corner fillets

A small proportion of particles is allowed to collect in corners or on the bottom in relatively stagnant regions to form fillets. This condition may offer advantages from the practical point of view because of the very large saving in power consumption compared with that required for complete suspension. This power saving may more than offset the loss of active solids.

16.2.4   Dispersion of floating solids

If the solids are less dense than the liquid, then they float on top. This condition has rarely been studied though quite clearly it is a situation that may well arise. When it has been studied, visual observation that stagnant zones of floating solids have been removed has been used as the criterion of adequate agitation, i.e. a broadly equivalent criterion to N_{JS}.

16.3   Mechanisms and models of particle suspension

For particles to be lifted from the bottom of the vessel, the local hydro-dynamics, i.e. velocities and turbulence levels must be suitable in that region. Though no completely satisfactory mechanism has yet been proposed, a combination of two seem most probable.

Firstly, there is a drag forcing arising from the fluid moving in a complex boundary layer flow across the bottom similar to the dynamics of channel flow[1]. Secondly, turbulent bursts propagate intermittently through the boundary layer and if the frequency and energy levels are high enough, suspension will be caused. Once lifted from the bottom of the vessel, the tendency to return there due to the action of gravity is counteracted by the upward drag due to the strong vertical flows which are always induced in vessels designed for particle suspension.

Considering first models based on turbulence, Kolar[2] initiated this approach but the most successful has been that developed by Baldi *et al.*[3,4]. They suggested that an energy balance could be made between the critical eddies (assumed of size of the order of the particles, d_p) and the height to which a particle must be lifted to become entrained (also assumed of order d_p). Thus

$$\rho_L u'^2 \propto d_p \Delta\rho g \tag{16.1}$$

In general, these eddies would be much larger than the dissipation scale of turbulence so that their fluctuating velocities, u', could be expressed as

$$u' \propto (\varepsilon_T d_p)^{1/3} \tag{16.2}$$

where ε_T is the local energy dissipation rate[5]. Assuming $\varepsilon_T \propto \bar\varepsilon_T$ and knowing that (see Chapter 8)

$$\bar\varepsilon_T = \frac{4\,Po\,N^3 D^5}{\pi T^3} \tag{16.3}$$

for tanks of geometry $H = T$, then at the agitator speed to just cause suspension so that $N = N_{JS}$,

$$\sqrt{\left(\frac{\Delta\rho g}{\rho_L}\right)} \cdot \frac{T d_p^{1/6}}{Po^{1/3} D^{5/3} N_{JS}} = \text{const} = Z \tag{16.4}$$

Alternatively

$$N_{JS} = \left(\frac{g\Delta\rho}{\rho_L}\right)^{1/2} \cdot \frac{1}{Po^{1/3}} \cdot \left(\frac{T}{D}\right) \cdot \frac{d_p^{1/6}}{D^{2/3}} \cdot \frac{1}{Z} \tag{16.5}$$

Baldi *et al.*[3] then argued that the assumption of ε_T at the base being proportional only to $\bar\varepsilon_T$ should be modified to allow for the effect of viscosity and energy dissipation between the impeller region and the base. With this reasoning they suggested that

$$Z = f(Re^*; T/D; C/D) \tag{16.6}$$

where Re^* was a Reynolds number for suspension defined by

$$Re^* = \frac{\rho_L D^3 N_{JS}}{\mu_L T} \tag{16.7}$$

As will be shown later, functional relationships like equation (16.5) are in reasonable agreement with experimental results.

Consider next the mean and fluctuating velocities near the base of the vessel. It has been shown[6-8] that the mean linear velocity, $\bar u$ (i.e. ignoring the direction of the velocity vector), and the mean r.m.s. fluctuating velocity, $\bar u'$, at different levels in the tank, are both given by

$$\bar u \propto \bar u' \propto \frac{N D^2}{(T^2 H)^{1/3}} \tag{16.8}$$

or re-writing in terms of the mean energy dissipation rate (or power input/unit mass), $\bar{\varepsilon}_T$, from equation (16.3), then

$$\bar{u} \propto \bar{u}' \propto \bar{\varepsilon}_T^{1/3} D^{1/3} \qquad (16.9)$$

At constant $\bar{\varepsilon}_T$,

$$\bar{u} \propto \bar{u}' \propto D^{1/3} \qquad (16.10)$$

so that for this condition, higher values of $\bar{u}$ and $\bar{u}'$ are obtained on scale-up and for large diameter impellers in a fixed size of vessel. In addition, the constants of proportionality in equation (16.8) decrease rapidly with increasing distance from the impeller. Thus, $\bar{u}$ and $\bar{u}'$ are greater at the base when the impeller is placed close to it. Again, these predicted effects of scale and impeller size and position agree quite well with experimental findings[9].

A number of other theoretical models have been proposed. Of these, Narayanan et al.[24] put forward a model with predictions markedly different from experimental findings; the formulation of another[25] has been severely criticized[26] and in any case, it only applies to conical-bottom vessels in the apex of which some solids always remain; and the quantitative application of laminar boundary layer theory was shown to give unrealistic numerical predictions[43].

16.4   Experimental measurement of particle suspension

16.4.1   Complete suspension speed, N_{JS}

This is the easiest condition to define and measure, particularly in the laboratory. A well-illuminated transparent tank and careful observation of the base to detect when no particle remains stationary upon it for more than 1–2 s is all that is required[10,11]. Reproducibility is helped by the fact that the last particles to be suspended come consistently from either directly beneath the impeller, or the base of the baffles or midway between the two depending on the flow pattern and impeller size. As a result, between 3–5% reproducibility can be achieved.

By using a vertical collimated light beam, a thin strip of the tank can be illuminated and photographs used to determine the particle distributions[9,11] at low concentrations. Alternatively, conductivity[13] or optical methods[14,15] can be used. At just complete suspension there is normally a decrease in concentration with tank height and often a clear liquid zone at the top with both phenomenon increasing rapidly with increasing particle size and density difference. These effects can be roughly quantified by the still-fluid terminal velocity which if less than 5 mm/s characterizes an easy suspension problem and if greater than 30 mm/s a difficult one[12]. If the particle size and density difference are sufficiently great, both the concentration gradient and the depth of the clear fluid zone are reduced by increases in N above N_{JS}. These observations are the basis of two other methods of determining N_{JS}.

The first method[16] offers distinct advantages where direct observation is

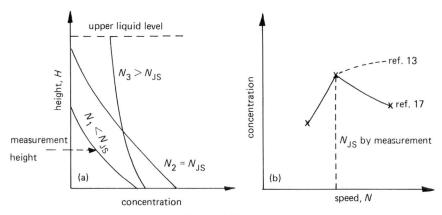

Figure 16.1(a)   Concentration profiles at different agitator speeds; (b) measured concentrations indicating N_{JS}

impossible. The concentration of the solid is determined at about 1/5 liquid depth either by sampling or by optical or conductivity methods. As the speed increases from zero, so does the local concentration until N_{JS} is reached. However, above this point the concentration falls again at a rate which depends on the concentration gradient at N_{JS} (see *Figure 16.1*). If sampling is used, the absolute value of the concentration depends on the rate and position but the break point occurs at the same value of N[17]. The break is generally a maximum but sometimes only a change of slope occurs[13]. The sampling method has also been shown to work quite well as a means of determining N_{JS} in three-phase mixing, i.e. gas dispersion and solid dispersion. Again, though in most cases peaks are found[18] in others there is only a change of slope. Though a very close correspondence between N at the peak or kink concentration and the visual N_{JS}[16-18] has been widely reported, recent work has indicated that the visual N_{JS} is usually just a little higher[48].

Einenkel and Mersmann[19] argued that since at N_{JS} there is often a clear zone at the top of the vessel, the quality of mixing at that point is not good enough. Therefore, they determined visually the speed at which the clear liquid at the top of the vessel was reduced to 1/10th of the total liquid height. Of course, for small particles, those that are suspended come to the top of the tank before the last particles lift off[48]. It does not appear to be a superior suspension criterion and it is perhaps noteworthy that their results differ significantly from the findings of others for particles less than about 300 μm.

16.4.2   Homogeneous suspension speed

Homogeneous suspension is very easy to define but extremely difficult to measure. Visual observation or photography, particularly where very large numbers of tiny particles are present, can be difficult and misleading; the disappearance of the clear liquid zone, together with measurement of the variation of pressure with depth, has been used but is a rather insensitive technique[20]. An optical or conductivity method appears most suitable, but when the former was used no attempt was made to detect the presence of the clear liquid interface[30]. Sampling can be used but this brings one to the heart of the problem.

The sample must be truly representative and this is extremely difficult to achieve. Sampling should be iso-kinetic, i.e. it should be in a region where the fluid and particles are moving at a velocity of the same magnitude and direction, and the withdrawal velocity should itself be at this same magnitude and direction. There are very few regions in a tank where the flow is of sufficiently low turbulence intensity and is sufficiently well defined to be sure that iso-kinetic sampling is being used. Indeed, it is quite difficult to find one point where the requirements can be properly met; and, even if one believes they are, an independent check that the sample taken does truly represent the condition at that point within the tank should, if possible, be made.

Homogeneous suspension is generally only important in continuous processing. To produce a continuous stirred tank reactor for particles of a wide size distribution equivalent to that for a fluid, the mean residence time and residence time distribution must be the same for all particle sizes and for the fluid. In addition, homogeneous suspension is required. However, deviations from homogeneous suspension (provided the residence time requirements are met) will have little effect on any process being carried out because the fluid is so well mixed that all particles will experience the same average environment. (On the large scale, this statement may not be true. In crystallizers, for instance, unacceptably high supersaturation levels may exist locally. If this type of problem arises, the agitation level required to overcome it will certainly ensure good suspension). Therefore, since homogeneous suspension is so difficult to measure with certainty and is intimately linked with solid withdrawal, this requirement can be modified as follows. The combination of agitation and withdrawal method must ensure that the particle size distribution and concentration in the vessel and the discharge (and feed if there is no particle growth or reduction or formation within the vessel) are all the same.

Experiments which demonstrate this have been called wash-out tests[16]. A mass of solids of known size distribution is placed in a tank and the concentration and distribution in the discharge are compared with that theoretically expected from a perfectly stirred tank model. *Figure 16.2* shows examples of good wash-out and bad wash-out curves, but experiments show quite conclusively[16,17,21] that a good wash-out, i.e. the ability to remove solids of identical size distribution and concentration as that in the tank, is as dependent on the method of withdrawal as on allowing homogeneous suspension.

16.4.3   Bottom or corner fillets

Whilst the above approximation to homogeneous suspension is especially useful for modelling systems where for example dissolution[22] or crystallization[23] occurs, in other cases conditions may be acceptable in which the particle size distribution and concentration in the tank are different from those in the exit stream. This condition may correspond to N_{JS} for the largest particles, or indeed a lower speed still may be tolerated with the build-up of fillets provided a steady state can be achieved. In this case, the average solid concentration and the proportion of the large particles in the tank may be considerably higher than in the discharge. In addition, the mean residence time of the solid may be a function of particle size and will generally be greater than that of the liquid. Of course, if all the solid in the tank comes from the feed, then at steady-state the inlet and outlet concentrations will be the same.

By using a combination of the wash-out technique and careful observation of fillet sizes, it was found[17] that in flat-bottomed, propeller-agitated, draught-tube baffled vessels the mass in the bottom fillet was inversely proportional to $N^{2.3}$. The maximum mass in the fillets was only 5% of the total solids present at speeds down to $0.6N_{JS}$. With $P \propto N^3$, this represents only 20% of the power at N_{JS}, a very considerable saving.

16.5   Experimental results and correlations for N_{JS}

16.5.1   Introduction

The factors which need to be considered and quantified are:

(a) Particle–fluid properties

1. *Particle–liquid density difference*[10,11,19,20,24,27-30,35].
2. *Fluid viscosity*[3,10,19,27,35]. For high viscosity, 8000 to 80000 m Pas, in the laminar flow region[31].
3. *Particle size*[3,10,11,15,19,20,27-30,34,35].
4. *Particle shape.* Its effect has not been quantified but appears to be small, provided the particles are approximately iso-dimensional[11]. However, small platelets of anthracite require much higher speeds than granular material of the same density[48].
5. *Particle size distribution.* Bourne[17] found that for a mixture of particle sizes N_{JS} was less than that required for the largest when only particles of that size were present. Baldi *et al.*[3,4] found that for binary mixtures, a mass mean diameter, $(d_p)_{4,3}$, enabled N_{JS} to be correlated with that from closely-sieved sizes and this has recently been confirmed for up to five sizes[48].
6. *The still-fluid terminal velocity.* Some workers have replaced each of the above properties with the still-fluid terminal velocity of the individual particle[20,30]. However, since the agitated vessel terminal velocity is considerably less than and difficult to relate to the still-fluid terminal velocity[6], this approach seems to offer no significant advantage except as a guide to difficult or easy suspension problems[12]. Einenkel and Mersmann[19] have introduced an additional complexity by considering the

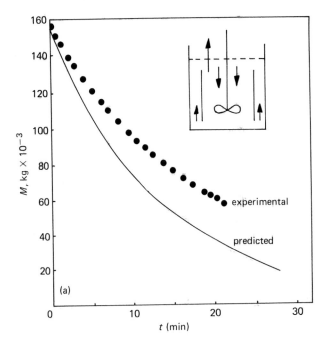

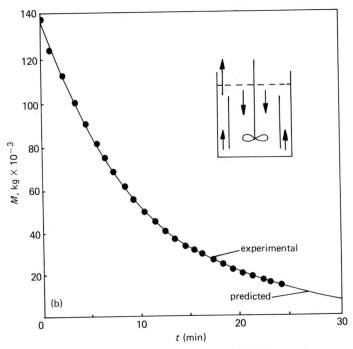

Figure 16.2   Examples of washout curves: (a) bad; (b) good

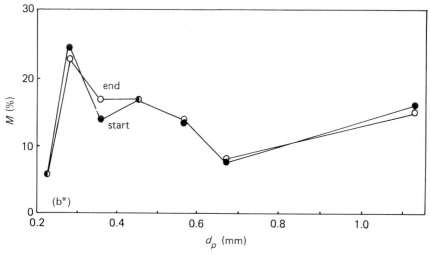

Figure 16.2(b)*   *Showing the unchanged size distribution with time[17]*

still-fluid terminal velocity of solids at the concentration of interest and the above comments again apply.

(b) *Solids concentration*

See refs. 3, 4, 10, 11, 15, 19, 24, 29, 30, 34 and 35 for details.

(c) *Geometry*

Most studies have used vessels with geometries which correspond closely to Rushton dimensions, i.e. $H = T$, with flat or dished bottoms with either radial impellers, axial flow impellers (propellers) or mixed flow (axial flow turbines). In general, four strip baffles of 1/10 to 1/12 their tank diameter to induce strong vertical flows have been employed though a small clearance at the wall or the base is sometimes used to prevent dead zones. Categories 1 to 3 below refer broadly to this geometry.

1. Comparison of impeller types[9,10,29,30,35].
2. Impeller elevation above the base[3,4,9-11,15,20,35].
3. Impeller to vessel size ratio[3,4,9 11,20,24].
4. Liquid depth[20,48].
5. Multiple impellers[20,48].
6. Other dimensional changes[13,15,16,20,21,25].

(d) *Scale-up*

See refs. 3, 4, 9, 10, 32, 33 and 48 for details.

16.5.2 Correlations for N_{JS}

Most workers have developed empirical correlations or graphical presentations to enable N_{JS} to be calculated. In general all are different; but the correlation of Zweitering[10] is based on experiments covering by far the widest range of the variables listed above and in an independent study, Nienow[11] found an almost identical relationship. Starting from a dimensional analysis, the correlation can be written as:

$$N_{JS} = \frac{S\nu^{0.1}d_p^{0.2}\left(\dfrac{g\Delta\rho}{\rho}\right)^{0.45}X^{0.13}}{D^{0.85}} \qquad (16.11)$$

where S is the dimensionless number accounting for the effect of geometry within categories 1, 2 and 3 above. The exponents in equation (16.11) are seen to be in good agreement with those in the theoretical development of Baldi *et al.* (equation 16.5) and also with those found by other workers as set out in *Table 16.1.* It is therefore recommended that Zweitering's correlation be used for small scale predictions except (a) where special geometries are involved and (b) where the correlations of others have been based on experimental conditions well away from those covered by Zweitering. The range of variables covered and special conditions are discussed below. For scale-up recommendations, see section 16.5.6.

16.5.3 Particle and fluid properties

The low exponent on d_p (complete range studied 63 μm to 7 mm) may seem a little surprising but can be accounted for by considering that both lift and drag are required to cause particle suspension. They in turn will depend on the linear velocity which the particle sees. As this velocity falls rapidly across the boundary layer to zero at the base (approximating to a turbulent boundary layer profile), large particles experience relatively large forces and small particles, small ones. Thus the effect of size nearly cancels out[1]. In addition, the imparted-turbulent energy theory predicts a low exponent[3,4]. For particles below about 200 μm, larger exponents are found[19,34]. However, it has been suggested that since N_{JS} falls away more rapidly than Zweitering's correlation would imply, its use therefore errs on the safe side[4].

The low exponent on ν (biggest range from 10^{-6} to 10^{-3} m^2/s (ref. 35)) is not at all surprising in such a turbulent flow system. However, there is rather a paucity of data for the effect of this parameter. The exponent of -0.33 for very tiny particles from the work of Einenkel and Mersmann[19] may be due to their experimental technique for determining N_{JS} as discussed earlier. Additional data in this region would be valuable.

The density difference (range 500 to 7500 kg/m^3) exponent is approximately the same as that for the terminal velocity of a particle. This seems quite reasonable[1] even though the drag coefficient in the Newton's law region (i.e. turbulent drag) is enhanced when the main flow is turbulent[6]. It also fits the

Table 16.1 Measurement of N_{JS}; scale and exponents on particle-fluid parameters

Author	Agitator* type	Exponent on				Scale of vessels (T m)	Baffles Y-yes: N-no
		X	v	$\Delta\rho$	d_p		
Zweitering[10]	R, A	0.13	0.1	0.45	0.2	0.15–0.7	Y
Nienow[11]	R	0.12	—	0.43	0.21	0.14	Y
Pavlushenko[27]	A	—	0.2	—	0.4	0.30	N
Kneule[29]	45°	0.17	—	0.5	0.17	0.15–0.4	Y, N
Narayanan[24]	R	0.22	—	0.5	<0.5	0.11	Y
Baldi[3]	R	0.13	0–0.23†	0.38–0.5†	0.13–0.17†	0.12–0.23	Y
Einenkel[19]	A	0.20●	0.1‡ / −0.33	0.5‡ / 0.57	0.17‡ / 0.67	0.14–0.79	Y
Herrindge[34]	R	0.18	—	0.42	0.3	0.15–1.0	Y
Rieger[35]	45°, R	0.13	0.16	0.42	−0.1	0.15–0.4	Y
Chapman[48]	A, 45°, R	0.12	—	0.40	0.15	0.29–1.8	Y
Weisman[20]	R	0.17		u_T		0.14–0.3	Y, N
Kolar[30]	45°, A	0.10		u_T		0.17–0.35	Y

* R—radial flow agitators; A—axial (propeller); 45°—45°-pitched blade turbine.
† Depending on clearance, C/H; ●, independent of conc. if >17% by volume.
‡ Upper figure for large particles; lower figure for small.

theory of Baldi *et al.*[3]. Obviously, density difference is the most important physical property causing difficulty for particle suspension.

16.5.4   Solid concentration

The effect of solid concentration X (range 0.1 to 20 wt %[10]) is covered by this exponent down to very small particle numbers. This is probably due to the fact that even two particles on the base will migrate towards each other and interact quite strongly. It has been suggested that around 17% by volume of solids, an effect similar to that found in hindered settling, should occur[19]. Above this concentration, no further increase in N_{JS} should be found. Evidence for this is given by Einenkel and Mersmann, the only workers to reach such high concentrations. However, as before, this result must be treated with caution. More work at high volume concentrations would be valuable.

16.5.5   Geometry

Graphs of S against T/D with T/C as a parameter for five impellers[10,11] are given in *Figure 16.3*. In general, the speed required decreases rapidly with an increase in D/T ratio ($N_{JS} \propto D^{-2}$ as implied by equation (16.8)[9] and slowly with a reduction in T/C. In general, for a particular value of T/C,

$$S \propto (T/D)^a \tag{16.12}$$

where $a = 0.82$ for propellers[10] and 1.3 for flat paddles[10] and disc turbines[11]. Therefore, from equation (16.11):

$$N_{JS} \propto SD^{-0.85} \propto T^a D^{-(0.85+a)} \tag{16.13}$$

and for one size of vessel

$$P_{JS} \propto N_{JS}^3 D^5 \propto D^{2.45-3a} \tag{16.14}$$

Thus for flat paddles and disc-turbines,

$$P_{JS} \propto D^{-1.45} \tag{16.15}$$

i.e. a relatively large, low speed agitator requires a lower power input. However, for propellers, the ratio of the agitator to vessel diameter is unimportant.

These general observations break down for most relatively large impellers ($D/T \geqslant 1/2$) close to the base of the vessel ($C/T \leqslant 1/6$)[9-11]. In all these cases a large stagnant zone directly beneath the agitator may have particles within it when all the remainder are very well suspended. The flat paddle is much less prone to this problem[9,10] presumably because its active blade length extends to the central shaft and very low S values are reported at T/D of 1.5.

A comparison of the relative efficiencies of different agitators in one size of vessel can also be made based on S values and power numbers. Since

$$N_{JS} \propto SD^{-0.85} \tag{16.16}$$

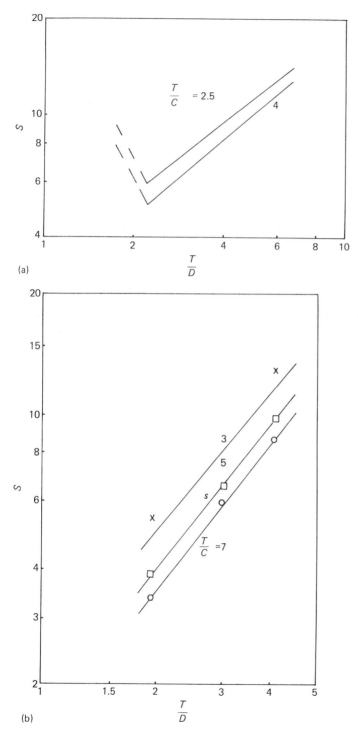

Figure 16.3   Five values for different impellers: (a) propeller[10]; (b) disc turbine[11];

(contd. overleaf)

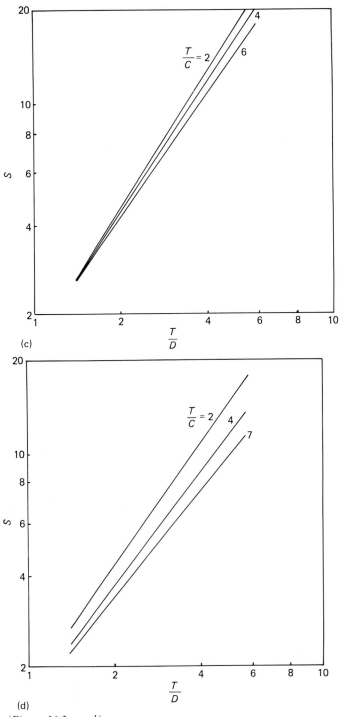

(c)

(d)

(*Figure 16.3 contd.*)
(c) *vaned disc*[10]; (d) *flat paddles (two blades, width* 0.5D)[10];

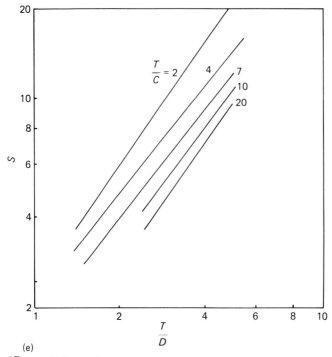

(e)

(*Figure 16.3 contd.*)
(e) *flat paddles (two blades, width 0.25D)*[10]

then for a particular size, position and type of agitator,

$$P_{JS} \propto P_0 N_{JS}^3 D^5 \propto P_0 S^3 D^{2.45} \tag{16.17}$$

Table 16.2 compares these relative efficiencies for the most common size $(D/T = 1/3)$ and clearance $(C/H = 1/3)$ and, as can be seen, the propeller requires by far the lowest power. If the optimum (i.e. minimum power) geometry for the different impellers is chosen for comparison, the low clearance flat paddle referred to above still consumes three times as much power as the propeller even at its optimum geometry of large size and low clearance (see *Table 16.3*).

Table 16.2   Relative power P_{JS} at $D/T = 1/3$ and $C/T = 1/4$

	S	Po	PoS³	$\dfrac{(PoS^3)}{(PoS^3)_{\text{propeller}}}$
Propeller	6.6	0.5	143	1
Vaned disc	8	4.6	2360	16.5
Disc turbine	8*	5†	2560	18
Paddle (2 blades, 1/4D)	8	2.5	1280	9
Paddle (2 blades, 1/2D)	6.3	5.9	1475	10

Data from ref. 10; except *ref. 11 and †ref. 36.

Table 16.3 Relative power P_{JS} at optimum geometry

	S	Po	D/T	C/T	$PoS^3D^{2.45}$	$\dfrac{(PoS^3D^{2.45})}{(PoS^3D^{2.45})_{propeller}}$
Propeller	6.6	0.5	1/3‡	1/4	9.7	1
Vaned disc	2.5	4.6	2/3	1/6	26.6	2.7
Paddle (2 blade, 1/2D)	2.3	5.9	2/3	1/7	26.6	2.7
Disc turbine	3.9*	5†	1/2	1/6	54	5.6

Data from ref. 10 except *ref. 11 and †ref. 36.
‡ At fixed clearance, $P_{JS} \neq f(D)$ for propeller.

Direct measurement of P_{JS} is in general agreement with the above trends but with less advantage to the propeller mainly due to variations in Po with geometry. The propeller is the most efficient[15] and relatively insensitive to position and relative size. The angle blade turbine is considerably more efficient than radial flow impellers[9,35] but its maximum efficiency is particularly sensitive to its positioning[9,15] with a clearance above the base (C/T) of 1/4 recommended. This sensitivity arises because of the particular flow pattern that the angle-blade turbine exhibits, i.e. it does not give a truly axial flow. At this position, its efficiency is comparable with that of the propeller[15].

Multiple impellers have little effect on the impeller speed required for complete suspension[20,48]. They produce a better distribution of solids, of course, but consume approximately n times the power where n is the number of impellers. Total liquid depth also has little effect[20,48]. To use S values from *Figure 16.3* for vessels in which $H \neq T$, it is tentatively recommended that C/T should still be used.

The base of a vessel is usually either flat, dished or conical, the latter especially in crystallizers. Dished- or flat-bottomed vessels have been found to have almost the same N_{JS} values[10,15]. However, if the dish approaches hemispherical, the proximity of the base to the impeller can lead to a significant enhancement in $(\bar{\varepsilon}_T)_{JS}$[15]. With conical bottoms, some solids always remain in the apex and correlations have been given allowing for this[13,25].

The use of a draught tube generally leads to a reduction in both N_{JS} and $(\bar{\varepsilon}_T)_{JS}$ when used in conjunction with either a propeller or an angle-blade turbine[17]. However, care should be taken with the position and size of the draught tube. For instance, if the bottom of the tube restricts the impeller outflow, the local head loss can cause a marked enhancement of $(\bar{\varepsilon}_T)_{JS}$[15]. Ideally, the flow area in the core, in the annulus, between the bottom of the tube and the base and between the top and the liquid surface should all be equal[17]. Finally, the use of a specially contoured base, maintaining the flow area constant (*Figure 16.4*) leads to an even further reduction in N_{JS} and $(\bar{\varepsilon}_T)_{JS}$[16] by eliminating all dead spots.

With a contoured bottom, the circulation velocity, v_c, is constant at all points. It was found that v_c for just complete suspensions was of the order of the still fluid terminal velocity, u_T, with (v_c/u_T) decreasing from 2 to 1 as the

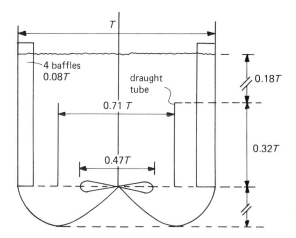

Figure 16.4   Geometry for minimizing $(\bar{\varepsilon}_T)_{JS}$ *(ref. 16)*

particle size increased. The lower values of (v_c/u_T) for larger particles is probably due to their greater protuberance into the main flow[1]. With a well-sized draught tube and a propeller, N_{JS} is lower by a factor of about 1.7 with a contoured bottom and by about 1.3 with a flat bottom[16,17]. Since Po is unaffected by such a draught tube[15], power reductions of about five-fold and two-fold are achieved respectively.

All these results indicate that using impeller power input to produce bulk flow is the best way of suspending particles. Propellers and angle blade turbines are 'high flow impellers' (see Chapter 8) and the addition of the draught tube further enhances this tendency. Radial flow impellers on the other hand are high shear types, best suited to gas or immiscible liquid dispersion. However, simple experiments using a downward jet of fluid have shown that flow alone is not sufficient to cause suspension and that jets require significantly higher power inputs than with rotating impellers[47].

16.5.6   Scale-up

For geometrically-similar systems,

$$N_{JS} \propto (\text{scale-factor})^{-b} \tag{16.18}$$

where the scale-factor may be impeller size or vessel size, e.g. from equation (16.11) $b = 0.85$. Since if Po is assumed constant, independent of scale,

$$\bar{\varepsilon}_T \propto N^3 \, (\text{scale-factor})^2 \tag{16.19}$$

then from equation (16.18)

$$(\bar{\varepsilon}_T)_{JS} \propto (\text{scale-factor})^{2-3b} \tag{16.20}$$

i.e. power/unit mass increases, remains constant or decreases depending on whether $b < 0.67$, $b = 0.67$ or $b > 0.67$ respectively and is extremely sensitive to the experimentally determined values of b. *Figure 16.5* shows the specific

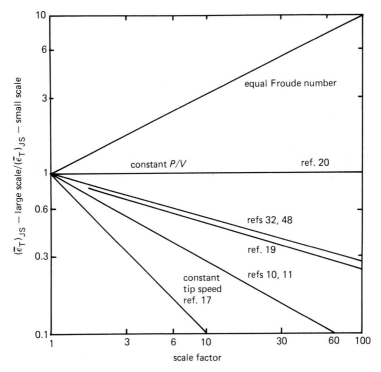

Figure 16.5   The dramatic effect of different scale-up relationships on $(\bar{\varepsilon}_T)_{JS}$

power, $(\bar{\varepsilon}_T)_{JS}$, changes as scale-up, developed from the literature by equations (16.18) to (16.20) where the work reported covered a wide range of vessel sizes. The large variation arises because of the extreme sensitivity to the exponent b. Also shown is the change in $(\bar{\varepsilon}_T)_{JS}$ when the scale-up rule of constant impeller tip speed, i.e. $N_{JS}D =$ constant, is used. This rule has been found to apply to the scale-up of draught tube, contoured-bottom vessels[17].

   Figure 16.5 also includes a manufacturer's scale-up rule[32]. A recently completed experimental study[48], measuring $(\bar{\varepsilon}_T)_{JS}$ and covering vessels from 0.3 to 1.83 m diameter, gave an almost identical result, i.e. $(\bar{\varepsilon}_{TJS}) \propto (\text{scale-factor})^{-0.28}$. Indeed, where $(\bar{\varepsilon}_T)_{JS}$ has been actually measured rather than inferred, a reduction on scale-up has always been reported[9,47,48]. Finally, the simple relationship (equation 16.10) would also suggest such a reduction.

16.6 Selection of geometry and the scale-up rule for $(\bar{\varepsilon}_T)_{JS}$

The propeller requires the lowest power and in addition, because changes of size have little or no effect on P_{JS} at any one operational scale, direct drive from the motor may well be possible. However, in this case, the fabrication costs of large propellers and, because of their weight, the cost of the shaft etc. becomes important. The 45° angle blade turbine is cheaper and provided the geometry is carefully chosen can be almost as efficient. Further reduction in power can be obtained by the inclusion of a draught tube but this implies greater fabrication costs. Draught tubes are, of course, used extensively in crystallizers.

For radial flow impellers, large D/T ratios in many cases offer a considerable reduction in P_{JS} at any one operational scale. However, any saving thus made may be offset by the increased cost of the gearbox required to produce the compatible low speed, particularly on the large scale. For particle suspension alone, they are clearly inferior but for three-phase agitation (gas dispersion and solid suspension) they are relatively insensitive to the effect of the gas flow[48].

The very special geometry proposed by Bourne *et al.* offers three important advantages over flat-bottomed tanks[21],

1. More certain scale-up is claimed;
2. Significantly lower impeller power and tip speeds;
3. Because of the lower speed, much less damage to suspended particles, which may be important in certain cases such as crystallization.

Of course, whether the manufacturer of such a special base could be justified economically is not certain. However, bottom entry impellers with an internal stuffing box are now being employed more frequently on the large scale for mechanical reasons. A simple shrouding might then be used to approximate the idealized shape, always taking care to avoid throttling of the flow.

On scale-up, there is considerable evidence that $(\bar{\varepsilon}_T)_{JS}$ reduces. However, the precise rate of reduction is still not firmly established. Clearly, the long standing rule-of-thumb of equal power/unit volume (or constant $(\bar{\varepsilon}_T)_{JS}$) leaves a margin of safety.

16.7 Homogeneous suspension and solid withdrawal

16.7.2 Homogeneous suspension

As already pointed out, the determination of the speed for homogeneous suspension is complicated because of the intimate link with solid withdrawal, whether continuously or in samples. In batch experiments, Weisman and Efferding[20] found that, above N_{JS}, the slurry-clear liquid interface height increased with and could be correlated with $\bar{\varepsilon}_T$. Lifting the baffles off the base, using multiple agitators and reducing the clearance above the base of the lowest agitator, all gave a reduction in the power necessary to achieve a given suspension height.

Measurement of the static head in the tank as a function of height indicated a constant concentration of solids over the depth of the slurry and it was concluded[20] that when the interface reached the upper surface a homogeneous suspension had been achieved. Many other workers[9,10-12,17], however, report a concentration gradient, especially for larger particles, and sediment transport theory[37] might lead one to expect this. Weisman and Efferding used rather small particles (10–180 μm) and perhaps this is the explanation. The matter is not resolved but the major problem for chemical processing is one of solid withdrawal during continuous operation.

16.7.2   Solid withdrawal

Satisfactory withdrawal requires iso-kinetic sampling or at least a good approximation to it. Draught-tube, baffled tanks have a relatively well-defined vertical flow, particularly around the mid-depth of the draught-tube. A withdrawal pipe then can be aligned with the flow[17] but the velocity must be greater than $5u_T$ for satisfactory transport[38] and at the same time it must satisfy the stoichiometry. These two requirements cause problems in small scale systems but intermittent withdrawal[16] at the point in the annulus of maximum velocity[38] may provide an answer.

If only the co-linear requirement is met, then solids are preferentially withdrawn if the withdrawal velocity is less than the fluid velocity and *vice versa*. This leads respectively to solids having a smaller or greater residence time than that of the fluid and the effect has been quantified for draught-tube, baffled, flat-bottomed tanks[39] and for similar vessels without a draught tube[40,41].

For turbine agitation, neither iso-kinetic or co-linear withdrawal is easy. The horizontal discharge leaves at about 60° to the radius close to the impeller blades and this changes to about 45° further out. Elsewhere the flow is truly three dimensional and of high turbulence intensity. Rushton[42] was able to recommend a radial withdrawal tube protruding 1/40th of a tank diameter into the vessel and Zacek *et al.*[45] achieved K values (see below) of 1 on a very small scale using intermittent withdrawal.

The simplest way of quantifying the deviation from the idealized CSTR solids residence time distribution is to measure the steady-state separation coefficient K defined as

$$K = \frac{C_{out}}{\bar{C}} \tag{16.21}$$

where C_{out} is the solids concentration in the outlet flow and $\bar{C}$ is the mean concentration in the vessel. K itself depends on the geometry, the solid and liquid properties and the withdrawal method. For a particular wash-out experiment,

$$K = \frac{C_{out}(t)}{\bar{C}(t)} = \text{constant} \tag{16.22}$$

if it is assumed that K is independent of the change in solid concentration as wash-out occurs[40,41]. A mass (or volume) balance leads to the following expression for the wash-out:

$$C_{out}(t)/\bar{C}(0) = K \exp\{-t/(\tau_L/K)\} \tag{16.23}$$

or

$$C_{out}(t)/C_{out}(0) = \exp(-t/\tau_S) \tag{16.24}$$

where τ_L and τ_S are the mean residence time of the liquid and solid respectively. Thus K can be obtained from the intercept or slope of the experimental data plotted on semi-log co-ordinates from equation (16.23) and τ_S can be obtained similarly from the slope from equation (16.24), see *Figure 16.26*. If the initial mass of solids added is M, then $M = \int_0^\infty C_{out}(t)\, dt$ if all the solids are washed out. Thus the technique also enables the detection of any trapped material. The use of K has been demonstrated in the design of solid–liquid chemical reactors[22] and of crystallizers[43,44].

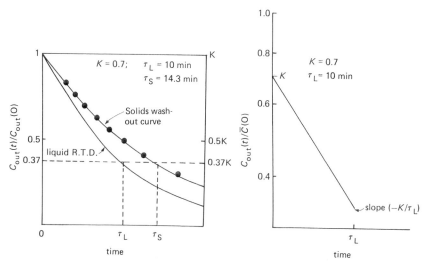

Figure 16.6   The steady-state separation coefficient, K, and its relationship to τ_L and τ_S

16.8   The dispersion of floating solids

This problem has rarely been studied. This process is generally more energy-intensive than solids suspension and is complicated by the fact that if the solids are fine, they may contain large amounts of entrapped air thereby reducing their effective density. This latter difficulty is further enhanced if the solids cannot be wetted or only wetted with difficulty.

Joosten *et al.*[46] found that a central vortex, obtained without baffles was not as effective at sucking the solids down as an off-centre vortex obtained by means of a short baffle (see *Figure 16.7*). They found a large ($D/T = 0.6$) four-bladed, 45°-pitch impeller placed near the base of the vessel required the minimum power and that for vessels from 0.27 to 1.8 m diameter, the minimum speed required, N_{DF}, could be determined from the equation

$$(Fr)_{DF} = \frac{N_{DF}^2 D}{g} = 3.6 \times 10^{-2} \left(\frac{D}{T}\right)^{-3.65} \left(\frac{\Delta\rho}{\rho_L}\right)^{0.42} \tag{16.25}$$

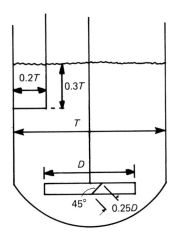

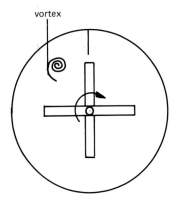

Figure 16.7   The preferred baffle configuration and the position of the vortex[46] (approximately to scale)

with particle concentration and size having a negligible effect. The particle size was between 2–10 mm and therefore air entrapment was not a problem.

A very recent study[47] used much finer plastic particles from 200 to 500 μm. Quantitative conclusions were marked by problems of air entrapment but the work showed that if a standard tank geometry was used, i.e. $H = T$ with four $0.1T$ baffles, an upward pumping propeller with a clearance above the base of 2/3rd the liquid height required least energy.

16.9   Conclusions

Any design for solids suspension or continuous withdrawal must inevitably be a compromise. Propellers (or 45°-angle blade turbines) offer advantages in that less power is required to completely suspend the particles than in other systems and the addition of a draught-tube reduces it further. Iso-kinetic withdrawal (or at least co-linear withdrawal) is also made much easier. If, as a further complexity, a contoured-bottom is introduced, lower power still is required and scale-up should be simplified.

If for other reasons, e.g. in a three-phase system, radial flow agitators are preferred, then least power is required if these are of relatively large D/T ratio and placed close to the vessel base (particularly if flat-bottomed). To prevent problems with a stagnant region at the centre, the turbine blades should extend in to the agitator shaft. Large agitators operate at low speed and the power saved here must be balanced against the initial high cost of gearboxes. With radial flow agitators, it is difficult to design for iso-kinetic withdrawal.

The physical property parameter causing the most difficulty in achieving particle suspension is density difference. Though a number of correlations have been put forward for calculating the agitator speed required to achieve complete suspension, on balance that proposed by Zweitering in 1958 is still recommended. This is because it covers the widest range of variables, is dimensionally homogeneous and the dependency of N_{JS} on physical properties and geometry are generally very similar to those reported by other workers.

All workers who have measured the specific power on different scales have shown it to fall. However, the precise relationship has not been established. Recent experimental work suggests and one manufacturer recommends

$$(\bar{\varepsilon}_T)_{JS} \propto (\text{scale-factor})^{-0.28}$$

Clearly the old rule of constant power/unit volume is conservative.

Little has been done on dispersing floating solids though the work of Joosten *et al.* represents a significant contribution. This is an area in which much work of practical value could be done, especially with fine, air-entrapping and non-wetting solids. It appears that the process is very energy intensive.

Notation

a	exponent
b	exponent
C	agitator clearance above base (m)
C_{out}	solids concentration in outlet flow (kg/m^3)
$\bar{C}$	solids concentration in vessel (kg/m^3)
D	agitator diameter (m)
d_p	particle size (m)
Fr	Froude number, dimensionless
g	gravitational constant (9.81 m/s^2)
H	liquid height in vessel (m)
K	steady state separation coefficient, dimensionless
M	initial mass of solids in vessel (kg)
N	agitator speed (rev/s)
P	power input to the liquid (W)
Po	power number, dimensionless
Re	Reynolds number for agitation, dimensionless
Re^*	Reynolds number for suspension (see equation 16.7), dimensionless
S	dimensionless constant, see equation (16.11)

T	vessel diameter (m)
t	time (s)
$\bar{u}$	the mean velocity (modulus) at a level in the vessel[6] (m/s)
$\bar{u}'$	mean r.m.s. velocity at a level in the vessel[6] (m/s)
u_T	still fluid terminal velocity of a particle (m/s)
V	vessel volume (m^3)
v_c	mean linear velocity in the annulus of a draught-tube (m/s)
X	% by wt of solids in the suspension, dimensionless
Z	a suspension parameter, see equation (16.4)
$\bar{\varepsilon}_T$	mean energy dissipation rate throughout the vessel $(=P/\rho V)$ (W/kg)
ε_T	local energy dissipation rate (W/kg)
v	kinematic viscosity (m^2/s)
μ_L	fluid dynamic viscosity (Pa s)
ρ_L	fluid density (kg/m^3)
$\Delta\rho$	density difference between particle and fluid (kg/m^3)
τ_L	liquid mean residence time (s)
τ_S	solid mean residence time (s)

Subscripts

JS   the condition of agitation at which particles just become completely suspended

DF   the condition of agitation at which floating particles are all just dispersed

References

1 GONCHAROV, V. N. (1962) *Dynamics of Channel Flow*, Leningrad.
2 KOLAR, V. (1967) *Coll. Czech. Chem. Comm.*, **32**, 526.
3 BALDI, G., CONTI, R. and ALARIA, E. (1978) *Chem. Eng. Sci.*, **33**, 21.
4 CONTI, R. and BALDI, G. (1978) *Proc. Int. Symp. on Mixing, Mons, Belgium*, Paper B2.
5 LEVICH, V. G. (1962) *Physico-Chemical Hydrodynamics*, Prentice-Hall, New Jersey.
6 NIENOW, A. W. and BARTLETT, R. (1975) *Proc. of 1st European Conf. on Mixing and Separation, Cambridge 1974*, BHRA, Cranfield, pp. B1–15.
7 LEVINS, D. M. and GLASTONBURY, J. (1972) *Trans. I. Chem. E.*, **50**, 32.
8 SCHWARTBERG, H. G. and TREYBAL, R. E. (1968) *Ind. Eng. Chem. (Funds.)*, 7, 1.
9 NIENOW, A. W. and MILES, D. (1978) *Chem. Eng. J.*, **15**, 13.
10 ZWEITERING, T. N. (1958) *Chem. Eng. Sci.*, **8**, 244.
11 NIENOW, A. W. (1968) *Chem. Eng. Sci.*, **23**, 1453.
12 OLDSHUE, J. Y. (1969) *Ind. Eng. Chem.*, **61** (9), 79.
13 MUSIL, L. (1976) *Coll. Czech. Chem. Comm.*, **41**, 839.
14 CALDERBANK, P. H. and JONES, S. J. R. (1961) *Trans. I. Chem. E.*, **39**, 363.
15 CLIFF, M. H., EDWARDS, M. E. and OHIAERI, I. (1981) *Proc. Conf. on Fluid Mixing, Bradford, 1981*, I. Chem. E. Symp. Ser. No. 64, pp. M1–M11.
16 AESBACH, S. and BOURNE, J. R. (1972) *Chem. Eng. J.*, **4**, 234.
17 BOURNE, J. R. and SHARMA, R. N. (1975) *Proc. of the 1st European Conf. on Mixing and Separation, Cambridge, 1974*, BHRA, Cranfield, pp. B3–25 to B3–39.
18 CHAPMAN, C. M., NIENOW, A. W. and MIDDLETON, J. C. (1981) *Trans. I. Chem. E.*, **59**, 134.
19 EINENKEL, W. D. and MERSMANN, A. (1977) *Verfahrenstechnik*, **11**, 90 (in German).
20 WEISMAN, J. and EFFERDING, L. E. (1960) *A. I. Ch. E. J.*, **6**, 419.
21 BOURNE, J. and SHARMA, R. N. (1974) *Chem. Eng. J.*, **8**, 243.

22 MATTERN, R. U., BILOUS, O. and PIRET, E. J. (1957) *A. I. Ch. E. J.*, **3**, 497.
23 RANDOLPH, A. D. and LARSON, M. A. (1971) *Theory of Particulate Processes*, Academic Press, New York.
24 NARAYANAN, S., BHATIA, V. K., GUHA, D. K. and RAO, M. B. (1969) *Chem. Eng. Sci.*, **24**, 223.
25 MUSIL, L. and VLK, J. (1978) *Chem. Eng. Sci.*, **33**, 1123.
26 DITL, P. and RIEGER, F. (1980) *Chem. Eng. Sci.*, **35**, 763.
27 PAVLUSHENKO, I. S., KOSTIN, N. M. and MATVEEV, S. E. (1957) *J. Appl. Chem. (USSR)*, **30**, 1235.
28 OYAMA, Y. and ENDOH, K. (1956) *Chem. Eng. (Tokyo)*, **20**, 66.
29 KNEULE, F. (1956) *Chemie Ing. Tech.*, **28**, 221.
30 KOLAR, V. (1961) *Coll. Czech. Chem. Comm.*, **26**, 613.
31 HIRKESORN, F. S. and MILLER, S. A. (1953) *Chem. Eng. Prog.*, **49**, 459.
32 GATES, L. E., MORTON, J. R. and FONDY, P. L. (1976) *Chem. Engng.*, **83**, 102.
33 MERSMANN, A., EINENKEL, W. D. and KAPPEL, M. (1976) *Int. Chem. Eng.*, **16**, 590.
34 HERRINDGE, R. A. (1979) *Proc. 3rd European Mixing Conf., York, 1979*, BHRA, Cranfield.
35 RIEGER, F., DITL, P. and NOVAK, V. (1978) *CHISA Congress, Prague*, Paper A5.3.
36 BATES, R. L., FONDY, P. L. and CORPSTEIN, R. R. (1963) *Ind. Eng. Chem. (Proc. Des. and Dev.)*, **2**, 310.
37 BATCHELOR, G. K. (1968) *Proc. 2nd Australasian Conf. on Hydraulics and Fluid Mechanics, Auckland, New Zealand*.
38 JAHNSE, A. M. and DE JONG, E. J. (1974) *Industria Chemica and Petrolifera*, **11**, 26.
39 STEVENS, J. D. and DAVITT, J. P. (1974) *Ind. Eng. Chem. (Fundls.)*, **13**, 263.
40 BALDI, G. and CONTI, R. (1978) *Proc. Int. Symp. on Mixing, Mons, Belgium, 1978*, Paper B5-1.
41 MACHON, V., KUDRNA, V. and HUDCOVA, V. (1980) *Coll. Czech. Chem. Comm.*, **45**, 2152.
42 RUSHTON, J. H. (1965) *A. I. Ch. E.—I. Chem. Eng. Symp.*, No. 10, p. 1.
43 BOURNE, J. R. and ZABELKA, M. (1980) *Chem. Eng. Sci.*, **35**, 533.
44 JAHNSE, A. K. and DE JONG, E. J. (1977) in *Industrial Crystallisation* (Mullin, J. W., ed.), Plenum Press, p. 403.
45 ZACEK, S., NYVELT, J., GARSIDE, J. and NIENOW, A. W. (1982) *Chem. Eng. J.*, **23**, 111.
46 JOOSTEN, G. E. H., SCHILDER, J. G. M. and BROERE, A. M. (1977) *Trans. I. Chem. E.*, **55**, 220.
47 OHIAERI, I. (1981) PhD thesis, University of Bradford.
48 CHAPMAN, C. M. (1981) PhD thesis, University of London.

Chapter 17

Gas–liquid dispersion and mixing

J C Middleton

Imperial Chemical Industries, p.l.c., New Science Group, Runcorn

17.1 Introduction—classification of gas–liquid mixing problems

Many examples of gas–liquid contacting operations are found in the process industries, often involving gas incorporation or absorption into liquid, perhaps with chemical reaction in the liquid, washing or humidifying a gas stream, removal of gas from liquid, and so forth.

Different contexts have different priorities for design. For example in effluent aeration or many fermentations, systems are dilute and reactions slow, so large vessels are used and energy efficiency is important but mass transfer intensity requirements are modest. In chlorination and sulphonation, the reactions are fast and the gases quite soluble, so high intensity mass and heat transfer are essential but contact times may be short. With many oxidations and hydrogenations, the gases are less soluble and longer contact time, or gas recycling, is needed; and often mixing pattern and solid suspension are also important. For food batters and creams, the rheological limitations overrule and control of local shear rate and temperature are all-important.

For the purpose of this chapter, the discussion will be centred on gas–liquid contacting in mixers, with emphasis on the mixing vessels commonly used in the chemical industry. Other types of mixer are used (for example 'motionless' mixers in pipes, jets, ejectors, beaters) and of course some degree of mixing does occur in other gas–liquid devices (bubble columns, plate columns, spray towers) not designated as 'mixers'.

The first classification of gas–liquid mixing problems is according to liquid viscosity, since this dictates the basic mixing mechanics and types of equipment applicable in practice.

For low viscosities (say $\leqslant 0.2$ Pa s), turbulence can be used to obtain good mixing, high interfacial area and high mass and heat transfer coefficients, and degassing can be achieved by gravity. Suitable devices include (see *Figure 17.1*):

Simple bubble columns. In which gas is sparged into the liquid at the bottom of the contactor;
Plate columns. Gas is redispersed and disengaged at each plate up the column;
Mechanically agitated vessels. In which an impeller rotates in a tank (usually baffled) to give enhanced rates of mass transfer;

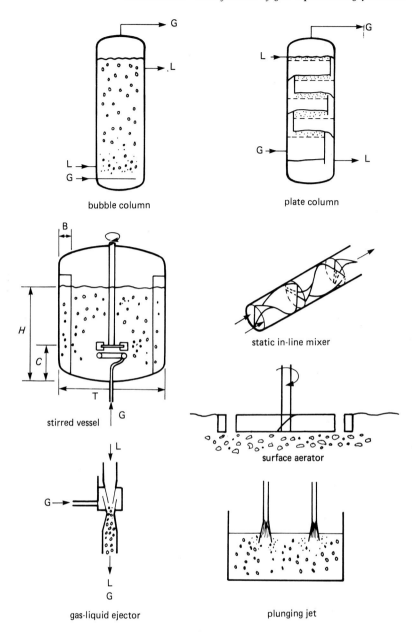

Figure 17.1   Gas–liquid contacting devices for low-viscosity systems

In-line static mixers[1,2]. Mixing energy is derived from the flow itself;
Jet ejectors. Gas is sucked into a liquid flow in a tube and dispersed by intense turbulence;
Plunging jets. Gas is entrained into the liquid surface by a liquid jet;
Surface aerators. Agitators at the liquid surface entrain gas from the head space into the liquid.

The choice of a particular type of equipment for a given duty will depend upon many factors such as the scale of operation and required rate of mass transfer. A comparison of the operating characteristics of some gas–liquid contacting devices is presented in *Table 17.1.*

For high viscosities, turbulence cannot be achieved in practice and a mechanism which is a combination of distributive and laminar shear or elongational mixing has to be used for incorporation of gas. Suitable devices include (see *Figure 17.2*):

Dough mixers. Bubbles are incorporated into the liquid surface;
Dynamic in-line mixers. These are capable of generating high shear rates suitable for producing a high gas content;
Scraped-film devices. A thin film is generated by a blade moving near a surface.

Highly viscous systems are difficult to de-gas under gravity, so degassing is best done in a thin-film device.

Further factors which should be taken into account in equipment selection are:

The rheology of the liquid;
The presence of solid particles;
The location of the main mass transfer resistance (in the gas or the liquid);
The optimization of heat transfer requirements;
The desired flow pattern for each phase (especially for reactors);
The importance of gas disengagement and foaming.

Unfortunately, tidy, well-established design methods to cover these factors are not yet available but research is actively improving the situation.

In the process industries, low viscosity Newtonian liquids predominate so this chapter concentrates on these. Even within this category, the performance of gas–liquid contactors can be influenced by poorly understood surface phenomena, which particularly affect bubble size and coalescence, so is not yet predictable from first principles.

This problem is usually overcome by basing designs on *scale-up* from laboratory or pilot-plant experiments. As will be seen later, even if this is done, scale-up is fraught with difficulties because of the empirical and incomplete nature of the available correlations which often have only a tenuous link to the basic mechanisms. Hence the designers' tendency to proliferate the 'standard' configurations, optimal or not.

In this discussion, attention is focussed upon gas dispersion in mechanically-agitated vessels which are generally the most effective and flexible contactors for many duties. The following sections will also be primarily concerned with situations in which it is necessary to absorb some of the dispersed gas into the liquid phase. However, the fundamental concepts set down for these systems also have some relevance to the other gas dispersing devices.

Table 17.1 Comparison of different gas–liquid contacting devices

Device	$k_L a$ (s^{-1})	V (m^3)	$k_L a V$ (m^3/s) ('duty')	a (m^{-1})	ε_L	Liquid mixing	Gas mixing	Power per unit volume (kW/m^3)
Baffled agitated tank	0.02–0.2	0.002–100	10^{-4}–20	~200	0.9	~Backmixed	Intermediate	0.5–10
Bubble column	0.05–0.01	0.002–300	10^{-5}–3	~20	0.95	~Plug	Plug	0.01–1
Packed tower	0.005–0.02	0.005–300	10^{-5}–6	~200	0.05	Plug	~Plug	0.01–0.2
Plate tower	0.01–0.05	0.005–300	10^{-5}–15	~150	0.15	Intermediate	~Plug	0.01–0.2
Static mixer (bubble flow)	0.1–2	Up to 10	1–20	~1000	0.5	~Plug	Plug	10–500

These are orders of magnitude for comparison, *not* design figures. They are based on air and water physical properties. Other considerations such as heat transfer, materials of construction, suspended solids, hazardous liquids or gases (requiring low inventory) must obviously be taken into account.

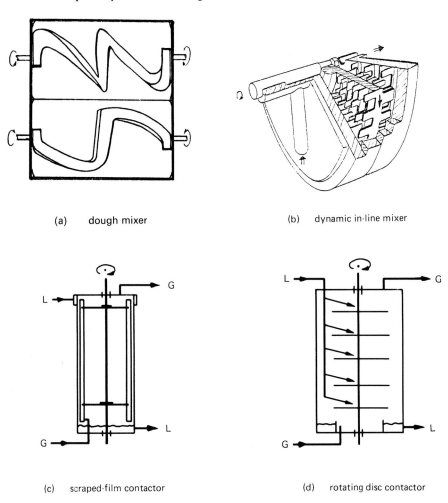

(a)   dough mixer

(b)   dynamic in-line mixer

(c)   scraped-film contactor

(d)   rotating disc contactor

Figure 17.2   Gas–liquid contacting devices for high-viscosity systems

17.2   Types and configurations of turbulent gas–liquid stirred vessels

The design of agitator and vessel shown in *Figure 17.1*, with a 6-blade 'Rushton' disc turbine, is now traditional and has been arrived at after many years of experience. However, recent research on the hydrodynamics of the system (e.g. Bruijn *et al.*[3], van't Riet[4]) has shown that other disc turbine agitators (with 12 or 18 blades, and/or with concave rather than flat blades (see *Figure 17.3*), have advantages. Probably even better designs will emerge from future work of this kind.

A suggested configuration for agitated vessels for gas–liquid contacting is, therefore, cylindrical (with vessel ends appropriate to the pressure), fitted with full wall baffles, width $T/10$. Typically in industrial practice, four baffles are

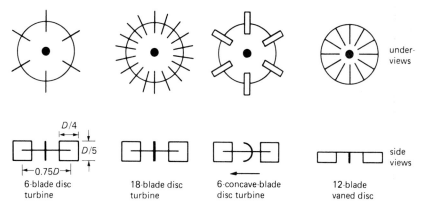

Figure 17.3   Agitators for gas–liquid dispersion

used for vessels less than 3 m diameter and six or eight in larger vessels. This number should be maintained even if other obstacles, e.g. coils etc., are present. The gas sparge ring is placed below and nearer than $D/2$ to the agitator. The height of the dispersion should be T, $2T$ or $3T$. If $H = T$, one agitator is required; if $H = 2T$ or $3T$, additional agitators should be mounted on the shaft, separated by a distance T. Recommended agitators are disc turbines with 18 flat blades or six concave blades, diameter $D = T/4$ to $T/2$, mounted a distance $C = T/4$ to $T/3$ from the vessel bottom.

The general range of application of such vessels is:

Liquid viscosity up to about 0.2 Pa s (Reynolds Number $\rho N D^2/\mu \geqslant 10^4$, i.e. the flow should be turbulent);
Gas superficial velocity up to about 0.2 m s^{-1} (possibly higher with the new agitator designs);
Shaft power input from 1 to 7 kW/m^3 (occasionally up to 15 kW/m^3 in small vessels);
Volume up to 100 m^3;
Small particle concentrations up to about 30% by wt.

These typical operating conditions should lead to specific interfacial areas of 100–500 m^2/m^3 and gas hold-up fractions up to about 0.4. (The absolute values depend on liquid properties and may be much greater in boiling or foaming systems.) They should also ensure good gas dispersion (see section 17.4).

Some alternative agitators are available which can be useful in certain circumstances but for which much less design information is published. For example, the vaned disc (*Figure 17.3*) is satisfactory where no recirculation of gas from above the agitator is required. 'Self aspirating' agitators have been on the market for many years[34-36]. These communicate the low pressure region behind the blades with the head space or a gas supply source via a hollow shaft and therefore use some of the input shaft energy to draw in the gas. Often they have multiple plates, sometimes curved like a centrifugal pump impeller, and a close-clearance stator around the agitator. They eliminate the need for recycle compressors, but are not capable of drawing in very large gas flowrates and are, of course, inflexible in that gassing rate depends upon agitator speed.

The roles of the agitator in a gas–liquid system are to:

(a) Break the gas into small bubbles for high interfacial area;
(b) Disperse the bubbles throughout the liquid;
(c) Keep the bubbles in the liquid (i.e. recirculate) for sufficient time;
(d) Mix the liquid throughout the vessel;
(e) Provide turbulent eddies to feed liquid to and from the interfaces;

and possibly

(f) Move the liquid past heat exchange surfaces and maximize heat transfer coefficients;
(g) Maintain particles in suspension.

High energy turbulence is required for (a) and high flow for (b, c, d, f, g). With turbines, high turbulence is created by the energetic roll vortex field which forms behind the blades especially if they have unobstructed inside edges as in the disc configurations (see *Figure 17.3*). For all the gas to flow through this region, it must enter the vessel close to and preferably underneath the disc. Once near the blade vortex (within say $D/50$ away from the blade edge), the centrifugal force will draw the gas in. To ensure sufficient turbulence is created, a power input > 100 W/m^3 must be added to the gas–liquid mixture by the agitator. Alternatively a criterion of a tip speed, $\pi ND > 1.5$ m/s or a Froude number, $N^2D/g > 0.1$ are often quoted.

To maximize the gas handling capacity of the impeller, many blades should be used on the agitator and they should preferably have the concave shape shown in *Figure 17.3*. For mechanical stability, an even number of blades, not less than six, should be used.

The nature of the gas inlet device is of only secondary importance if the design is such that gas is effectively captured and dispersed by the agitator. For efficient mass transfer (see section 17.8), a multiple orifice ring sparger is recommended of gas outflow diameter $= 3D/4$. However, it is only marginally better than a single open pipe sparging centrally beneath the disc. Spargers should always be nearer than a distance $D/2$ below the agitator line.

Surface entrainment of gas with the recommended configuration will be slight except at very high speeds. With sparged gas passing through, the extent of surface entrainment is very much reduced (Nienow et al.[5]), so it will be ignored in the following discussion.

17.3 A design basis for gas–liquid agitated vessels

A design will typically be required to meet specified rates of gas–liquid mass transfer and heat transfer and achieve suitable mixing of the liquid and gas phase at a certain throughput (or batch size). The designer will need to know the power consumption, gas holdup (voidage) fraction, ε_G, and foam height associated with such duty.

These factors depend upon many variables, of which the basic ones are:

System variables. Viscosity, density and thermal conductivity of the liquid, interfacial tension, diffusion coefficients, chemical reaction rate constants;
Operating variables. Impeller speed, gas flow rate, liquid volume;
Equipment variables. Impeller type and diameter, geometry of the equipment.

The most complex specification that the designer must be able to meet is the rate of gas–liquid mass transfer. Mass transfer rates are calculated from the basic equation

$$J = k_L a V \overline{(C_{LA}^* - C_{LA})} \tag{17.1}$$

Its use requires knowledge of the interphase mass transfer coefficient, k_L, the interphase surface area, a (or their product $k_L a$), and the appropriate average mass transfer driving force $\overline{(^*_{LA} - C_{LA})}$ and the dispersion volume, V, which is related to the gas hold-up in the vessel, ε_G. Sections 17.4 to 17.8 consider the evaluation of these quantities in some detail.

Unfortunately much of the published work on the dispersion of gases is restricted to air–water systems and it is difficult to extend the results of such work to other systems of industrial importance. Differences in behaviour are generally observed between pure liquids and solutions. Often in the latter case, coalescence of bubbles is greatly reduced ('non-coalescing' systems) leading to smaller bubble sizes, larger interfacial areas and greater gas hold-ups than with pure liquids ('coalescing' systems). This is discussed further in section 17.5. Small bubbles also follow the fluid more readily and therefore they recirculate more easily back to the impeller region. There is thus a distinct difference in behaviour between a pure and impure liquid, often even when the impurity is present in very small quantities.

As a result of these complexities, design from first principles is generally not possible. The designer must avoid this problem by characterizing his system from careful small scale experiment and scaling-up by using semi-empirical relationships such as those given below. In scaling-up, consideration must be given to which are the most important parameters in the process in question.

The experiments must be done, preferably at steady state, at the same conditions of temperature, pressure and concentrations (even of minor components) as in the intended process, to avoid unpredictable changes in coalescence. They must be carried out in a system of the recommended configuration, preferably at similar power per unit volume and superficial gas velocity as may be used at full scale. The agitator speed should be varied to check whether mass transfer is important and gas hold-up should be measured (e.g. by level probe or γ-ray density scan). In cases involving chemical reactions, further tests are required as described in section 17.10.

17.4   Power consumption

Knowledge of the power, P_g, absorbed by the gas–liquid dispersion from the agitator is required for the determination of mass transfer rates, $k_L a$, gas hold-up and interfacial area on the large scale from small scale tests and, of course, for mechanical design. The ungassed power P is readily obtained from:

$$P = Po\rho N^3 D^5 \tag{17.2}$$

where the ungassed power number Po may be obtained from *Table 17.2*.

These values are for Newtonian liquids with $Re > 10^4$ with no surface aeration. They refer, of course, to 'agitator shaft' power, so allowance must be made for electrical and friction losses when calculating motor power input.

The sparged gas stream also imparts some power to the system but this is often negligible compared with the mechanical power, P_g. However, strictly the combined power input should be included in correlations involving the energy dissipation rate per unit volume, for example, in predictions of bubble size, gas hold-up and mass transfer coefficients. These are considered in subsequent sections.

When gas is sparged into these agitators at a given speed, N, the power decreases (see *Figure 17.4*) because of the formation of gas cavities behind the

Table 17.2   Recommended unaerated power numbers

Geometry	Po
4 flat blade disc turbine	4.0
6 flat blade disc turbine	5.5
12 flat blade disc turbine	8.7
18 flat blade disc turbine	9.5
6-concave blade disc turbine	4.0
16 blade vaned disc ($W/D = 0.1$)	4.0

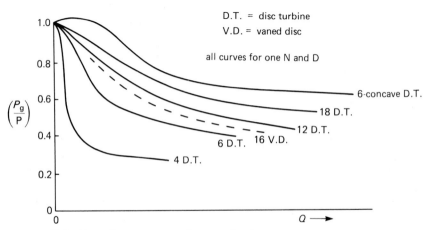

Figure 17.4   Typical power curves for gassed agitators

blades (*Figure 17.5*). The extent of this decrease is not yet fully predictable from basic principles as discussed below. Several workers have attempted to correlate P_g as a function of the system and operating variables. These included empirical relationships (e.g. Calderbank[6], Michel and Miller[11], Pharamond[12], Hassan and Robinson[13]), and graphical techniques (e.g. Nagata[14]) etc. but these take no account of the fluid mechanics in the impeller region. Such investigations have been mainly based on small scale vessels using water and are found to be inapplicable for large vessels and for impure and rheologically-complex fluids.

A more useful approach for general prediction is to investigate the separate factors contributing to the behaviour of P_g. The key to the prediction of P_g is understanding the formation of the captive 'cavities' of gas which are drawn by centrifugal force into the roll vortices formed behind the blades (see *Figure 17.5*). The gas drawn into the cavities can consist of freshly sparged gas and that recirculated from the dispersion back into the cavity. Since this recirculation depends on the bubble size and hence upon the coalescence properties of the dispersion, it cannot be predicted with a high degree of precision from any general correlation.

Greaves and Kobbacy[17] have produced empirical correlations which do start to accommodate the various flow regimes and coalescence classifications. However, they were derived from very small scale work with a limited range of configurations and are not widely tested or always accurate at large scales.

Three basic types of cavity shape have been distinguished[3,4,37], and these are illustrated in *Figure 17.5*. The 'large cavities' are the chief cause of the reduction in P_g with increasing Q_G. However, it is the total amount of gas entering the cavities which controls their size and therefore P_g. The gas flow number, Q_G/ND^3, although at first sight a sensible correlating group, does not allow for gas recirculation, only that from the sparger. Therefore Q_G/ND^3 cannot in general bring together all the data for various N and D in a given tank (see *Figure 17.6*), or for the same N in different tanks (*Figure 17.7*), or for different bubble sizes in coalescing (large bubbles) and non-coalescing (small bubbles) systems (*Figure 17.6*).

The total gas entering the agitator region is $(Q_G + Q_R)$, where Q_R is the volumetric flow rate of recirculated gas. Q_R will consist in general of some recirculated sparger gas plus at very high speeds some gas entrained from the surface though this is normally negligible[5]. Bruijn *et al.*[3] have demonstrated this by showing that if recirculation and surface entrainment are eliminated by using large vessels with very small impellers, Q_G/ND^3 does correlate P_g for one size of agitator at different speeds. This is shown in *Figure 17.7*, where curves A_1, B_1 and C_1, are for a 'no-recirculation' geometry and are coincident, in contrast to the separate curves, A_2, B_2 and C_2, for different speeds in a standard vessel with recirculation. *Figure 17.8* shows again that Q_G/ND^3 is unsatisfactory as a direct scale-up criterion for gassed power requirements.

In more recent work, P_g/P has been related to the different gas flow patterns with changing speed as identified by Nienow *et al.*[15]. These are shown in *Figure 17.9*. These flow patterns can be associated closely with a graph of

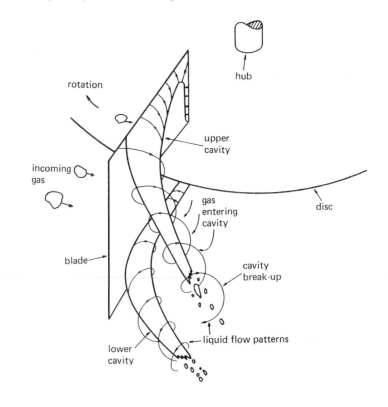

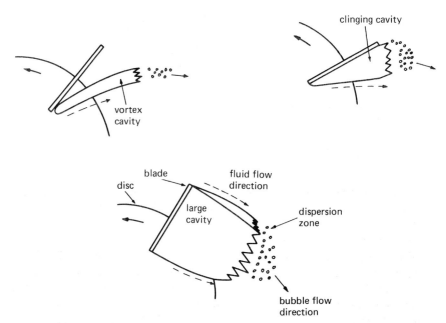

Figure 17.5   Cavity formation behind blades

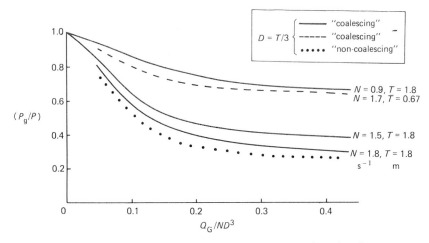

Figure 17.6   Power curves for 6-blade turbines. Various speeds and scales

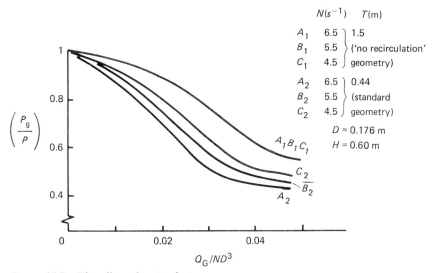

Figure 17.7   The effect of recirculation on power

(P_g/P) against gas flow number, Q_G/ND^3, with increasing N when Q_G is held constant. A characteristic example is shown in *Figure 17.10* and explained below.

At the lowest N (region *a*) the gas passes mainly through the agitator without dispersion and the liquid flows around the outer part of the blades undisturbed by the gas. Thus, the gassed power is not much less than the ungassed. As N increases, gas is captured by the vortices behind the agitator blades and dispersed and P_g first decreases as captured gas 'streamlines' the blades, forming 'large cavities'. Further increase of N diminishes the cavities and changes their form to 'vortex' cavities. The curve passes through a

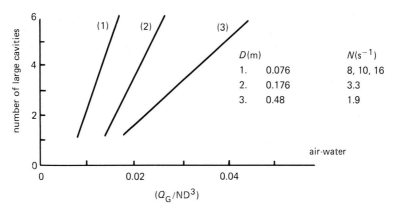

Figure 17.8   The effect of scale on aeration number and number of large cavities for 6-blade disc turbines[4]

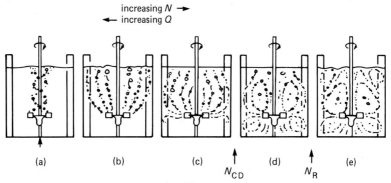

Figure 17.9   Regimes of bubble flow[15]

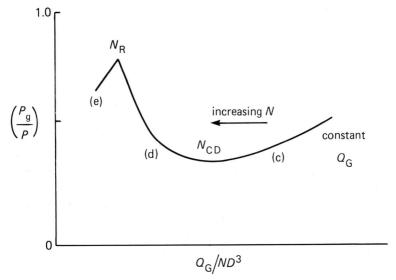

Figure 17.10   Power curves at constant gas rate[15]

minimum, at a speed, N_{CD}, the minimum speed to just completely disperse the gas. It then rises while small recirculation patterns start to emerge. The maximum corresponds to the speed, N_R, at which gross recirculation of gas back into the agitator sets in.

The speed N_{CD} can be predicted from this minimum and this is valuable since it marks the speed below which gas does not penetrate beneath the agitator and the whole tank is not effectively used. In *Figure 17.9* this is the transition from state (c) to state (d).

Nienow *et al.*[15] have given the following empirical correlation (in SI units) for coalescing systems for tanks up to 1.8 m diameter ($H = T$, $C = T/4$) for 6-blade disc turbines:

$$N_{CD} = (4Q_G^{0.5} T^{0.25})/D^2 \tag{17.3}$$

for pipe spargers. However, for ring spargers,

$$N_{CD} = (3Q_G^{0.5} T^{0.25})/D^2 \tag{17.4}$$

N_R is given by

$$N_R = (1.5 Q_G^{0.2} T/D^2) \tag{17.5}$$

These values will be somewhat lower for 'non-coalescing' systems and for turbines with more than six blades.

Smith and Warmoeskerken[16] show how different expressions for recirculation in each regime of flow behaviour can be worked out from power data, and future work connecting this with correlations of recirculation measurements (see section 17.7.2) seems a promising way forward.

To summarize, further research work is required on the prediction of recirculation before gassed power consumption and hence mass transfer rates can be confidently predicted. Meanwhile, the use of agitators with either many flat blades (12 to 18) or with six concave blades (see *Figure 17.3*) is recommended. This choice minimizes the effect of the unpredictability of P_g, since these impellers have a comparatively flat gassed power characteristic (see *Figure 17.4*).

17.5 Bubble size and coalescence

Many small bubbles are required in a dispersion to give a large gas–liquid interfacial area and thus effective mass transfer. Small bubbles are more readily entrained into the circulating liquid stream and so give more gas backmixing than larger bubbles under similar hydrodynamic conditions.

Using 6-blade disc turbines in baffled tanks with pure liquids, i.e. coalescing systems, Calderbank[6] measured interfacial area and hold-up and obtained an equation for the surface/volume mean bubble diameter (in SI units):

$$d_B = 4.15 \left[\frac{\sigma^{0.6}}{(P_g/V)^{0.4} \rho^{0.2}} \right] \varepsilon_G^{0.5} \left(\frac{\mu_g}{\mu} \right)^{0.25} + 0.0009 \tag{17.6}$$

for situations where bubbles do not grow or shrink during their passage.

However, it is found that small concentrations of electrolytes, surfactants, alcohols, oils, etc., in agitated tanks or bubble columns of water have a profound effect on the bubble size and gas hold-up, ε_G (see section 17.6) and change the empirical constants in equation (17.6). This effect is quite distinct from the effects of changes in agitation or gas velocity. It is related to bubble coalescence after the initial break-up. In pure water, d_B is typically ~ 5 mm and $\varepsilon_G \sim 0.1$, whereas in a 'non-coalescing' (e.g. electrolyte) solution under the same agitation conditions, d_B might be ~ 0.5 mm and $\varepsilon_G \sim 0.25$. The effect is caused by solute concentration or temperature gradients near the interface setting up local surface tension gradients which oppose the drainage and stretching of the liquid film between approaching bubbles and hence hinder coalescence[38].

The decrease in d_B or increase in ε_G or surface area, a, levels out as electrolyte concentration increases[38]. Marucci and Nicodemo[7] also indicate this for various solutes in bubble columns. Smith et al.[8] and Zlokarnic[9] have studied the effect of electrolytes on $k_L a$ and a in stirred tanks. Since $a = 6\varepsilon_G / d_B$, with the typical figures quoted above, a will increase by a factor of about 25. However, Cooke[41] and Smith et al.[8] found that $k_L a$ increased only by about twice for standard stirred tanks whilst Zlokarnik[9] found an increase of about five times for some self-aspiring agitators. Smith et al.[8] showed that approximately the same scale-up relationship for $k_L a$ can be used for both non-coalescing and coalescing systems within certain limits. However, more recently in an extensive review of the earlier literature (some of which may require correction (see section 17.8)), van't Riet[39] suggested two distinct correlations for $k_L a$ in water depending on whether electrolytes were present or not.

In 'real' industrial systems where rapid mass transfer and/or reaction occurs, even larger effects on d_B, ε_G and $k_L a$ can be found, caused by local surface tension gradients (the 'Marangoni' effect).

With large solute molecules (surfactants, proteins), mutual electrostatic repulsion of interfaces may also have an effect. Solid particles can also have a substantial influence. In some reacting, boiling or biological systems, foaming can readily occur.

17.6 Gas hold-up fraction

Most published correlations for gas hold-up are derived from experiments with either pure liquids ('coalescing' systems) or aqueous solutions of electrolytes ('non-coalescing') and are of the form

$$\varepsilon_G \propto (P_g/V_L)^A (v_s)^B \tag{17.7}$$

Values of A range from 0.2 to 0.7 and of B from 0.2 to 0.7, with a tendency for A to be higher for non-coalescing systems than for coalescing. With the current state of knowledge, it is best to take the correlation covering the widest range, particularly of scale. On these grounds, the correlation of Smith et al.[8] is selected in which $A = 0.48$ and $B = 0.4$ for both systems for $v_s = 0.005$ to

0.05 m/s and $(P_g/V_L) = 1$ to 5 kW/m³. For 'real' systems, the values of ε_G can be very different from and generally much higher than those for pure liquids, as mentioned in section 17.5. Also bubble diameter may change with time because of absorption, desorption and evaporation. These factors mean that such correlations are not, in general, universally applicable. However, since hold-up is controlled by bubble size and the amount of gas rising and recirculating, all which are functions of (P_g/V_L) or v_s for a given system, the basic form of the correlation seems reasonable.

17.7 Concentration driving force

The remaining parameter to be evaluated for the determination of $k_L a$ from experiments and for the design of equipment is the correct concentration driving force, $\overline{(C_{LA}^* - C_{LA})}$. This in turn depends on the extent of gas mixing which determines C_{LA}^* throughout the vessel and the liquid mixing which determines C_{LA}.

17.7.1 Liquid mixing

There is very little information available on the degree of liquid mixing in gas–liquid agitated vessels. The intensity of agitation required for effective gas dispersion is usually sufficient to provide a high degree of mixing in the liquid, provided that the liquid pumping capabilities of the agitator are not impaired by large gas cavities, i.e. $N > N_{CD}$. In general therefore, it is reasonable to assume that C_{LA} is constant throughout the vessel.

It has been shown by Middleton[22] that for gas hold-up fractions up to at least 0.1, liquid circulation time (related to mixing time) is changed only slightly by the presence of the gas, so correlations for liquid only may be used. However, above about 20% hold-up, circulation time is increased[42], i.e. mixing is worse than without gas. At much higher gas hold-ups ($> \sim 0.7$) as encountered in some foamy boiling systems, the liquid is mostly in the form of films between bubbles, with very restricted mobility and the agitated tank cannot provide good liquid mixing in such cases.

17.7.2 Gas mixing

The degree of gas backmixing is mostly related to the proportion of the gas flow which recirculates through the agitator. As has already been seen in section 17.4, this recirculation also influences the power consumption. Gas is mainly mixed with other gas only at the agitator in the gas cavities behind the blades and minimal mixing occurs elsewhere in the vessel. If bubbles do coalesce after formation, they do so very rapidly and very near the agitator, where the gas has just been mixed anyway so bubble–bubble coalescence has a negligible influence on gas backmixing except that it controls average bubble size which influences gas recirculation. Thus, there is an apparent anomaly in that 'non-coalescing' systems have a greater amount of gas backmixing than 'coalescing'.

The degree of gas backmixing therefore depends upon the amount of gas recirculating (which is a function of local liquid velocity and turbulence, flow pattern and bubble size) as compared with the amount fed to the vessel. Of the gas generally recirculating, that which backmixes does so by being drawn into the cavities. There is very little information available on the amount of recirculation[4] but some preliminary data[21] for 6-blade disc turbines ($C = H/3$, $H = T$) is presented in *Figure 17.11*.

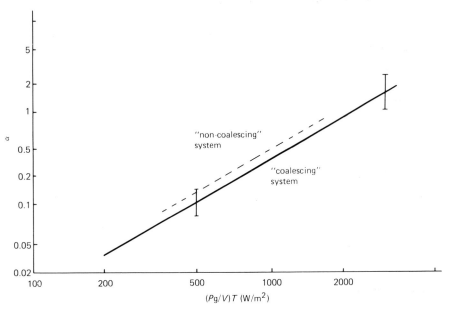

Figure 17.11   Gas recirculation ratio (α) correlation. Data covers: vessels 0.6–1.8 m diameter, $H = T$; $C = T/3$; $D = T/3$ and $T/2$; tip speed 1.6–4.7 m/s; $v_s = 5.40 \times 10^{-3}$ m/s

Here the data are expressed as a recirculation coefficient, α, where

$$\alpha = \frac{\text{Volumetric flow of gas recirculating into agitator}}{\text{Volumetric flow of gas fed to vessel}} \qquad (17.8)$$

Measurements[5] were only made on water and electrolyte solutions and may not apply to other systems and agitators.

α can be used to calculate the mean gas concentration driving force for simple absorption, or incorporated in reactor models for more complex situations. For $\alpha \simeq 0$, the gas is in plug flow and the mean value of the driving force is $\overline{(C^*_{LA} - C_{LA})}_{\text{log mean}}$ where

$$\overline{(C^*_{LA} - C_{LA})}_{\text{log mean}} = (\Delta C_{\text{in}} - \Delta C_{\text{out}})/\ln(\Delta C_{\text{in}}/\Delta C_{\text{out}}) \qquad (17.9)$$

where 'in' and 'out' refer to the concentration driving force of the inlet and outlet respectively. C_{LA} is constant throughout the vessel and C_{LA}^* is the liquid phase saturation concentration of the soluble gas.

For $\alpha > 100$, the gas is approximately fully mixed and

$$\Delta C_{\text{mean}} = \Delta C_{\text{in}} = \Delta C_{\text{out}} \tag{17.10}$$

For intermediate values of α,

$$\Delta C_{\text{mean}} = (\Delta C_{\text{in}} - \Delta C_{\text{out}})/(\alpha+1) \ln\left[\frac{\Delta C_{\text{in}} + \alpha\Delta C_{\text{out}}}{(\alpha+1)\Delta C_{\text{out}}}\right] \tag{17.11}$$

This is a simplified expression[46] which is strictly only applicable to small α since it neglects mass transfer from the recirculating gas but it is a reasonable approximation. This treatment does not account for bubble shrinkage or growth effects on α.

17.8   Gas–liquid mass transfer

In general, the mass transfer between gas and liquid is described by

$$J = K_L a\overline{(C_{LA}^* - C_{LA})} \tag{17.12}$$

where J is in (moles transferred/s/m³ of dispersion), K_L is the overall mass transfer coefficient, a is the gas–liquid interfacial area per unit of volume dispersion and $\overline{(C_{LA}^* - C_{LA})}$ is the mean concentration driving force. The significance of a and $\overline{(C_{LA}^* - C_{LA})}$ have been discussed in the previous section. $(1/K_L)$ is the overall mass transfer, envisaged as being composed of resistances on the gas side of the interface and on the liquid side, in series, i.e.

$$\frac{1}{K_L} = \frac{1}{Ek_G} + \frac{1}{k_L} \tag{17.13}$$

where E is the equilibrium constant.

For small bubbles (say <0.2 mm)[44], k_G is governed by diffusion only since no convection occurs. An approximate value[48] may be obtained from:

$$k_G = \frac{2\pi^2 D_{AG}}{3d_B} = \frac{6.6 D_{AG}}{d_B} \tag{17.14}$$

However, if the contact time is short ($<40\%$ approach to equilibrium)

$$k_G = 2\sqrt{D_{AG}/\pi t} \tag{17.15}$$

where t is the contact time or lifetime of the bubble. With larger bubbles ($d_B \geqslant 2$ mm), k_G is increased by internal convection up to 2.25 times the above values[44]. Thus k_G controls the rate if $k_G E \ll k_L$. (Care must be taken with the units in this inequality because of the wide range of E definitions employed.) Usually $k_L \ll k_G E$, so that the liquid side resistance predominates and therefore

$K_L = k_L$. In practice, it is important to check this assumption since it affects the choice of equipment and scale-up method. `

The gas–liquid mass transfer factor $k_L a$ for standard geometries has been extensively measured and correlated for air–water (coalescing) and air–electrolyte (non-coalescing) solutions. The most successful correlations are of the form

$$k_L a = X(P_g/V_L)^Y (v_s)^Z \qquad (17.15)$$

Over the years the reported values of Y have ranged over 0.4 to 0.95 and of Z, from 0 to 1. Some of this variation is due in earlier work to uncertain behaviour of the air/sulphite solution system used, friction in power measurements, non-uniform geometry etc. Dissolved oxygen meters have now been developed which enable physical oxygen absorption to be measured simply, without the complications of a reaction[8,49].

Recently it has been demonstrated[41,45] that the assumed gas flow pattern has a dramatic influence on the value of $k_L a$ (except at the lowest values) derived from concentration measurements, and that the flow pattern is neither approximately plug flow nor backmixed. Unfortunately this means that most of the published $k_L a$ data is suspect since either an ideal gas flow pattern has been implicitly assumed, or the depletion of transferring gas in the gas phase has been ignored.

Using a new technique (involving transient liquid and exit gas concentration measurements) which eliminates the need to explicitly describe the gas flow pattern, the data shown in *Figure 17.12* and the following correlations[41] have been generated for air–water ('coalescing system') at 20°C:

$$k_L a = 1.2(P_g/V_L)^{0.7}(v_s)^{0.6} \qquad (\pm 10\%) \qquad (17.16a)$$

and for air–electrolyte solution ('non-coalescing') at 20°C:

$$k_L a = 2.3(P_g/V_L)^{0.7}(v_s)^{0.6} \qquad (\pm 20\%) \qquad (17.16b)$$

from experiments covering a range of gas rates and agitator speeds but one agitator (6-blade disc turbine, $D = T/2$) and one vessel size ($T = 0.3$ m).

Smith *et al.*[8] showed (albeit without correction for the gas phase flow pattern) that $k_L a$ data from various scales ($T = 0.4$ to 1.8 m), geometries ($T/D = 2$ to 3, $H/T = 1$), and agitator types (6, 12, 18 flat-blade, 6-concave-blade, 6-perforated blade disc turbines) could all be correlated by an equation of the form of equation (17.16), so equation (17.16) may apply to other scales and configurations, though this has yet to be demonstrated experimentally. A more recent review[39], however, presents different values for the index Y for 'coalescing' and 'non-coalescing' systems. It should be borne in mind that this was based on earlier data without correction for the effect of gas flow pattern.

The addition of anti-foam[43] or a few ppm of oil causes a drastic reduction in $k_L a$ and the effect seems to be time dependent[10]. Evidently oil leaks must be avoided!

For these reasons and those quoted earlier, it is not possible to predict $k_L a$ for other systems with confidence, but the constancy of Y and Z above implies

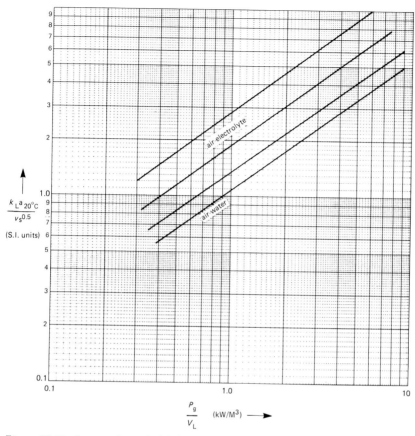

Figure 17.12 k_L *correlation (6-blade disc turbine)*

that it might at least be possible to use

$$k_L a \propto (P_g/V_L)^{0.7}(v_s)^{0.6} \qquad (17.17)$$

for scale up.

In some cases, design for 'a' alone is required (see section 17.11). Calderbank[18] gives for $d_B < 2.5$ mm:

$$k_L = 0.31 \left[\frac{(\rho - \rho_g)\mu g}{\rho^2} \right]^{1/3} \left[\frac{\mu}{\rho D_{AL}} \right]^{-2/3} \qquad (17.18)$$

or for $d_B > 2.5$ mm:

$$k_L = 0.42 \left[\frac{(\rho - \rho_g)\mu g}{\rho^2} \right]^{1/2} \left[\frac{\mu}{\rho D_{AL}} \right]^{-1/2} \qquad (17.19)$$

These k_L values are for pure turbulent liquids only, but the equations show that, assuming d_B does not change across the 2.5 mm value, k_L is independent of agitation variables and therefore,

$$a \propto k_L a \propto (P_g/V_L)^{0.7}(v_s)^{0.6} \qquad (17.20)$$

for scale-up. However, some workers find that k_L is a continuous function of d_B.

Finally, it should be noted that dissolved additives, e.g. electrolytes, polymers, antifoams, oils, alcohols etc. and small particles can grossly affect k_L, often reducing it. In general though, they increase a by increasing ε_G and decreasing d_B by an even larger factor so that $k_L a$ usually increases, though occasionally it has been found to decrease.

17.9   Heat transfer

In many instance when gas is dispersed into a liquid it is necessary to add or remove heat at some stage during the process. In mechanically agitated vessels this can be achieved using either jacketted vessels or helical or baffle coils immersed in the vessel. Edney and Edwards[19] indicate that, for holdups $\leqslant 15\%$, the rates of heat transfer with gas addition are very close to the values obtained without gas addition, i.e. with the single phase liquid only being agitated. Thus it is possible to estimate rates of heat transfer using the extensive literature on heat transfer for agitated single phase liquids. This literature has been reviewed by Edwards and Wilkinson[20] for both Newtonian and non-Newtonian systems.

Frequently heat transfer by boiling is encountered in agitated gas–liquid reactors, usually with a reflux condenser, for temperature control of exothermic reactions. In many systems boiling can increase the gas hold-up to 0.7 or more. The vapour may form new bubbles and/or enlarge existing bubbles of sparged gas. The vapour will dilute the sparged gas and the mass transfer driving force will therefore be reduced, usually by more than the surface area is increased. Other factors being equal, gas recirculation should be minimized in such cases to reduce this dilution effect. It is safest with boiling systems to use the maximum (i.e. the exit) gas flowrate for hydrodynamic calculations to avoid, for example, unexpected flooding of the agitator with gas and vapour.

Obviously the design of boiling sparged agitated vessels has to be a compromise between free escape of vapour for temperature control, i.e. using a large free surface area with low H/T, and minimizing gas recirculation, e.g. by using a large H/T.

17.10   Gas–liquid mixers as reactors

The preceding sections cover the physical process of contacting gas with liquid in stirred vessels. This section shows how this knowledge is applied to the selection and design of gas–liquid reactors in which the absorbed gas reacts with a liquid component. It concentrates on mixing vessels as gas–liquid reactors, although the degree and mode of mixing can have an important effect on the conversion and yield of a reaction and in some cases other devices, e.g. packed columns, bubble columns, liquid film contactors, with one or both phases in plug flow are more appropriate.

The treatment centres around the relative rates of gas–liquid mass transfer and reaction. The reaction rate controls the overall rate if it is very slow; but often it is fast, so that the overall rate is mass transfer controlled. Very fast reactions can influence the diffusion process, causing enhancement of mass transfer above the purely physical rate. This enhancement is itself a function of the reaction rate. In extreme cases ('instantaneous' reaction), mass transfer again controls the overall rate of the process.

Ideally the reactor design requires quantitative data for both the chemical kinetics and the mixing and mass transfer. The difficulties of predicting the physical effects have been outlined in the earlier sections. Often the chemistry is also difficult to analyse, and simplification must be made. In any case, small scale experiments must be carried out to determine relevant rates and other necessary reaction characteristics such as heat of reaction, degree of foaming and bubble coalescence etc. These experiments then allow scale-up to the final plant design. If time and techniques permit, this experimentation also guides the choice of reactor type and indicates whether and how complex mathematical modelling should be carried out.

Generally the difficulties of chemical analysis together with the uncertainties in gas–liquid fluid dynamics render scale-up of gas–liquid reactors a hazardous procedure. Experiments therefore must be done under the same conditions of temperature, pressure, catalyst type etc. as it is intended to use on the final plant. Allowances for uncertainty must be made in scaling-up to the final plant because of the difficulty and expense of experiments. In the past, designs have been rather arbitrary and therefore often sub-optimal and the large stirred pot has been very popular, mainly because it resembles a chemist's flask whilst also providing reasonable heat transfer, particle suspension and flexibility. This section aims to assist in deciding whether these expensive vessels are really the most efficient and economic for a given process.

It covers briefly the theory of interacting chemical rate processes with physical ones and the possible influence of mixing on reaction yields. Some suggestions are given on experimentation to elucidate the relative importance of the two rate processes and finally a summary of reactor modelling techniques suitable for gas–liquid systems.

17.10.1   Theory of mass transfer with reaction

This topic is covered fully in many works[23][25] and only a summary of the important relationships is presented here.

Consider a reaction between a dissolving gas A and a liquid phase reactant B, with q moles of B reacting per mole of A so that

$$nA + mB \rightarrow \text{Products} \qquad (17.21)$$

and $q = m/n$. Let C_{LA} and C_{LB} represent molar concentrations of A and B respectively in the liquid. The rate of reaction of A, J_L, is then given by

$$J_L = k_{nm} C_{LA}^n C_{LB}^m \tag{17.22}$$

where J_L has the units, moles/s/unit volume of liquid. Alternatively,

$$J = k_{nm} C_{LA}^n C_{LB}^m \varepsilon_L \tag{17.23}$$

when J has the units moles/s/unit volume of reactor. n and m are the orders of reaction in A and B and ε_L is the liquid hold-up fraction. A 'reaction time' t_R can be defined as

$$t_R = (1/k_{nm} C_{LA}^{(n-1)} C_{LB}^m)\left(\frac{2}{n+1}\right) \tag{17.24}$$

The mass transfer of A into the liquid is given by

$$J = k_L a \overline{(C_{LA}^* - C_{LA})} \tag{17.25}$$

where J is in moles/s/vol of reactor. A mass transfer 'diffusion time', t_D, can also be defined as

$$t_D = D_{AL}/k_L^2 \tag{17.26}$$

where D_{AL} is the diffusivity of A in the liquid.

If a fast reaction is occurring near the interface within the 'diffusion film', it will enhance the mass transfer rate and equation (17.25) becomes

$$J = k_L^* a \overline{(C_{LA}^* - C_{LA})} \tag{17.27}$$

where

$$k_L^* = \left[\frac{2D_{AL} k_{nm}(C_{LA}^* - C_{LA})^{(n-1)}(C_{LB}^m)}{n+1}\right]^{1/2} \tag{17.28}$$

It is important to appreciate that k_L^* depends only on the reaction rate and not on the hydrodynamics, which only affect k_L. Also $C_{AL} \approx 0$ for these fast reactions since gas is reacted as soon as it is transferred. Sometimes this enhancement factor must be modified for heat of reaction effects[26].

Five main regimes[23-25] are identified depending on the ratio of reaction time to diffusion time. These are shown in *Table 17.3*, with diagrams of the concentration profiles near the interface and notes on the parameters of importance.

The table hints that for Regime I, a is unimportant provided it is adequate but high liquid hold-up is required, so a bubble column is appropriate. For Regimes II, IV or V, high a and k_L are needed, so a stirred tank, for example, would be suitable though a high pressure is required for II. For Regime III, all the liquid needs to be in the 'film', so a thin film device such as a packed column should be used. The choice depends also on the mixing mode, the heat transfer requirements etc.

17.10.2   Locale of diffusion limitation

An enhancement of k_L in Regime III can mean that the gas film transfer coefficient k_G becomes important, so this must be checked (see section 17.8).

Table 17.3   The various gas–liquid reaction regimes and parameters of importance

REGIME		CONDITIONS	IMPORTANT VARIABLES	CONCENTRATION PROFILES
I	Kinetic control Slow reaction	$\sqrt{\dfrac{t_D}{t_R}} < 0.02$	Rate α ϵ_L α k_{nm} α $(C_{AL}^*)^n$ α $(C_{BL})^m$ Independent of a (if a adequate) Independent of k_L	
II	Diffusion control Moderately fast reaction in bulk of liquid $C_{AL} \approx 0$	$0.02 < \sqrt{\dfrac{t_D}{t_R}} < 2$ Design so that $\dfrac{\epsilon_L}{a} > 100 \dfrac{D_{AL}}{k_L}$	Rate α a α k_L α C_{AL}^* Independent of k_{nm} Independent of ϵ_L (if ϵ_L adequate)	
III	Fast reaction Reaction in film $C_{AL} \approx 0$ (pseudo first order in A')	$2 < \sqrt{\dfrac{t_D}{t_R}} < \dfrac{C_{BL}}{q\,C_{AL}^*}$ $C_{BL} \gg C_{AL}^*$	Rate α a α $\sqrt{k_{nm}}$ α $\left(C_{AL}^*\right)^{(n+1)/2}$ Independent of k_L Independent of ϵ_L	
IV	Very fast reaction General case of III	$2 < \sqrt{\dfrac{t_D}{t_R}}$ $C_{BL} \sim C_{AL}^*$	Rate α a depends on $k_L, k_{nm}, C_{AL}^*, C_{BL}$ Independent of ϵ_L	
V	Instantaneous reaction Reaction 'at interface'. Controlled by transfer of B to interface from bulk. $J \alpha k_L a$	$\sqrt{\dfrac{t_D}{t_R}} \gg \dfrac{C_{BL}}{q\,C_{AL}^*}$	Rate α a α k_L Independent of C_{AL}^* Independent of k_{nm} Independent of ϵ_L	

17.10.3 Mixing mode and reaction

Two types of 'longitudinal mixing' (mixing in the general flow direction or 'residence time distribution') are distinguishable for 'ideal' continuous flow reactors. These are:

(a) *Plug flow.* All the reactants are in the reactor for the same period, so that the mean concentration driving force is the log-mean of the initial and final driving forces. Tubular reactors in turbulent flow (including in-line mixers), or several stirred tanks in series, approximate to this.

(b) *Backmixed.* A reactant has a wide distribution of residence times in a given phase within the reactor and the mean concentration equals the outlet concentration. Continuous stirred tanks, or plug flow reactors with large recycle loop, aim to approximate to this.

Longitudinal mixing should not be confused with 'radial' or 'cross' mixing, which is uniformity across the general flow direction and is required for all reactors.

In addition, of course, there are batch reactors where all the reactants are in the reactor for the same period so that the mean concentration is the log mean of initial and final conditions.

For reactions of order $n \geqslant 1$, the required reactor volume is greater for a backmixed reactor than for a plug flow one, e.g. by a factor of 3.91 for 90% conversion and 21.5 for 99% conversion, for a first order reaction. This is also true for gas–liquid mass transfer controlled systems, where in the main it is the gas phase residence time distribution which affects the driving force and hence reactor size for a given throughput.

For certain more complicated reactions, the yield of the reaction is also influenced by the mode of mixing in a similar fashion to the effects discussed in Chapter 10. Full discussion of this is outside the scope of this chapter and ref. 25 gives a good introduction. Briefly, for consecutive reactions,

$$A \rightarrow B \rightarrow C$$

a high yield of B (compared with C) is favoured by plug flow, but if C is the desired product, a backmixed reactor should be chosen. For parallel reactions

$$A \rightarrow B \text{ (product) (of order } p \text{ in } A)$$

$$A \rightarrow C \qquad \text{(of order } r \text{ in } A)$$

then if $p > r$, a high yield of B requires high C_{LA} and hence plug flow; conversely if $p < r$, high C_{LA} must be avoided, so a backmixed reactor is preferred. If $p = r$, the yield is unaffected by the mixing mode.

17.10.4   Scale of mixing

To see the above effects, obviously the timescale of mixing must be comparable with the reaction time. The backmixing found in stirred tanks may be described by a seconds-to-minutes timescale (often called 'macromixing'). If, however, the reaction is mixing sensitive and is substantially complete within milliseconds, any yield effects will arise from mixing on the microsecond to millisecond timescale ('micromixing'). These reactions tend to occur either around the bubbles or locally at the reactant feedpoints, and overall residence time distribution is irrelevant[27], cf. Chapter 10. In such cases, a highly turbulent static mixer reactor of the type described in Chapter 13 might be more appropriate than a stirred tank.

17.10.5   Experimental work for classification of reaction regimes

The ideal design arises from measurements of chemical kinetics unhindered by mixing or mass transfer effects. This is possible for reaction times greater than about 10 milliseconds for some liquid systems, using rapid mixers[28] or by mixing at low temperature then suddenly increasing the temperature to cause reaction. This is the 'temperature jump' technique[28]. Such instruments are available commercially.

For gas–liquid reactions, these methods can be used if the gas is predissolved in a liquid solvent. If this is not possible, a high intensity gas–liquid contactor is required to give $t_D \ll t_R$, to enable t_R to be studied. A small continuous gas–liquid stirred tank could be used to measure t_R of the order of 10 s or greater directly without mass transfer effects (i.e. experimenting within Regime I). If the reaction is simplified to one of known order, the stirred tank can be used to infer reaction rate constants for faster reactions by utilizing the Regime III and IV rate expressions in *Table 17.3*. A suggested course of experiments would be

(a) Changing C_{LA}^* by altering pressure or solvent concentration to distinguish the regimes, unless $n = 1$.
(b) If $n = 1$, the agitator speed should be varied and if this has no effect, Regime I is indicated.
(c) If agitation has an effect, the temperature should be increased by, say, 10°C. If the rate increases by a factor of 1.2 or 1.3 Regime II is indicated, but if it increases by two to five times, Regime III is indicated. (This experiment decides whether the agitator speed effect was due to an influence on $k_L a$ or just a.) Special devices, e.g. a laminar jet[24], can also be used to distinguish between Regimes II and III. A very small temperature effect might indicate gas film diffusion controlled mass transfer especially if the solubility is high or the gas diluted with inerts. Large changes in C_{LA}^* often occur with changes in temperature. Such changes can confuse these general trends and boiling conditions should be avoided if possible.

The above experiments, if done under conditions equivalent to full scale ones with a well-mixed stirred tank reactor at steady state, give the basic rate of overall reaction plus information on what influences it. These can be used for scale-up calculations, either keeping to a stirred tank, or where appropriate, scaling up a different type of reactor, e.g. a bubble column for Regime I, a cascade of stirred tanks if plug flow is required in Regime II, or a packed tower or gas–liquid annular flow tubular reactor for Regime III or for gas-film controlled mass transfer.

A different reactor can change the regime. A good aim is to look for a reactor which operates in Regime I, i.e. high $k_L a$ where the chemistry is not restrained by mass transfer. The final choice must obviously take account of requirements for heat transfer, particle suspension, foam control, materials of construction and a feasible size for the full scale equipment. Only in the simplest cases are there sufficient degrees of freedom to allow economic optimization of reactor type.

Most practical reactions are more complicated than has been inferred here, but with the aid of careful experiments this can often be simplified to one controlling step. The rapid mixer plus plug flow reactor is useful for unravelling the details. If such a simplification cannot be made, the results of the kinetics experiments will have to be fed into a computer reactor model (see section 17.10.6) for interpretation and the model used for scale-up.

17.10.6   Gas–liquid reactor modelling

More complicated approaches are required when the reaction yield is sensitive to the mode of mixing and the process is carried out in a reactor which is neither in ideal plug flow, nor perfectly mixed, but somewhere between. The mixing is then described by a model of the flow pattern, whose equations must be solved together with those describing the reaction kinetics. The solution of the equations is iterative and generally requires a substantial computer programme.

There are three types of mixing model for continuous reactors, based on:

(a) Arbitrary zones of mixedness and segregation (or dispersion and recoalescence) of elements of fluid;
(b) Measured flow patterns;
(c) Calculated flow patterns.

Much effort has been put into developing type (a) models[29], but they remain unsuitable for scale-up because they require fitted parameters which are not available. In other words, they can describe mixing but not predict it. They have not been applied to gas–liquid systems.

Type (b) models are at present the most useful, though still requiring more hydrodynamic data. A successful example[30] considers the liquid and gas circulations in a stirred tank as loops of connected mixed zones (see *Figure 17.13*) with feed and offtake where appropriate. The zones can have different $k_L a$ and gas hold-ups if these have been determined experimentally. The mean circulation flow and relevant number of zones have to be supplied from experiments, so the model is again not in itself predictive. However, the parameters have a physical meaning and so they are more amenable to scale-up.

A measurement technique for determining mean circulation times and their distribution, is available[22]. A scale-up correlation for mean circulation time, θ_c, applicable to zero or low gas hold-up with the standard vessel configuration, is

$$\theta_c = 0.5 V^{0.3} \left(\frac{1}{N}\right)\left(\frac{T}{D}\right)^3 \tag{17.29}$$

The correlation for the distribution of circulation times can be converted to the number of zones, X, required in a loop by

$$X = 6.4 V^{0.3} N^{-1.3} \tag{17.30}$$

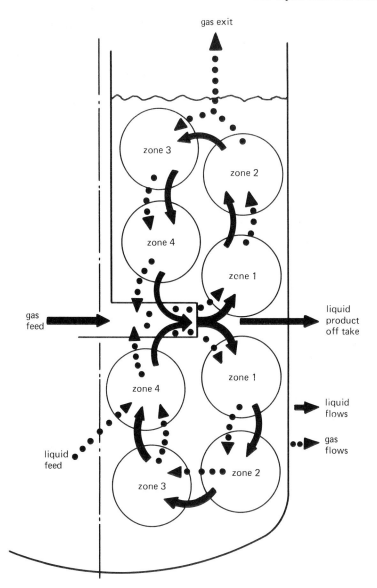

Figure 17.13 The mixed zones in loops' model for stirred reactors

A more detailed modelling technique using the same information has also been proposed[31].

A reactor model[32] is also available which uses actual local fluid velocity and turbulence measurements throughout a vessel to describe the flow pattern (simplified to a rectangular grid of 'cells'). This is getting even nearer the true hydrodynamics, but does not yet have sufficient velocity data behind it for reliable application to large vessels or gas–liquid systems.

The ultimate model is type (c), which involves a complete solution of the

'Navier–Stokes' equations using a turbulence model and the reaction equations using finite difference techniques. So far this has been done only with simple, single phase reactions in two-dimensional flows (e.g. combustors). In principle, it is capable of extension to more general situations, although with present computers the size of the program would make such models very expensive to run.

17.11   Example of scale-up

The experiments outlined above indicate whether and to what extent k_G, k_L, a, ε_L, feed concentrations, pressure, mixing and reaction constants will be important for scale-up. In this section, the example of scale-up of $k_L a$ (or a Regime II reaction) in a continuous stirred tank will be considered. The approach can obviously be extended to other regimes.

Assume that the plant vessel must react L m³/hr of liquid to give the same conversion on the large scale as in the experiments. This duty means that

$$\frac{\text{Gas transferred}}{\text{Mole throughput}} = \text{constant for all scales}$$

i.e.

$$\frac{k_L a V_L (C_{LA}^* - C_{LA})}{C_{LBF} L} = \text{constant} \tag{17.31}$$

where C_{LBF} is the molar concentration of B in the feed. Assuming $C_{LA} = 0$, and degree of backmixing and C_{LBF} are constant, then

$$k_L a V_L (C_{LA}^*)_{\text{out}} \propto L \tag{17.32}$$

Since $V_L / L \propto \theta_L / \varepsilon_L$, this implies that

$$\frac{k_L a \theta_L (C_{LA}^*)_{\text{out}}}{\varepsilon_L} = \text{constant} \tag{17.33}$$

where θ_L is liquid mean residence time.

Provided Henry's Law is obeyed, then $(C_{LA}^*)_{\text{out}} \propto (C_{GA})_{\text{out}}$ at a given pressure. $(C_{GA})_{\text{out}}$ must often be kept constant, e.g. to prevent explosive gas/vapour mixtures in the reactor head space. Assuming this is so, a mass balance on the gas gives:

$$Q_G \{ (C_{GA})_{\text{in}} - (C_{GA})_{\text{out}} \} \propto k_L a V_L (C_{GA})_{\text{out}} \tag{17.34}$$

If $(C_{GA})_{\text{in}}$ is fixed by the process, then for constant $(C_{GA})_{\text{out}}$:

$$k_L a V_L / Q_G = \text{constant} \tag{17.35}$$

From equation (17.16)

$$k_L a \propto (P_g / V_L)^{0.7} (Q_G / T^2)^{0.6}$$

For the sake of this example, a correlation for P_g from ref. 17 is used though it has limitations (see section 17.4). Within the 'efficient dispersion' regime[17]

$$P_g \propto N^{3.3} D^{6.3} Q_G^{-0.4} \qquad (17.36)$$

If geometrical similarity is now assumed, i.e. $D \propto T$, $H \propto T$, $V \propto T^3$, then

$$k_L a \propto N^{2.31} Q_G^{0.32} T^{1.11} \qquad (17.37)$$

Now from equations (17.32) and (17.35) and allowing that $(C_{LA}^*)_{out} = $ constant, then

$$Q_G \propto L \qquad (17.38)$$

and from equation (17.32), putting $V_L \propto T^3$, and equation (17.37),

$$N^{2.31} Q_G^{0.32} T^{4.11} \propto L \qquad (17.39)$$

or with equation (17.38)

$$N^{3.4} T^{6.0} \propto L \qquad (17.40)$$

Equations (17.38) and (17.40) are therefore the scale-up relationships.

The design may be fixed by imposing one further constraint, which will depend upon the process. Suppose, for example, particles must be kept in suspension, then $NT^{0.76}$ should also be constant on scale-up with geometric similarity (from $\varepsilon_{JS} \propto D^{-0.28}$ (Chapman)) (see Chapter 16). From equations (17.38), (17.40) and for $NT^{0.76} = $ constant, then

$$Q_G \propto L \propto (T^{-0.76})^{3.4} T^{6.0} \propto T^{3.4} \qquad (17.41)$$

or

$$T \propto L^{0.29} \qquad (17.42)$$

(*Note:* In passing, it should be noted that recent work[40] has indicated that $NT^{0.76} = $ constant, is probably not applicable when gassing. However, it is only used here to illustrate the approach and an alternative relationship could be selected as appropriate in order to produce definitive scale-up relationships.)

With scale-up to any set of rules, other aspects of mixing are directly affected. For instance, gassed power from equation (17.36) is

$$P_g \propto N^{3.3} T^{6.3} Q_G^{-0.4}$$

so that in this case from equation (17.41)

$$P_g \propto (T^{-0.76})^{3.3} T^{6.3} (T^{3.1})^{-0.4} \propto T^{2.4} \qquad (17.43)$$

This might imply a maximum permissible scale of operation. On the other hand power per unit volume is given by

$$(P_g/V_L) \propto T^{-0.6} \qquad (17.44)$$

suggesting a lower agitation intensity on scale-up.

Superficial gas velocity, v_s, is given by

$$v_s \propto Q_G/T^2 \propto T^{3.4}/T^2 \propto T^{1.4} \qquad (17.45)$$

so that foaming or liquid entrainment is enhanced on scale-up. Similarly gas hold-up, ε_G, from equation (17.7) is

$$\varepsilon_G \propto (T^{-0.6})^{0.48}(T^{1.4})^{0.4} \propto T^{0.27} \tag{17.46}$$

so that a need for more head space is indicated. This need is further enhanced by the reduction in surface area/unit gas flow, i.e.

$$T^2/Q_G \propto T^{-1.4} \tag{17.47}$$

which would imply especially difficulties for heat removal in boiling systems.

Heat transfer itself is approximately related to the agitation conditions by

$$Nu \propto Re^{0.67} \tag{17.48}$$

so that

$$h \propto \frac{1}{D}(ND^2)^{0.67} \tag{17.49}$$

or the heat flux per unit of throughput

$$\frac{hA}{L} \propto \frac{1}{T}(T^{-0.76}T^2)^{0.67}\left(\frac{T^2}{T^{3.4}}\right) \propto T^{-1.6} \tag{17.50}$$

Clearly, as is well known, heat transfer becomes a problem in large scale applications and external heat exchangers are often required.

Finally, some of the other hydrodynamic effects can be considered. For instance, liquid backmixing is affected since the ratio of the circulation time to the mean residence time changes. Thus the mean residence time, θ_L, is given by

$$\theta_L \propto \frac{V_L}{L} \propto \frac{T^3}{T^{3.4}} \propto T^{-0.4} = \text{constant} \tag{17.51}$$

and the circulation time, θ_c, by[22]

$$\theta_c \propto \frac{V_L^{0.3}}{N}\left(\frac{T}{D}\right)^3 \propto \frac{(T^3)^{0.3}}{T^{-0.76}} \propto T^{1.7} \tag{17.52}$$

so that

$$\theta_c/\theta_L \propto T^{2.1} \tag{17.53}$$

which implies progressively increasing deviations from a backmixed reactor residence time distribution. Gas backmixing can be obtained from *Figure 17.11* for constant Q_G/V, i.e. $Q_G \propto T^3$ (equation (17.41), which is approximately the case here, so that from equation (17.44)

$$\alpha \propto (P_g/V_L)(T)^{1.4} \propto T^{0.6} \tag{17.54}$$

Thus whilst liquid mixing gets poorer gas mixing improves. Considering the ratio of the speed to the speed required for complete dispersion from equation (17.3) gives

$$N/N_{CD} \propto T^{-0.76}/(T^{3.4})^{0.5}T^{0.2}T^{-2} \propto T^{-0.6} \tag{17.55}$$

so that dispersion of gas throughout the vessel deteriorates. Finally the maximum shear in the impeller vortex region[4] is proportional to N so that

$$\text{Maximum shear rate} \propto T^{-0.76} \tag{17.56}$$

Thus shear sensitive materials are less likely to be damaged on scale-up.

Some of the deteriorations on scale-up can be alleviated by changes in geometry, e.g. larger impeller to tank diameter or multiple agitators to improve backmixing; smaller tank height to diameter ratio to increase surface area for boil-off and gas release and hence reduced foaming. The techniques outlined above can be simply modified to accommodate these variations.

Notation

a gas–liquid interface area/unit dispersion volume (m^{-1})

C agitator clearance from base (m)

C_{GA} concentration of soluble component in gas stream $(\text{moles}/\text{m}^3)$

C_{LA} dissolved gas concentration in liquid bulk $(\text{moles}/\text{m}^3)$

C_{LA}^* dissolved gas concentration at interface $(\text{moles}/\text{m}^3)$

C_{LB} liquid reactant concentration $(\text{moles}/\text{m}^3)$

d_B bubble diameter (surface-volume mean) (m)

D agitator overall diameter (m)

D_A diffusivity of dissolved gas A (m^2/s)

D_{AG} diffusivity of A in the gas phase (m^2/s)

E gas–liquid equilibrium constant $(=C_G^*/C_L^*)$

g gravitational acceleration (m/s^2)

h heat transfer coefficient $(\text{W}/\text{m}^2\,\text{K})$

H height of liquid (m)

J reaction rate per unit reactor volume $(\text{moles}/\text{s}/\text{m}^3)$

k thermal conductivity $(\text{W}/\text{m}\,\text{K})$

k_{nm} reaction velocity constant $(\text{m}^3/\text{mole})^{m+n-1}\,\text{s}^{-1}$

k_L liquid film mass transfer coefficient (m/s)

k_L^* enhanced liquid film mass transfer coefficient (m/s)

k_G gas film mass transfer coefficient (m/s)

K_L overall mass transfer coefficient (m/s)

L mean liquid volumetric flow rate (m^3/s)

n order of reaction

N agitator speed (rev s^{-1})

N_{CD} agitator speed to just completely disperse the gas (rev s^{-1})

N_R agitator speed at which gross recirculation starts (rev s^{-1})

p pressure (N/m^2)

P ungassed agitator power consumed by liquid (W)

P_g agitator power consumed by gassed liquid (W)

Po ungassed power number

Q_G mean gas volumetric throughput rate (m^3/s)

Q_R volumetric flow rate of recirculating gas (m^3/s)

t_D diffusion 'time' $= D_{AL}/k_L^2$ (s)

t_R reaction 'time', see equation (17.24) (s)

T vessel diameter (m)

v_s superficial gas velocity (m/s)

V dispersion volume (m^3)

V_L liquid volume without gassing (m^3)

X number of mixed zones in loop

ΔC $C_{LA}^* - C_{LA}$ (moles/m^3)

ε_G gas hold-up per unit dispersion volume

ε_L liquid hold-up per unit dispersion volume

α gas volumetric recirculation rate per unit gas feed rate

θ_c mean liquid circulation time (s)

θ_L mean liquid residence time (s)

μ liquid viscosity (Pa s)

μ_g gas viscosity (Pa s)

ρ liquid density (kg/m^3)

ρ_g gas density (kg/m^3)

σ surface tension (N/m)

Nu Nusselt number (hD/k)

Re Reynolds number $(ND^2\rho/\mu)$

References

1 TAUSCHER, W. and MATHYS, P. (1975) *Proc. 1st Eur. Conf. on Mixing and Centrifugal Separation, September, 1974*, BHRA Fluid Engineering, Bedford, pp. D3-25 to D3-32.

2 CHEN, S. J. (1975) *Proc. 1st Eur. Conf. on Mixing and Centrifugal Separation, September, 1974*, BHRA Fluid Engineering, Bedford, pp. D2-13 to D2-24.

3 BRUIJN, W., VAN'T RIET, K. and SMITH, J. M. (1974) *Trans. I. Chem. E.*, **52**, 88.

4 VAN'T RIET, K. (1975) PhD thesis, Delft Technical University.

5 NIENOW, A. W., MIDDLETON, J. C. and CHAPMAN, C. M. (1979) *Chem. Eng. J.*, **17**, 111.

6 CALDERBANK, P. H. (1958) *Trans. I. Chem. E.*, **36**, 443.

7 MARUCCI, G. and NICODEMO, L. (1967) *Chem. Eng. Sci.*, **22**, 1257.

8 SMITH, J. M., MIDDLETON, J. C. and VAN'T RIET, K. (1978) *Proc. 2nd Eur. Conf. on Mixing, April, 1977*, BHRA Fluid Engineering, Cranfield, pp. F4-51 to F4-66.

9 ZLOKARNIC, M. (1978) *Adv. in Biochem. Engrg*, **8**, 134.

10 WARMOESKERKEN, M. and SMITH, J. M. (1978) *Proc. Int. Symp. on Mixing, Fac. Poly de Mons, 1978*, pp. C13-1 to C13-6.

11 MICHEL, B. J. and MILLER, S. A. (1962) *A. I. Ch. E. J.*, **8**, 262.

12 PHARAMOND, J. C., ROUSTAN, M. and ROQUES, H. (1975) *Chem. Eng. Sci.*, **30**, 907.

13 HASSAN, I. T. M. and ROBINSON, C. W. (1977) *A. I. Ch. E. J.*, **23**, 48.

14 NAGATA, S. (1975) *Mixing: Principles and Applications*, Halstead–John Wiley, New York.

15 NIENOW, A. W., WISDOM, D. J. and MIDDLETON, J. C. (1978) *Proc. 2nd Eur. Conf. Mixing, April, 1977*, BHRA Fluid Engineering, Cranfield, pp. F1-1 to F1-16 and X54.

16 SMITH, J. M. and WARMOESKERKEN, M. (1981) *I. Chem. E. Symp. Ser.*, No. 64, J1.

17 GREAVES, M. and KOBBACY, K. H. (1981) *I. Chem. E. Symp. Ser.*, No. 64, L1.

18 CALDERBANK, P. H. and MOO-YOUNG, M. B. (1961) *Chem. Eng. Sci.*, **16**, 39.

19 EDNEY, H. G. and EDWARDS, M. F. (1976) *Trans. I. Chem. E.*, **54**, 160.

20 EDWARDS, M. F. and WILKINSON, W. L. (1972) *The Chem. Engr.*, August 1972, p. 310 and September 1972, p. 328.

21 NIENOW, A. W., MIDDLETON, J. C. and CHAPMAN, C. M. (1980) *Paper at EFCE Symp. Gas–Liq. Mixing, Cologne, 1980*.

22 MIDDLETON, J. C. (1979) *Proc. 3rd Eur. Conf. on Mixing, April, 1979*, BHRA Fluid Engineering, Cranfield, pp. 15–36.

23   ASTARITA, G. (1967) *Mass Transfer with Chemical Reaction*, Elsevier, New York.

24   DANCKWERTS, P. V. (1970) *Gas–Liquid Reactions*, McGraw-Hill, New York.

25   COULSON, J. M. and RICHARDSON, J. F. (1971) *Chemical Engineering*, Pergamon Press, London, vol. 3.

26   MANN, R. and MOYES, M. (1977) *A. I. Ch. E. J.*, **23**, 17.

27   BOURNE, J. R., MOERGELI, V. and RYS, P. (1978) *Proc. 2nd Eur. Conf. on Mixing, April, 1977*, BHRA Fluid Engineering, Cranfield, pp. B3-41 to B3-54.

28   ROUGHTON, F. (1963) *Techniques of Organic Chemistry* (Freiss, J., ed.), Interscience, New York, vol. VIII, Part 2.

29   RAO, D. P. and EDWARDS, L. L. (1973) *Chem. Eng. Sci.*, **28**, 1179.

30   MANN, R., MIDDLETON, J. C. and PARKER, I. B. (1978) *Proc. 2nd Eur. Conf. on Mixing, April, 1977*, BHRA Fluid Engineering, Cranfield, pp. F3-35 to F3-50.

31   MANN, R., MAVROS, P. P. and MIDDLETON, J. C. (1981) *I. Chem. E. Symp. Ser.*, No. 64, G1.

32   PATTERSON, G. K. (1975) in *Turbulence in Mixing Operations* (Brodkey, R. S., ed.), Academic Press, New York, Chapter 5.

33   VAN'T RIET, K., BOOM, J. and SMITH, J. M. (1976) *Trans. I. Chem. E.*, **54**, 124.

34   ZUNDELEVICH, Y. (1979) *A. I. Ch. E. J.*, **25**, 763.

35   ZLOKARNIC, M. (1966) *Chem. Ing. Tech.*, **38**, 357.

36   KOEN, C. and PINGAUD, B. (1978) *Proc. 2nd Eur. Conf. on Mixing, April, 1977*, BHRA Fluid Engineering, Cranfield, pp. F5-67 to F5-82.

37   NIENOW, A. W. and WISDOM, D. J. (1974) *Chem. Eng. Sci.*, **29**, 1994.

38   LEE, J. C. and MEYRICK, D. L. (1970) *Trans. I. Chem. E.*, **48**, T37.

39   VAN'T RIET, K. (1979) *Ind. Eng. Chem. (Proc. Des. Dev.)*, **18**, 357.

40   CHAPMAN, C. M., NIENOW, A. W., MIDDLETON, J. C. and COOKE, M. (1983) *Chem. Eng. Res. Des.*, **61**, 71, 82, 167 and 182.

41   COOKE, M. (1982).

42   BRYANT, J. and SADEGZADEH, S. (1979) *Proc. 3rd Eur. Conf. on Mixing, April, 1979*, BHRA Fluid Engineering, Cranfield, pp. 325–336.

43   ANDREW, S. P. S. (1981) *Paper to NW Branch, I. Chem. E.*

44   MOTARJEMI, M. and JAMESON, G. J. (1978) *Chem. Eng. Sci.*, **33**, 1415.

45   CHAPMAN, C. M., GIBILARO, L. G. and NIENOW, A. W. (1982) *Chem. Eng. Sci.*, **37**, 891.

46   LEVENSPIEL, O. (1972) *Chemical Reaction Engineering*, John Wiley, New York.

47   AKITA, K. and YOSHIDA, F. (1974) *Ind. Eng. Chem. (Proc. Des. Dev.)*, **13**, 84.

48   CALDERBANK, P. H. (1967) in *Mixing* (Uhl, V. and Gray, J., eds.), Academic Press, New York, vol. II.

49   NIENOW, A. W. (1980) *Proc. Conf. Profitable Aeration of Wastewater, April, 1980*, BHRA Fluid Engineering, Cranfield.

Chapter 18

The mixer as a reactor: liquid/solid systems

A W Nienow

Department of Chemical Engineering, University of Birmingham

In this chapter the range of agitation conditions and solid–liquid characteristics is identical to that set out in Chapter 16.

18.1 Reactor types

The agitated vessel containing solid particles will generally be operated as a reactor in one of three ways indicated in *Figure 18.1*. First, it may be operated as a batch reactor (*Figure 18.1a*); secondly, it may be part of a continuous process but only the fluid may pass through with the solids remaining in the vessel (*Figure 18.1b*), though within this category the particle size may remain fixed (for instance, if the solid is a catalyst) or they may cover a wide size range (for instance, where the mixer is acting as a dissolver); thirdly, both the solids and liquid may be flowing through continuously as in a crystallizer.

In the first two cases, the agitation condition which just completely suspends the solids is normally sufficiently vigorous to ensure that the composition of the liquid phase is approximately homogeneous throughout the vessel. Thus, the reactor model must allow for particles to grow, shrink or remain of constant size, with, in the first case, a fluid which is spacially homogeneous but of time-varying composition, and, in the second case, a fluid composition constant in both time and space.

In the third case, the residence time distribution (RTD) of the solid becomes an important factor. Though the liquid RTD will again approximate closely to the perfectly mixed condition required for a continuous stirred tank reactor model, generally the solid will not. Therefore the actual solid RTD must be determined as set out in Chapter 16 for a satisfactory reactor design to be made.

The modelling of all these three cases has been well set out by Mattern, Bilous and Piret[1] for dissolving solids. Bourne and Zabelka[2] have done the same for continuous crystallizers with non-ideal solids RTD. No great difficulty should be found in applying these models to reactor design, provided 'reaction rates' are known, and the same is true when the particles do not change in size. Of course, the 'reaction rates' are in general a function of the

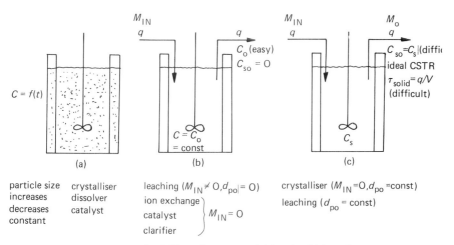

Figure 18.1   Some types of solid/liquid reactors: (a) batch; (b) liquid continuous; (c) both continuous

agitation conditions and the remainder of this chapter will concentrate on this relationship.

For crystallizers, the problem becomes considerably more complex because not only are 'reaction rates' (crystal growth rates in this case) functions of the agitation conditions but so is the rate of secondary nucleation, i.e. the number of tiny crystals of the order of magnitude of a micron that come from the impact of large crystals with themselves and with the walls and agitator. It is these tiny nuclei which grow and determine (along with a number of other factors) the mean crystal size and size distribution.

A detailed discussion of crystallizer design and secondary nucleation is beyond the scope of this chapter but the problem has been considered for a wide range of conditions by Randolph and Larson[3] and some simple and idealized cases have been treated in great detail by Larson and Garside[4]. However, the points to be made in this lecture about 'reaction rates' apply also to crystal growth rates and in addition a few comments on the effect of agitation on secondary nucleation and particle abrasion are included.

18.2   Reaction rates

For the purposes of this description, the reaction mechanism is considered to consist of two steps. The first of these takes place entirely in the liquid phase and is a bulk diffusion step. It is characterized by a mass transfer coefficient k and is a function of the hydrodynamics of the system, i.e. in this case, the agitation conditions. If the process of interest is a simple physical dissolution, then indeed only this bulk diffusion step is important and the 'reaction rate' is the rate of dissolution. In general, however, there is also a reaction step which takes place at the 'surface' of the solid.

Again, for the purpose of this chapter, this step is defined as any process which is unaffected by the agitation conditions, e.g. the surface integration step in crystallization[5] and the combined in-pore diffusion and chemical reaction at the surface associated with such particles as ion-exchange resins and catalysts[6]. *Figure 18.2* shows diagrammatically this idealized two-step mechanism.

From this simple model it is possible to write equations (18.1) and (18.2) to describe the diffusion and surface integration steps respectively;

$$R_G = kA(C_\infty - C_i) \tag{18.1}$$

$$R_G = k_R A(C_i - C^*)^{n'} \tag{18.2}$$

The value of n' can be considered as the 'order of reaction' for the process in question. Generally C_i is unknown and may be eliminated to give

$$R_G = k_R A[(C_\infty - C^*) - R_G/Ak]^{n'} \tag{18.3}$$

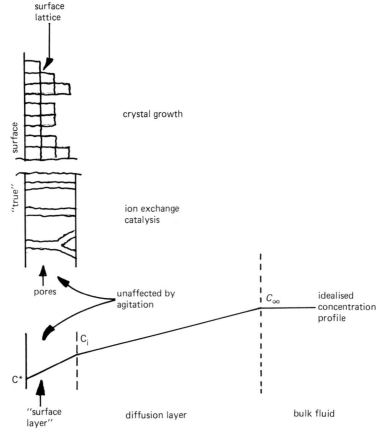

Figure 18.2   'Two-step' reaction model

When $n' = 1.0$ or 2.0, analytical solutions give

$$n' = 1.0; \qquad R_G = \{kk_R/k + k_R\}A(C_\infty - C^*) \qquad (18.4)$$

$$n' = 2.0; \qquad R_G = k\{1 + k/[2k_R(C_\infty - C^*)]$$
$$- [(1 + k/[2k_R(C_\infty - C^*)]) - 1]^{1/2}\}$$
$$\times A(C_\infty - C^*) \qquad (18.5)$$

For other values of n' numerical solutions must be used.

Writing equation (18.3) for the two limiting cases of diffusion and 'chemical reaction' control respectively gives

$$R_G = kA(C_\infty - C^*) \qquad (18.6)$$

$$R_G = k_R A(C_\infty - C^*)^{n'} \qquad (18.7)$$

For 'chemical reaction' to control, $k_R/k < \sim 0.01$ and, in that case, any increase in agitation over and above that necessary just to completely suspend the particles so that all of the surface area, A, is available for reaction fails to produce any further increases in reaction rate. Indeed, for an extremely slow 'reaction step', even complete suspension of the particles may be unnecessary with a general and gentle turnover of the solids to prevent stagnant pockets being all that is required.

Alternatively, for bulk diffusion to control $k_R/k > \sim 100$ for $n' = 1$ and $k_R(C_\infty - C^*)/k > \sim 1000$ for $n' = 2$. The case of bulk diffusion control corresponds to that where 'the reaction rate' is most sensitive to agitation, for $N < N_{JS}$ due to increases in the active surface area and k, and, for $N > N_{JS}$, due to increases in k alone.

Experiments to estimate n' and k_R usually rely on producing hydrodynamic conditions in which k is so great that equation (18.7) can be assumed to apply. In practice, this is often impossible and this problem and the evaluation of n' and k_R even when equation (18.7) is not applicable has been discussed in detail by Garside[7]. However, n' and k_R can only be obtained by experiments. These will usually be conducted on a small scale and the designer of large-scale plant must therefore either be able to obtain n' and k_R or alternatively be able to estimate R_G directly from them.

On the other hand, k is the mass transfer coefficient for diffusion and should be capable of being estimated from one of the many mass transfer relationships available. Thus the limiting case of diffusion control can be determined theoretically or experimentally. The main difficulty in the theoretical approach is in deciding which of the many mass transfer equations to use and what is the particle-fluid slip velocity for the complex hydrodynamics in the stirred tank.

Finally, the designer must be able to select a suitable value of agitation speed, N, and make some estimate of the relationship between N and R_G. Since in the range of conditions covered by this chapter $Re > 2 \times 10^4$ and $Po \doteqdot$ constant, then both

$$P \propto N^3 \qquad (18.8)$$

and

$$\varepsilon_T \propto N^3 \qquad (18.9)$$

Since the power requirement increases rapidly with speed, such increases must be justified by a suitable increase in R_G. Of course, the best that can be obtained will correspond to R_G under diffusion control. Therefore the remainder of this chapter will concentrate on the relationship between k and N and ε_T. (N.B. In this chapter $\varepsilon_T = P/\rho_L V$, the mean energy dissipation rate.)

18.3   Mass transfer

18.3.1   Measurement of the mass transfer coefficient, k

The diffusion mass transfer coefficient for stirred tanks has usually been measured in batch experiments with the rate of mass transfer being obtained from the dissolution of organic acids and electrolytes and also from ion-exchange and certain metal–metal solution reactions, both under bulk diffusion control[8]. Generally the change of particle size has been negligible; and in the latter two cases spherical particles giving an accurate surface area are obtainable, though the density range is then limited and there is always some doubt as to whether experiments are being done in a fully diffusion-controlled regime.

With dissolution experiments, there is either the problem of obtaining an accurate measure of surface area, particularly with small particles, or of only working with a limited size range if pellets are produced to overcome this problem. In nearly all cases k has been calculated from equation (18.6) with A being taken as the total surface area of the particles in the stirred tank whatever the condition of particle suspension.

In some experiments integral mass transfer coefficients from some initial size to zero have been calculated by measuring the total dissolution time θ of rapidly-dissolving particles[8,9]. In that case

$$\bar{k} = 3\rho_s / \theta a_p (C^* - C_\infty) \qquad (18.10)$$

where a_p is the specific surface area of the initial particle (though strictly some allowance for the convective flux should be made[10]). The advantage of this approach is the speed with which data are collected, making it possible to cover a large number of the variables which are inherent to particle–liquid systems in agitated vessels (see *Table 18.1*). The disadvantage is that it is difficult to assess the effect of particle size on k. However, most experimental studies have shown this to be small[8] and predictions based on a Froessling-type equation (see equation 18.13b) and a terminal velocity-slip velocity approach indicate this theoretically[13]. Thus, comparison of $\bar{k}$ values with k values is still of value. On balance, it seems a useful experimental technique.

Table 18.1 Parameters which may affect solid–liquid mass transfer in agitated vessels

Physical properties	Solid density; liquid density; density difference; viscosity; diffusivity.
Geometrical parameters	Particle size and shape; tank size and shape; impeller type, size and clearance; liquid depth; open or flooded; baffles and if so, baffle type and size.
Impeller– fluid–solid interactions	Impeller speed and power input; solid 'free' or fixed. If free, are the particles suspended and if so, what are their spatial and velocity distributions? What is the liquid flow pattern and the distribution of turbulence intensity throughout the vessel? Does surface aeration occur? What is the effect of solid inventory?

18.3.2 The relationship between k and impeller speed N

In general, experiments have shown that

$$k \propto N^a \tag{18.11}$$

where the value of a has varied from 0 to 1.4. It is probable that such a wide variation has been found because in some cases (i.e. $a \to 0$) the experiment was not conducted under diffusion-controlled conditions and in others ($a > 1.0$), $N < N_{JS}$.

Where care has been taken to make close observations of the impeller–fluid–particle interaction in the tank, from $N \to 0$ to $N \gg N_{JS}$, data typified by *Figure 18.3* are obtained[8,28]. This figure has been idealized but, basically, k increases rapidly with increasing N as increasing surface area is presented to the fluid for $N < N_{JS}$ and then more slowly for $N_S > N_{JS}$[11-13]. Above a certain speed, N_{SA}, surface aeration begins and rapidly increases, leading to a further reduction in the rate of increase of k with $N_S > N_{SA}$[8,28], mainly due to the blanketing of the particle surface with ingested air.

In practice, then, the desirable working range will probably be $N_{JS} < N_S < N_{SA}$ and in this region

$$k_S \propto N_S^b \tag{18.12}$$

where b generally lies between about 0.4 and 0.6 for efficient suspension geometries[8,9], i.e. ones with low values of $(\varepsilon_T)_{JS}$, and between about 0.65 and 0.9 for inefficient geometries[8,9]. In each case, the larger exponent is found with larger and/or denser particles[8,28].

In addition, it is worth noting that a survey of 14 experimental studies where sufficient data were presented to make the analysis showed that the ratio of k_{max}/k_{min}, the maximum to minimum measured mass transfer coefficient for a particular solid–liquid pair, respectively, was always less than 2 (ref. 14)!

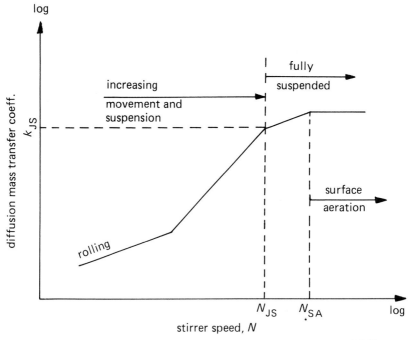

Figure 18.3   Mass transfer coefficient over wide range of stirrer speeds[8,28]

18.3.3 k_{JS} at N_{JS} and as a function of $(\varepsilon_T)_{JS}$

Experiments over a wide range of impeller-tank configurations[9] clearly indicated that the diffusion controlled mass transfer coefficient, k_{JS}, at the impeller speed just causing particle suspension, N_{JS}, was constant independent of configuration. Similar experiments with a wide variety of electrolytes and particle sizes showed that, though k_{JS} varied with particle size and electrolyte, it was not a function of configuration[8] (see *Figure 18.4*). Measured values of $(\varepsilon_T)_{JS}$ at N_{JS} also clearly indicated that, even with very wide variations in $(\varepsilon_T)_{JS}$, k_{JS} remained constant (*Figure 18.5*). Recent experimental studies have confirmed the approximate constancy of k_{JS}[15,16].

The practical implications of the findings of these last two sections relate to the following:

(a) *The choice of impeller-tank configuration.* Since k_{JS} is constant at N_{JS} regardless of $(\varepsilon_T)_{JS}$, an impeller-tank configuration requiring the least power input for particle suspension is the most economic in terms of power.

(b) *The choice of impeller speed.* If such an impeller-tank configuration is chosen, then

$$k_S \propto N_S^b \tag{18.12}$$

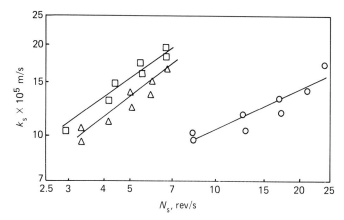

Figure 18.4   Mass transfer coefficient v. stirrer speed for 2230 μm NaCl crystals[9].
□, *disc turbine:* $D = T/2$; △, *two-blade paddle;* $D = 0.75T$; ○, *45° pitch blade turbine;* $D = T/4$

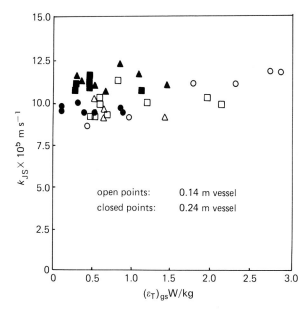

Figure 18.5 k_{JS} v. $(\varepsilon_T)_{JS}$ for 2230 μm NaCl crystals in two sizes of vessel for 40 impeller-vessel combinations[9] (●, 45° pitch turbine; ▲, flat paddle; ■, disc turbine)

where $b = 0.4$–0.6. Since

$$(\varepsilon_T)_S \propto N_S^3 \tag{18.9}$$

a tenfold increase in power/volume (or mass) only leads a 35–55% increase in k_S. Remember, firstly, that the achievement of the just fully suspended condition often requires a fairly high power/mass[14] and, secondly, that this rate of increase of k_S above k_{JS} (and therefore in R_G) is the best possible since only when the overall rate is totally diffusion-controlled will the dependence on agitation be so great. This, in general, designing for $N_S > N_{JS}$ is often uneconomic.

18.3.4   Prediction of the bulk diffusion mass transfer coefficient, k

The most logical method of predicting k is, firstly, to choose one of the many mass transfer relationships for particle–fluid systems, and, secondly, to estimate a slip (or relative) velocity u_S (for $N_S > N_{JS}$) for its application to the agitated vessel, though exactly what u_S is in such complex particle–fluid dynamics is not clear[17,18].

18.3.4.1 The mass transfer equation: The most general form of mass transfer equation is

$$Sh = A + B\,Re_p^n\,Sc^m \tag{18.13a}$$

Because experiments are very difficult to conduct with a high degree of accuracy over a wide range of conditions and diffusivity data is notoriously unreliable[19] it is not easy to test the accuracy of different theoretical models by comparison of the predicted and experimental values of the correlating constants, particularly since there are four of them. However, the basic Froessling equation

$$Sh = 2 + B\,Re_p^{1/2}\,Sc^{1/3} \tag{18.13b}$$

though derived for laminar bulk flow of fluid around a single particle, has been found to be capable of empirical modification to fit a wide range of configurations[20] and turbulent bulk flows[21]. In addition, extensive experimental work[22] has indicated that a value of $B = 0.72$ gives a better fit to the data than the theoretical value of 0.6. Additional complexity does not seem warranted because of the imprecise nature of the slip (or relative) velocity term in the Reynolds number. Therefore the mass transfer equation chosen is

$$Sh = 2 + 0.72\,Re_p^{1/2}\,Sc^{1/3} \tag{18.13c}$$

18.3.4.2 The slip velocity u_s[14]: Broadly, two approaches have been adopted. One is based on Kolmogoroff's theory of local isotropic turbulence. The outcome of this theory is that the slip velocity u_{JS} at N_{JS} for any solid–liquid pair is directly related to $(\varepsilon_T)_{JS}$. However, as shown in Chapter 16, $(\varepsilon_T)_{JS}$ varies markedly from one impeller to another, with changes in D/T and with changes of scale. Thus k_{JS} should also vary with these changes. *Figure 18.5* and the results discussed in section 18.2 show that this is not so.

In addition, recent experiments have confirmed that different agitators give very different k_s values at equal $(\varepsilon_T)_s$[15,23]. Also there are serious conceptual anomalies in applying the Kolmogoroff theory to solid–liquid mass transfer in stirred tanks[17,18]. Nevertheless, the Kolmogoroff theory has often been used leading to the Reynolds number, Re_p, being written as $(\varepsilon_T d_p^4/\nu^3)$. *Figure 18.6* shows data correlated on this basis[14]. A point to note is the facility that this type of graph has for apparently correlating disparate data. The two lines for 2230 μm NaCl particles actually start at k_{JS} with $(\varepsilon_T)_{JS}$ values differing by a factor of 30 (ref. 14). Thus, this form of presentation can be rather misleading.

The other approach is the terminal velocity-slip velocity theory[8,9,13,17]. It is postulated that the slip velocity will be of the order of the particle terminal

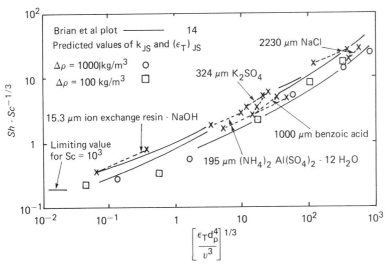

Figure 18.6   Correlation of stirred tank data based on Kolmogoroff's theory[14,24]

velocity u_T when the agitator speed is N_{JS}. Substituting u_T in equation (18.13c), a mass transfer coefficient k_T can be calculated and if the postulate is correct k_{JS} determined experimentally should be constant and of the order of k_T. As already indicated, k_{JS} is found to be constant and in addition k_{JS}/k_T is found to lie between 1 and 2. *Figure 18.7* shows the variation of k_{JS}/k_T (which has been called E, the enhancement factor) with particle size[14].

In estimating k_T, u_T is calculated from

$$u_T = 0.152 g^{0.71} d_p^{1.14} \Delta\rho^{0.71} / \rho_L^{0.29} \mu^{0.43}; \qquad d_p < 500 \ \mu m \qquad (18.14)$$

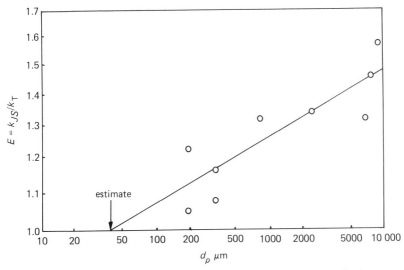

Figure 18.7   The dependence of the enhancement factor E *on particle size*

and

$$u_T = (\tfrac{4}{3}gd_p\Delta\rho/\rho_L)^{1/2}; \qquad\qquad d_p > 1500\ \mu m \qquad\qquad (18.15)$$

which is the usual terminal velocity equation for the Newton's Law region but with $C_D = 1.0$ to allow for the enhanced turbulent drag coefficient found in stirred tanks[17,25,26].

Figure 18.7 can also be expressed as

$$k_{JS}/k_T = E = (d_p/(d_p)_{40\,\mu m})^{0.08} \qquad\qquad\qquad (18.16)$$

since it can be seen that $E = 1$ at $d_p = 40\ \mu m$ and increases to 1.5 at 10 000 μm, i.e. ~ 10 mm diameter. It is postulated that the enhancement factor E is brought about by the high turbulence level in stirred tanks having a progressively greater effect at the surface of the particle as it gets larger[8]. However, at around 40 μm enhancement becomes negligible since the mass transfer contribution from the 2 term in equation (18.13) begins to swamp that from the Reynolds–Schmidt group in typical solid–liquid systems.

In passing, it should be noted that predictions based on the terminal velocity-slip velocity theory will also fit data plots based on the Kolmogoroff approach (see *Figure 18.6*). More recently, another attempt to bring the two approaches together has been proposed by Conti and Sicardi[27]. They plotted the enhancement factor E from data treatment based on the terminal velocity-slip velocity theory against $(\varepsilon_T d_p^4/v^3)$, i.e. the Reynolds number based on the Kolmogoroff theory. *Figure 18.8* shows the outcome and comparison of *Figures 18.7* and *18.8* indicate that an apparently improved correlation is obtained. However, it is difficult to determine whether this is actually the case or just an artefact arising from the use of d_p^4 in the abscissa when d_p varies very considerably whilst the ranges of v and ε_T are rather limited.

The mass transfer coefficient, k_S, and slip velocity, u_S, for $N_S > N_{JS}$: The terminal velocity-slip velocity theory gives no indication from first principles

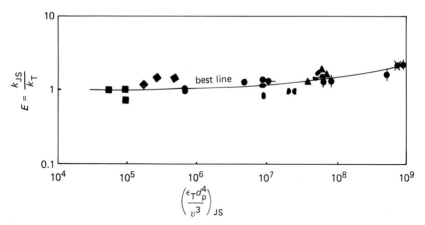

Figure 18.8   The enhancement factor E against a Kolmogoroff Reynolds Number[27]

what value the exponent on N_S should be in the relationship

$$k_S \propto N_S^b \tag{18.12}$$

for $N_{JS} < N_S < N_{AS}$ nor of the exponent c in the same range of impeller speeds in the relationship

$$u_S \propto N_S^c \tag{18.17}$$

The Kolmogoroff theory is specifically designed to give values for the exponent c by considering the relationships between u_S and $(\varepsilon_T)_{JS}$ and between $(\varepsilon_T)_{JS}$ and N_S[17]. However, it generally implies that the exponent c is constant and therefore through equation (18.13) for all but the smallest particle sizes, so is b. The value of b generally predicted is 0.75. Experimentally b is found neither to be constant nor equal to 0.75.

Finally, attempts to measure u_S have generally proved unsuccessful because of the phase angle relationship between the turbulent fluid and the turbulent motion of particle within the fluid[17,25,26].

The practical recommendation for estimating k_S in this region (assuming that working at $N_S > N_{JS}$ is necessary or justified) is to estimate k_{JS} as outlined above using an efficient suspension geometry, i.e. one for which $(\varepsilon_T)_{JS}$ is low. k_S can then be determined from equation (18.12) with $b = 0.4$ (to be on the safe side).

Since neither experiment nor either of the theories can give reliable values of u_S in this range and since, as will be seen later, knowledge of u_S may offer some distinct practical benefit, it is possible to develop a simple means of predicting u_S from the terminal velocity-slip velocity theory[29].

From equation (18.13) and neglecting the constant 2 which is usually unimportant for particles $> \sim 100 \ \mu m$,

$$k_T \propto u_T^{1/2} \tag{18.18}$$

for one particle size of a solid–liquid pair. In addition

$$k_{JS} = E k_T \tag{18.16}$$

and if it is assumed that k_{JS} arises from a slip velocity u_{JS} and k_S from a slip velocity u_S, then

$$k_{JS} \propto u_{JS}^{1/2} \tag{18.19}$$

and

$$k_S \propto u_S^{1/2} \propto N_S^b \tag{18.20}$$

Thus from equations (18.16), (18.18) and (18.19)

$$(u_T/u_{JS}) = (k_T/k_{JS})^2 = (E)^{-2} \tag{18.21}$$

and from equation (18.20) since this applies for $N_{JS} \leqslant N_S \leqslant N_{SA}$

$$(u_{JS}/u_S) = (N_{JS}/N_S)^{2b} = (k_{JS}/k_S)^2 \tag{18.22}$$

Finally, combining equations (18.21) and (18.22) and eliminating u_{JS} gives

$$u_S = E^2 (N_S/N_{JS})^{2b} u_T \tag{18.23}$$

Again, in practice, an assumption of $b=0.4$ is recommended.

Obviously, if the particles are of neutral density, the slip velocity cannot be estimated from the terminal velocity. Only one study has been done with $\Delta\rho \to 0$ and from that work[26],

$$u_S = 0.93 (v/d_p)(d_p^{4/3}\varepsilon^{1/3}/v)^{1.23}(D/T)^{0.35} \tag{18.24}$$

This relationship is not entirely satisfactory because substitution in equation (18.13c) will indicate

$$k_S \propto N^{0.6} \tag{18.25}$$

when the constant 2 is negligible. As already discussed, other work[8,13,28] has shown that the exponent on N falls to around 0.3 to 0.4 as $\Delta\rho$ decreases. However, for the present, equation (8.24) is all that is available and theoretical justification has recently been published[30].

18.4   The use of the slip velocity equation

Two possible uses are suggested for equation (18.23). Firstly, stirred tank mass transfer experiments could be used to determine b for particles of a known size and density in a particular geometry. This information might then be used to estimate the increase at $N_S > N_{JS}$ in the overall reaction rate in a similar geometry for particles of the same size and density where both reaction steps are important and difficult to separate quantitatively.

Secondly, if reaction rate data as a function of velocity are available for other geometries, then determination of u_S enables an estimate of the 'reaction rate' to be expected in a stirred tank. This approach has been successfully applied to crystallization data in growth cells, fluidized beds and stirred tanks[29], as can be seen in *Figure 18.9*.

18.5   Particle impacts and abrasion

Three particle collision mechanisms can occur in an agitated vessel. These are (a) particle-impeller, (b) particle-vessel and (c) particle-particle. Most of the work on collisions has been related to secondary nucleation, but there are other systems where mechanical abrasion following impact may occur and may be undesirable, e.g. breakdown of friable catalysts, or desirable, e.g. removal of an impervious outer skin which forms on ore particles during some leaching processes.

When the particles are present in low concentration (number/volume), the most important of the three mechanisms is particle-impeller impaction, being the most energetic and most frequent. The rate of impaction, ω, is estimated

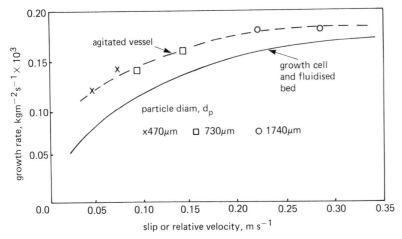

Figure 18.9   Comparison of single crystal growth cell, fluidized bed and agitated vessel growth rates for ammonium alum at 32°C (ref. 29)

from[31]

$$\omega \propto \quad \begin{array}{c}\text{The frequency of}\\\text{particle passage}\\\text{through the impeller}\end{array} \quad \times \quad \begin{array}{c}\text{Concentration}\\\text{of particles}\end{array} \quad \times \quad \begin{array}{c}\text{Probability of}\\\text{impacting during}\\\text{a passage}\end{array}$$

or

$$\omega \propto (ND^3/V)(C_S)(p) \tag{18.26}$$

and the impact probability, p, is given by[31]

$$p \propto (NDu_T/Lg) \propto Nu_T \tag{18.27}$$

for a particular solid–liquid pair.

If it is assumed that the amount of abrasion is proportional to the energy of the impact, e, and that

$$e \propto N^2 D^2 \tag{18.28}$$

then the abrasion rate J in (number/(volume)(time)) is given by

$$J \propto e\omega \propto (N^3 D^5/V)(C_S)(p) \propto \varepsilon_T N \tag{18.29}$$

Obviously no abrasion will occur unless the impact energy is greater than some threshold level. However, if it does, then it can be increased by increasing the power input, ε_T, and using small, high-speed, turbine-type impellers to do it as this increases the impact probability. However, since under most scale-up rules N decreases with increase of scale, abrasion will generally decrease on scale-up.

Conversely, abrasion can be reduced by minimizing the power input and also putting in that power with large, low-speed, propeller-type stirrers. For reacting particles, this means that the systems which are efficient for particle suspension and reaction are also the optimum for minimizing abrasion.

If the solids are present in high concentration, then particle-particle impaction occurs so much more frequently than particle-impeller that, provided the impacts are above the threshold level, this mechanism becomes dominant. In that case[32]

$$J \propto C_s^2 \varepsilon^{3/2} \qquad (18.30)$$

Again, the extreme sensitivity to ε_T suggests that, if abrasion is undesirable, ε_T must be minimized as much as possible.

18.6   Conclusions

The mixer as a solid–liquid reactor is generally most efficient, i.e. (R_G/ε_T is a maximum) at the just fully suspended condition for all impellers and for any particular solid–liquid pair, $R_{G_{JS}} \simeq$ constant. Thus the overall optimum is obtained when $(\varepsilon_T)_{JS}$ is lowest. Thus an impeller-tank configuration that is efficient for suspension is also efficient for reaction. In many cases this corresponds to a large low-speed impeller and in this respect the recent development of cheap and reliable gearboxes capable of handling the large torques that low speeds imply is noteworthy[33].

The overall 'reaction rate' requires knowledge of both the diffusion and surface reaction step. The diffusion step can be estimated from a slip velocity–terminal velocity theory but values for the surface reaction rate must be obtained by experiment and this may be difficult. Slip velocities can also be calculated from the terminal velocity–slip velocity theory and these may enable overall reaction rates in more easily defined hydrodynamic systems to be used to estimate the rates obtainable in a stirred vessel. For efficient suspension geometries,

$$(R_G)_s \propto (\varepsilon_T)_s^c \qquad (18.31)$$

where $c \not > \sim 0.13$ for fully diffusion-controlled processes and $c = 0$ for surface reaction control.

If particle abrasion is a problem, the use of large slow streamlined impellers giving low $(\varepsilon_T)_{JS}$ is again desirable. However, if abrasion (or particle breakdown) is required, then the amount of energy necessary to achieve it is minimized by the use of small high-speed turbine-type impellers. However, since particle suspension may then not be achieved, some optimum configuration must be sought.

Notation (see also Chapter 16 but note that ε_T is mean energy dissipation rate, $P/\rho_L V$ in this chapter)

a, b, c	exponents, dimensionless
a_p	particle specific surface (m^{-1})
A	the total surface area of particles (m^2)
A, B	constants in equation (18.13a), dimensionless
C	concentration of the reacting species (kg m^{-3})

C_S	solids concentration (number of particles/unit volume) (m^{-3})
D_L	diffusivity $(m^2\ s^{-1})$
e	energy of impacts (J)
E	enhancement factor, see equation (18.16), dimensionless
k	bulk diffusion mass transfer coefficient $(m\ s^{-1})$
k_R	'reaction rate' coefficient $(kg^{1-n'}\ m^{3n'-2}\ s^{-1})$
m, n	exponents in equation (18.13a), dimensionless
M	mass flow rate of solid particles, see *Figure 18.1* $(kg\ s^{-1})$
n'	exponent for 'reaction rate' step, dimensionless
p	probability of impaction of a particle on the impeller, dimensionless
q	liquid through flow rate, see *Figure 18.1* $(m^3\ s^{-1})$
Re	particle Reynolds number, $u_p d_p/v$, dimensionless
R_G	overall reaction rate $(kg\ s^{-1})$
Sc	Schmidt number, v/D_L, dimensionless
Sh	Sherwood number, kD_p/D_L, dimensionless
t	time (s)
u	particle–fluid slip velocity $(m\ s^{-1})$
ρ_s	solid density $(kg\ m^{-3})$
θ	total dissolution time for a particle size, d_p (s)
τ	mean residence time, see *Figure 18.1* (s)
ω	frequency of passage of a particle through the impeller (s^{-1})

Subscripts

i	at the 'surface' where the bulk diffusion and surface integration steps meet
∞	in the bulk solution
S	at an impeller speed $N_{JS} < N_S < N_{SA}$
SA	at the impeller speed giving rise to surface aeration
IN	at the inlet conditions of a continuous reactor
O	at the outlet conditions of a continuous reactor

Superscripts

*	refers to the concentration at the 'true' fluid–solid interface
$^-$	integral k value, see equation (18.10)

References

1 MATTERN, R. V., BILOUS, O. and PIRET, E. L. (1957) *A. I. Ch. E. J.*, **3**, 497.
2 BOURNE, J. R. and ZABELKA, M. (1980) *Chem. Eng. Sci.*, **35**, 533.
3 RANDOLPH, A. D. and LARSON, M. A. (1971) *Theory of Particulate Processes*, Academic Press.
4 LARSON, M. A. and GARSIDE, J. (June 1973) *The Chem. Engr.*, **274**, 318.
5 MULLIN, J. W. (1972) *Crystallisation*, Butterworths, 2nd ed.
6 LEVENSPIEL, O. (1962) *Chemical Reaction Engineering*, Wiley.
7 GARSIDE, J. (1971) *Chem. Eng. Sci.*, **26**, 1425.
8 NIENOW, A. W. (1969) *Can. J. Chem. Eng.*, **47**, 248.
9 NIENOW, A. W. and MILES, D. (1978) *Chem. Eng. J.*, **15**, 13.

10 NIENOW, A. W., UNAHABHOKHA, R. and MULLIN, J. W. (1969) *Chem. Eng. Sci.*, **24**, 357.
11 HIXSON, A. W. and BAUM, S. J. (1941) *Ind. Eng. Chem.*, **33**, 478.
12 KNEULE, F. (1956) *Chem. Ing. Tech.*, **28**, 221.
13 HARRIOTT, P. (1962) *A. I. Ch. E. J.*, **8**, 93.
14 NIENOW, A. W. (1975) *Chem. Eng. J.*, **9**, 153.
15 DITL, P. (1982) Private Communication.
16 OUSENIK, A. (1982) *Proc. 4th European Conf. on Mixing*, BHRA Fluid Engineering, Cranfield, pp. 463–470.
17 NIENOW, A. W. and BARTLETT, R. (1975) *Proc. 1st European Conf. on Mixing, Cambridge, 1974*, BHRA Fluid Engineering, Cranfield, pp. B1-1 to B1-95.
18 LEVINS, D. M. and GLASTONBURY, J. R. (1972) *Chem. Eng. Sci.*, **27**, 537.
19 NIENOW, A. W. (1965) *Brit. Chem. Eng.*, **10**, 827.
20 ROWE, P. N. and CLAXTON, K. T. (1965) *Trans. I. Chem. E.*, **43**, T321.
21 GALLOWAY, T. R. and SAGE, B. H. (1968) *Int. J. Heat Mass Transfer*, **11**, 539.
22 ROWE, P. N., CLAXTON, K. T. and LEWIS, J. B. (1965) *Trans. I. Chem. E.*, **43**, T14.
23 AUSSENAC, D., ALRAN, C. and COUDERC, J. P. (1982) *Proc. 4th European Conf. on Mixing*, BHRA Fluid Engineering, pp. 417–421.
24 BRIAN, P. L. T., HALES, H. B. and SHERWOOD, T. K. (1969) *A. I. Ch. E. J.*, **15**, 727.
25 SCHWARTZBERG, H. G. and TREYBAL, R. E. (1968) *Ind. Eng. Chem. (Fundls)*, **7**, 1.
26 LEVINS, D. M. and GLASTONBURY, J. R. (1972) *Trans. I. Chem. E.*, **50**, 32 and 132.
27 CONTI, R. and SICARDI, S. (1982) *Chem. Eng. Comm.*, **14**, 981.
28 NAGATA, S. (1975) *Mixing; Principles and Applications*, Kodansha, Tokyo, Chapter 6.
29 NIENOW, A. W., BUJAC, P. D. B. and MULLIN, J. W. (1972) *J. Crystal Growth*, **13/14**, 488.
30 BATCHELOR, G. K. (1980) *J. Fluid Mech.*, **98**, 609.
31 NIENOW, A. W. (1976) *Trans. I. Chem. E.*, **54**, 205.
32 NIENOW, A. W. and CONTI, R. (1978) *Chem. Eng. Sci.*, **35**, 543.
33 STRONG, M. D. (Cleveland Mixer Co., USA) (1977) *Energy Savings with Low Speed Mixers*, Lecture to Eng. Found. Conf. on Mixing, Rindge, New Hampshire.

Index